D0563881

Comets and the Origin and Evolution of Life

Springer
*New York
Berlin
Heidelberg
Barcelona
Budapest
Hong Kong
London
Milan
Paris
Santa Clara
Singapore
Tokyo*

Paul J. Thomas   Christopher F. Chyba
Christopher P. McKay
Editors

Comets and the Origin and Evolution of Life

With 47 Illustrations

Springer

99-59

Paul J. Thomas
Department of Physics and Astronomy
University of Wisconsin, Eau Claire
Eau Claire, WI 54702–4004
USA

Christopher F. Chyba
Department of Geosciences
Princeton University
Princeton, NJ 08544–1003
USA

Christopher P. McKay
NASA/Ames Research Center
Space Science Division
MS 245-3
Moffett Field, CA 94035
USA

Library of Congress Cataloging-in-Publication Data
Thomas, Paul J.
    Comets and the origin and evolution of life / Paul J. Thomas,
  Christopher F. Chyba, Christopher P. McKay.
       p.   cm.
    Includes bibliographical references and index.
    ISBN 0-387-94650-0 (alk. paper)
    1. Comets.   2. Life—Origin.   3. Exobiology.   I. Chyba,
  Christopher F.   II. McKay, Christopher P.   III. Title.
  QB721.T37   1996
  577—dc20   96-11740

Printed on acid-free paper.

Production managed by Bill Imbornoni; manufacturing supervised by Johanna Tschebull.
Camera-ready copy prepared from the editors' TeX files.
Printed and bound by Braun-Brumfield, Inc., Ann Arbor, MI.
Printed in the United States of America.

9 8 7 6 5 4 3 2 1

ISBN 0-387-94650-0 Springer-Verlag New York Berlin Heidelberg   SPIN 10425472

Contents

Contributors

D.C. Boice
>Southwest Research Institute
>6220 Culebra Road
>San Antonio, TX 78228
>USA

L. Brookshaw
>Department of Physics
>University of California
>Davis, CA 95616
>USA

C.F. Chyba
>Department of Geosciences
>Princeton University
>Princeton, NJ 08544-1003
>USA

A. Delsemme
>Department of Physics and Astronomy
>University of Toledo
>Toledo, OH 43606
>USA

W.F. Huebner
>Southwest Research Institute
>6220 Culebra Road
>San Antonio, TX 78228
>USA

J. Kissel
>Max Planck Institut fur Kernphysik
>Postfach 103980
>D-69029 Heidelberg
>GERMANY

F.R. Krueger
>   Ing.-Bureau Krueger
>   Messeler Str. 24
>   D-64291 Darmstadt
>   GERMANY

A. Lazcano
>   Departamento de Biología
>   Facultad de Ciencias
>   Universidad Nacional Autónoma de México
>   Apdo. Postal 70-407, Cd. Universitaria
>   México 04510, D.F.
>   MEXICO

C.P. McKay
>   NASA Ames Research Center
>   Moffett Field, CA 94035
>   USA

D. Morrison
>   NASA Ames Research Center
>   Moffett Field, CA 94035
>   USA

J. Oró
>   Department of Biochemical and Biophysical Sciences
>   University of Houston
>   Houston, TX 77204
>   USA

M. Podolak
>   Department of Geophysics and Planetary Sciences
>   Tel Aviv University
>   Ramat Aviv, 69 978
>   ISRAEL

D. Prialnik
>   Dept. of Geophysics and Planetary Sciences
>   Tel Aviv University
>   Ramat Aviv, 69 978
>   ISRAEL

K. Roessler

Institut fur Nuklearchemie
Forschungszentrum KFA
D-52425 Julich
GERMANY

C. Sagan

Laboratory for Planetary Studies
Cornell University
Ithaca, NY 14853
USA

N.H. Sleep

Department of Geophysics
Stanford University
Stanford, CA 94305
USA

D. Steel

Anglo–Australian Observatory
Coonabarabran, NSW 2357
AUSTRALIA
and
Department of Physics and Mathematical Physics
University of Adelaide, SA 5005
AUSTRALIA

P.J. Thomas

Department of Physics and Astronomy
University of Wisconsin
Eau Claire, WI 54702
USA

K.J. Zahnle

NASA Ames Research Center
Moffett Field, CA 94035
USA

Introduction: Comets and the Origin of Life

P.J. Thomas, C.F. Chyba, and C.P. McKay

Extraterrestrial objects such as comets and asteroids (and their associated dust) played a significant dual role in the early history of the Earth's biosphere: they were both deliverers of organic material and volatiles, and also destroyers of organic material, by the heat and shock of violent impact. The study of the origins of life on Earth attempts to uncover the physical processes operating during this earliest, very turbulent era in the history of our planet. There is still much to learn, despite recent progress in our knowledge of such fundamental issues as the nature of organic chemical processes in space, the chemistry of comets and asteroids and the nature of the early terrestrial environment.

To examine current developments in this dynamic field, a conference on "Comets and the Origins and Evolution of Life" was held in Eau Claire, Wisconsin, 30 September–2 October, 1991. The 55 participants included scientists from 13 countries, presenting 37 papers.

The focus of the conference was to consider the role of comets in the origins and evolution of life, in particular in light of recent improvements in our knowledge of the chemical nature of Comet Halley (particularly from the European Space Agency's *Giotto* and Soviet *Vega* spacecraft), the study of interplanetary dust particles (IDPs), an improved understanding of plausible mechanisms of organics synthesis inside comets, progress in numerical simulations of cometary orbital evolution and models of the impacts of comets and cometary dust on the Earth. In addition, given the prospect of upcoming spacecraft missions to comets such as NASA's Comet Rendezvous/Asteroid Flyby mission (now canceled) and the joint NASA/ESA *Rosetta* mission, it seemed timely to discuss the contribution that spacecraft observations could make to this topic. Current research presented at this confernce has already been published in a special edition of the journal *Origins of Life* (*Origins Life*, **21**, 265–440, 1991–1992).

This book has two purposes. First, it is a review of the history of our study of comets and the evolution of Earth's biosphere. The chapter by Oró and Lazcano reviews the development of our scientific research in this field since the germinal paper of Chamberlin and Chamberlin in 1908. Second, it is a summary of current research. On a broad scale, Delsemme estimates the volatile inventory brought to the early Earth by comets. Thomas and Brookshaw discuss the survivability of organic molecules in large comet and asteroid collisions on the early Earth, while

Chyba and Sagan describe a wide range of mechanisms for introducing organic molecules to the Earth, from IDP deposition to impact shock synthesis. Looking at the other—destructive—face of cometary bombardment, Zahnle reviews the disruptive influences of large impacts on the early terrestrial biosphere.

In chapters focusing on our knowledge of the chemical environments inside comets, Kissel, Krueger, and Roessler describe possible synthesis mechanisms for creating organic molecules while Huebner and Boice discuss evidence for polymers and other large organic molecules within comets. Podolak and Prialnik examine the possibility that early intense internal heating from radioactive ^{26}Al might have caused temporary liquid water environments within comets.

Turning to the present day environment of the Earth, the chapters by Steel and Morrison discuss the perturbations that comet impacts pose to the biosphere, a topic that has provoked considerable public interest in the years since the conference. Finally, McKay discusses the possibility that comets contain viable lifeforms, as opposed to nonliving organic molecules, and discusses the goals and prospects for spacecraft exploration of selected comets in the near future.

In October 1980, the late Cyril Ponnamperuma organized a colloquium on comets and the origin of life at the University of Maryland. The proceedings of the colloquium were edited by Professor Ponnamperuma and published as *Comets and the Origin of Life* (Reidel: Boston, 1981). That meeting was intended to review knowledge of the field in anticipation of the return of Halley's Comet in 1986, much as the Eau Claire conference and this present volume are intended to synthesize our progress subsequent to that apparition. The editors warmly acknowledge their debt to Professor Ponnamperuma's original efforts.

The editors wish to express appreciation to many who helped to organize the Eau Claire conference. The scientific organizing committee, comprising S. Chang, A. Delsemme, W. Huebner, J. Oró, D. Steel, and P. Weissman provided advice and helped to broaden the participation of scientists so that this conference was truly international.

At Eau Claire, many people provided enthusiastic assistance and advice. In particular, the editors wish to thank Mary Traynor and Sister Michaela Hedican of the St. Bede Center for providing the unforgettably beautiful and peaceful setting for the conference. Many University of Wisconsin, Eau Claire staff, in particular, Jim Simonson, Bob Elliott, and Vince Strief helped with many organizational issues that helped the conference to flow smoothly. In the preparation of this book, Cindy Friedman and especially Julie McLaughlan assisted with the typing and preparation of the manuscripts.

This conference was funded by NASA's Exobiology, Origins and Comet Rendezvous/Asteroid Flyby programs.

1
Comets and the Origin and Evolution of Life

J. Oró and A. Lazcano

ABSTRACT The historical development of the study of comets and the origins of life is reviewed.

1.1 Introduction

Did comets play a role in the origin and early evolution of life on Earth? This possibility, which has been suggested independently at least twice during this century (Chamberlin and Chamberlin, 1908; Oró, 1961) has recevied considerable attention in the past few years, and has been transformed into a highly interdisciplinary field that now includes speculations on the possibility that cometary impacts may have led to several major biological extinctions (Thomas, 1992). To a considerable extent, this renewed interest in the part that comets and other volatile-rich minor bodies may have played in the origin and evolution of life is one of the outcomes of the space exploration programs, which have provided striking images of the cratered surfaces of planets and their satellites.

A number of recent developments and hypotheses that have also reinforced the interest in this issue include: (a) the development of inhomogeneous accretion models for the formation of the Earth (Wetherill, 1990), and the single-impact theory of the origin of the Earth–Moon system (Cameron and Benz, 1989), both of which predict a volatile-depleted young Earth; (b) the current debate on the detailed chemical composition of the prebiotic terrestrial atmosphere (Oró et al., 1990); (c) the spectrometric observations of comet Halley obtained from the Vega and Giotto misions, which suggest that many different types of cyclic and acyclic organic compounds, which may perhaps include adenine, pyrimidines and their derivatives, are present in cometary nuclei (Kissel and Krueger, 1987); (d) the possibility that the Cretaceous–Tertiary (K/T) extinction was caused by a large meteorite (Alvarez et al., 1980) or a comet (Davis et al., 1984); and (e) the detection of extraterrestrial nonproteinic amino acids near the K/T 65-million year old boundary, which may be of ultimate cometary origin (Zhao and Bada, 1989; Zahnle and Grinspoon, 1990).

Since the reconstruction of the physical and chemical characteristics of the primitive environment is fraught with problems and pitfalls, it is difficult to evaluate in detail the cometary contribution to prebiological processes. This has led to rather extreme points of view, which range from claims that bacteria and viruses originating in the interstellar medium pass from comets to the Earth (Hoyle and Wickramasinghe, 1984) to a sobering challenge of the idea that the input of extraterrestrial organic molecules and volatiles played a major part in the origin of terrestrial life (Miller, 1991a, b). As argued throughout this book, the roles that comets may have actually played in prebiotic and biological evolution probably lie somewhere in the middle of this wide spectrum of possibilities.

As summarized in this introductory chapter, a widely held idea supported by different independent estimates is that cometary impacts represented a rich veneer of volatiles during the early stages of the Earth's history. As the solar system aged, such catastrophic events became increasingly rare, but nonetheless they may have had dramatic implications for the evolution of the biosphere (Steel, this volume), and may represent a potential threat for our own species (Morrison, this volume). The purpose of this chapter is to review the different (and sometimes conflicting) hypothesis on the role of comets in the origin and evolution of life, and to provide a brief summary of the historical background which led to the development of these different ideas.

1.2   Comets and the Origin of Life: An Idea with a Long History

There is well-established scientific tradition linking cometary phenomena with the appearance of life. Prompted and funded by Edmund Halley, Issac Newton published his *Principia* in 1686, a book which represents a major breakthrough in the development of Western thought. But not even the grave and pompous Sir Issac, for all his love for mathematical accuracy, was immune to the appeal of popular beliefs, even if they had not been experimentally proven. As Oparin (1938) has written, more than once Newton expressed his conviction that cometary emanations could lead to the spontaneous generation of plants — a phenomenon neither he nor Halley had the chance to corroborate when the latter's comet returned to perihelion in 1759.

A hundred years later the young field of evolutionary biology faced a major crisis. In 1859 Charles Darwin published the first edition of *The Origin of Species*, but that very same year Louis Pasteur began the experiments that would eventually lead to the discredit of the idea of spontaneous generation, the mechanism implicitly advocated by most nineteenth century scientists to explain the emergence of primary forms of life. As Lord Kelvin wrote, "the impossibility of spontaneous generation at any time whatever must be considered as firmly established as the law of universal gravitation." It was not easy to support the concerpts of natural selection and common ancestry of all living beings in the absence of a secular explanation for the origin of life — but comets and meteorites came to the rescue. As an attempt to find a way out of this scientific *cul de sac*, the German physiologist Hermann von Helmholtz (1871) wrote that "who could say whether the comets and meteors which swarm everywhere through space, may not scatter germs wherever a new world has reached the stage in which it is a suitable place for organic beings."

Von Helmholtz's attempts to revive the old idea of panspermia were supported by other distinguished scientists, including Lord Kelvin and Svante Arrhenius. By the turn of the twentieth century this hypothesis joined the large list of explanations that were being developed to explain the origin and early evolution of life on Earth, but not everyone was amused with it (Farley, 1977; Kamminga, 1988; Lazcano, 1992a, b). In 1911 the American geologist Thomas C. Chamberlin wrote against the increasing infatuation of some of his contemporaries with the idea that spores driven by light pressure could travel across the Universe, transporting life from planet to planet (Chamberlin, 1911). T.C. Chamberlin was no newcomer to the origin of life issue. A few years before he had published, with his son and colleague Rollin T. Chamberlin, an insightful paper titled "Early terrestrial conditions that may have favored organic synthesis." In this unique work it was suggested that "planetesimals," i.e., the small sized bodies from which the planets were formed, may have represented a significant source of organic compounds for the primitive Earth, influencing both abiotic synthesis and biological evolution (Chamberlin and Chamberlin, 1908).

What was the driving force behind the Chamberlin father-and-son team, that led them to such pioneering ideas? The senior Chamberlin studies on glaciations had led him to recognize several different early ice ages (Chamberlin, 1893, 1894, 1896; Alden, 1929) and, eventually, to the rejection of the widely held Laplacian nebular theory (MacMillan, 1929; Fenton and Fenton, 1952). According to the Laplacian orthodoxy, the Earth had originated as a molten hot sphere surrounded with a dense CO_2-rich primoridal atmosphere. As the Earth grew older and cooler, the atmosphere became less thick and less moist, leading to a progressive cooling of the environment and to the appearance of longer winters which eventually culminated in an ice age. However, T.C. Chamberlin had found geological evidence of long-term climate fluctuations, i.e., of alternative periods of warmth and coolness, and this discovery could not be reconciled with the idea of progressively cooling climates implied by Laplace's theory. Chamberlin was forced to look for other explanations (Fenton and Fenton, 1952).

With the help of the American astronomer Forest R. Moulton, T.C. Chamberlin began working on an alternative hypothesis, according to which the young, eruptive Sun had been approached by a smaller star with drew huge tides of solar matter, which eventually condensed into small, organic-rich solid bodies called planetesimals. According to this idea, which was developed independently of an equivalent scheme suggested by James Jeans, the cooling of this solar ejecta eventually led to the formation of a swarm of small, cold bodies from which the Earth accreted (Moulton and Chamberlin, 1900; Chamberlin, 1904).

The paper that T.C. Chamberlin coauthored with his son in 1908 was the natural outcome of his views on the origin of the Earth. The presence of meteoritic organic compounds had been firmly established since the mid-nineteenth century, when Jöns Jacob Berzelius, analyzed the Alais meteorite in 1834, a carbonaceous C1 chondrite, which fell in Alais, France, in 1806, and also when F. Wöhler found organic matter in the Kaba meteorite, a C2 carbonaceous chondrite (Berzelius, 1834; Wöhler, 1858; Wöhler and Hörnes, 1859). The Chamberlins were obviously well acquainted with such discoveries, and in a bold stroke of chemical insight, they incorporated them in their paper. "The planetesimals are assumed to have contained carbon, sulphur, phosphorus and all other elements found in organic matter," they argued in a qualitative description that predates contemporary models of the primitive Earth, "and as they impinged more or less violently upon the surface formed of previous accessions of similar matter, there should have been generated various compounds of these elements," and added that his process would generate "hydrocarbons, ammonia, hydrogen phosphide, and hydrogen sulphide gases mingled with the ordinary gases carried by the planetesimal furnishing rather remarkable conditions for interactions and combinations, among which unusual synthesis would not be improbable."

This remarkable work went unnoticed by those concerned at the time with the study of the origins of life. The Chamberlins had written their paper as part of a larger attempt to understand the origin of the solar system, but at that time very few scientists saw a direct, causal relationship between the formation of our planet and its primitive environment, and the origin of life. In fact, during that period the nonbiological synthesis of sugars (Butlerow, 1861) and of amino acids (Löb, 1913) had already been reported, but these experiments were not considered a simulation of the early Earth (Miller, 1974). Since it was generally assumed that the first living beings had been autotrophic, plant-like organisms, neither these abiotic synthesis nor an extraterrestrial source of organic material appeared to be a necessary prerequisite for the origin of life. As the Chamberlins themselves wrote, "... the purpose of this paper is not to meet a geologic necessity, but merely to consider those conditions in the early history of the globe which may be thought to have been specially favorable to organic synthesis, irrespective of the question whether the natural evolution of life was wholly dependent upon them or was merely facilitated by them" (Chamberlin and Chamberlin, 1908). In spite of their geological insight and deep comprehension of evolutionary theory, their ideas soon sank into an 80-year period of scientific oblivion.

1.3   Chemical Evolution of Cometary Nuclei

Nonetheless, comets kept a place in the modern theory of the origins of life. "Hydrocarbon lines have also been found in the spectra of comets, those heavenly bodies which from time to time pass through our solar system from interplanetary space," argued Alexander I. Oparin in his classical monograph on the appearance of life, "... thanks to the studies of several scientists it has been found that cyan (CN), a compound of carbon and nitrogen, and carbon monoxide are present in the gases which form the tail of comets ... thus, we can demonstrate beyond doubt the presence of hydrocarbons on a number of heavenly bodies. This fact gives full support to the conclusion we have already drawn. There came a time in the life of the Earth which the carbon had been set free from its combination with metal and had combined with hydrogen forming a number of hydrocarbons. These were the first 'organic' compounds on the Earth" (Oparin, 1924).

As summarized in this book, recent discoveries have shown that the organic chemistry of comets is much more complex than ever suspected by Oparin and his contemporaries. It is true that cometary nuclei contain the most pristine material in the solar system (Delsemme, this volume), but theoretical modeling, experimental simulations, and observational results suggest that in addition to refractory, dark carbonaceous components and simple molecules and radicals (i.e., C_2, C, CH, CN, HCN, CH_3CN, NH, NH_3, OH, H_2O), comets may also contain significant amounts of complex organic compounds of biochemical significance. These may have different sources, including inherited interstellar components (Greenberg, 1983; Greenberg and Grimn, 1986; Huebner and Boice, this volume; Delsemme, this volume), as well as those resulting from the interaction of the cometary nuclei volatile components with cosmic rays (Donn, 1976), solar radiation (Grün et al., 1991), and the decay products of imbedded presolar unstable isotopes such as ^{26}Al (Irvine et al., 1980; Podolak and Prialnik, this volume).

The above conclusion is supported by a series of simulation experiments carried out in different laboratories (Kissel, this volume; Huebner and Boice, this volume), in which solid-phase experiments with icy-mixtures of relative simple compounds (CH_4, C_2H_6, N_2, NH_3, H_2O, H_2CO, etc.) have been irradiated by high energy electrons, protons, and ions, as well as by UV light (Oró, 1963; Schutte et al., 1992; Briggs et al., 1992) under varying conditions including different ratios of ice precursors (Strazzula and Johnson, 1991), various degrees of irradiation (Grün et al., 1991), and a wide range of temperatures that include the possibility of liquid water (Lerner et al., 1991; Navarro-González et al., 1991; Negrón-Mendoza et al., 1992). The relevance of these simulations to cometary chemistry is a lingering problem. The widely different conditions under which these experiments have been performed make their direct exploration to cometary nuclei a controversial issue. However, some general trends are apparent, like the efficient free radical-mediated formation of large, nonvolatile organic polymers, which upon hydrolysis yield many smaller compounds like glycine, urea, glycerol, lactic acid, formamidine, and others (Briggs et al., 1992; Huebner and Boice, this volume).

There is a good correlation between current inventories of the molecules found in interstellar clouds and in comets and those used as precursors in experiments simulating the conditions of the primitive Earth (Oró et al., 1990). The former include hydrogen cyanide (HCN), water (H_2O), cyanamide (CN), hydrogen sulphide (H_2S), ammonia (NH_3), formic acid (HCOOH), H_2N, and formaldehyde (H_2CO), etc. As argued elsewhere, at least in one principle all major biochemical monomers could be synthesized from molecules present in the interstellar medium and in comets (Lazcano-Araujo and Oró, 1981). Complex nonvolatile polymers may represent as much as 30% by mass of cometary dust particles (Langevin et al., 1987; Huebner, 1987; Eberhardt et al., 1987; Korth et al., 1989; Encrenaz and Knacke, 1991; Briggs et al., 1992). Polyoxymethylene, a formaldehyde polymer previously used with hydroxylamine as a precursor in the abiotic synthesis of amino acids and hydroxyacids (Oró et al., 1959), has also been identified in cometary spectra (Mitchell et al., 1987; Huebner, 1987; Huebner and Boice, this volume). Polyaminoethylene and other dark HCN polymers are also likely to be present in the nonvolatile organic fraction of cometary dust (Matthews and Ludicky, 1986).

Given the relatively high abundance of chemical precursors used in the abiotic synthesis of amino acids, the absence of detectable levels of these biochemical compounds in Halley's spectra is rather surprising (Kissel and Krueger, 1987). However, as argued elsewhere (Oró and Mills, 1989), the final steps of abiotic formation of amino acids require a liquid phase, as in the Stecker-cyanohydrin synthesis (Miller, 1957). If this is the case, amino acids should not be expected to be present in comets nor in the interstellar environment. The fact that an extensive radioastronomical search for glycine has yielded negative results (Hollis et al., 1980) is consistent with this explanation. It is possible, however, that under certain conditions of heavy UV flux, that glycine, and simple amino acids, could be synthesized in the gas phase, as some laboratory experiments have indicated. (R.S. Becker, personal communication.) Thus, under extreme UV and dust protection, glycine may be present in certain regions of our galaxy, although the tentative evidence obtained thus far needs confirmation. (L. Snyder, personal communication.)

On the other hand, observations of comet Halley using the impact mass spectrometers on board the Vega 1 spacecraft provided evidence that cometary dust is composed of a chondritic core with an organic mantle composed mainly of highly unsaturated compounds. Many of these are cyclic and acyclic organic molecules such as genzene and toluene, and may also include pyrrole, pyridines, imidazol, adenine, xanthine, pyrimidines, and some of their derivatives like oxyimidazole and oxypyrimidines (Kissel and Krueger, 1987; Kissel, this volume). Life may not be present on cometary nuclei; but the presence of these molecules supports previous contentions that these minor bodies are a frozen sample of the organic chemistry synthetic processes going throughout the universe.

1.4   The Collisional History of the Early Solar System

It is generally accepted that the Earth was formed approximately 4.6×10^9 years ago as a result of the gravitational condensation of the solar nebula, a viscous disk of gas and dust circling the protoSun (Cameron, 1988; Wetherill, 1990). In spite of the many uncertainties haunting current descriptions of the Earth's formation, it is agreed that our planet was formed in a region of the solar nebula from solid material devoid of volatiles. Although some models of the early solar system predict that the loss of nebular gas at the region of the nebula where the Earth was formed took place after our planet was formed (Hayashi et al., 1985), most theoreticians believed that the Earth condensed in a region already depleted in gas.

Although radial mixing involving volatile-rich planetisimals may have taken place in late stages of planetary formation (Wetherill, 1990), it has been shown that solar type prestellar objects blow away their protostellar gas envelopes in a timescale of only 10^{6-7} years (Strom et al., 1989). These observations support the hypothesis that the concentration of nebular gas at the vicinity of the Earth must have been rather small, a conclusion in good agreement with the well-known terrestrial depletion of noble gases relative to solar abundances (Brown, 1952; Suess and Urey, 1956; Cameron, 1980). Thus, even though the solar system was formed from a cloud whose elementary abundances and chemical composition probably resembled the dense gas and dust regions rich in interstellar compounds, it is unlikely that interstellar organics are the direct precursors of the first living systems (Lazcano-Araujo and Oró, 1981).

The above conclusion is reinforced by models suggesting that the Earth–Moon system was formed due to a single giant impact. This idea, known as the single-impact hypothesis, has been developed by Cameron and his collaborators (Benz et al., 1986, 1987; Cameron and Benz, 1989) and argues that a near Mars-sized body collided with the protoEarth. Upon collision, all the iron from the impactor was injected into the core of the protoplanet, ejecting into an Earth-circling orbit a debris cloud which later, though quickly, aggregated into the Moon. If such a catastrophic encounter actually took place, it must have blasted away practically all the water and the volatiles that the Earth may have previously retained.

Thus, in order to explain the observed terrestrial abundances of water and the so-called biogenic elements, i.e., C, H, O, N, S, P, etc., an exogenous veneer of volatiles is required. Cometary nuclei appear to be the best candidates to fulfill this role. This possibility is completely consistent with the vestiges of the intense bombardment stage registered in the cratered surfaces of planets and their satellites. Although only three probable impact craters dating from Precambrian times have survived (Grieve and Robertson, 1979), the Earth was obviously not spared from the collisional process that characterized the early history of the solar system, and in which comets may have played a major role (Chyba and Sagan, this volume).

Recent spectroscopic studies of β Pictoris suggest that intense cometary collisions may be part of the early evolution of planetary systems throughout the whole Galaxy. Although most attempts to detect the existence of cometary nuclei orbiting other stars or planetary systems have failed (Chyba, 1993), there is increasing spectroscopic evidence of a swarm of small cometary-like bodies falling at high velocities towards β Pictoris, a relatively young star around which planet formation is either occurring now or has just been completed (Lagrange-Henri et al., 1988).

Originally classified by visible light measurements as an anomalous shell star (Slettebak, 1975), infrared observations showed that β Pictoris was surrounded by a circumstellar disk (Aumann et al., 1984). This was confirmed when visible light images of a highly flattened disk seen edge on, formed by solid particles in nearly coplanar, low-inclination orbits extending for more than 400 astronomical units from β Pictoris became available (Smith and Terrile, 1984). Since this star is known to be much younger than its 10^9 man-sequence lifetime, it has been assumed that the gas and dust disk that surround it corresponds to a planetary system in the early stages of formation (Smith and Terrile, 1984). Such objects may be extremely abundant. Recent observations of the Orion Nebula using the Hubble

Space Telescope suggest that approximately half of the stars seen on ground-based images are surrounded by protoplanetary disks (O'Dell et al., 1993).

Since the favorable edge-on orientation of the β Pictoris disk allows direct spectral observation of its gaseous components, a number of visible and UV studies have been made of the β Pictoris complex (Hobbs et al., 1985; Kondo and Bruhweiler, 1985; Lagrange et al., 1987; Lagrange-Henri et al., 1988; Vidal-Majdar et al., 1986).

These observations have shown that strong variations occur in the redshifted part of the Al III, Mg II, and Fe II resonance spectroscopic lines. Neither stellar photospheric phenomena nor a pulsating circumstellar shell can account for the observed temporal variations, and Langrange-Henri et al. (1988) have suggested that they may be due to the rapid vaporization of a large swarm of small 10^{13-17} gram cometary-like bodies falling at high velocities toward β Pictoris.

The above explanation would require up to 10^2 collisions per year (Lagrange-Henri et al., 1988; Beust et al., 1990). This figure is 10^{3-4} times greater than Joss' (1974) estimate of approximately 10^{-2} current impacts per year with the Sun, which is clearly a lower limit when considering primitive solar system conditions. The hypothesis that a large number of comet-like bodies are colliding with larger objects in the β Pictoris complex is clearly consistent with current models of the impact history of the early solar system (Chyba, 1993), and indirectly supports the contention that comets and other volatile-rich minor bodies contributed significantly to the formation of the terrestrial ocean and atmosphere.

1.5   A Cometary Origin for the Terrestrial Volatiles?

Six months before the Chamberlin and Chamberlin (1908) paper was published, the Tunguska explosion took place. As reviewed above, neither event had an impact in the ripening of the different theories that were being developed at the time to explain the origins of life. However, although the possibility that the Tunguska object was of cometary nature or not is still a matter of debate (Brown and Hughes, 1977; Kresák, 1978; Chyba et al., 1993), the original suggestion that cometary collisions may have played a role in the appearance of life on our planet was based on data from the 1908 Siberian explosion (Oró, 1961). This hypothesis was suggested within the framework of Oparin's (1924, 1938) theory on the origins of life, and followed closely the prebiotic synthesis of adenine, amino acids, and other biochemical monomers from HCN (Oró, 1960), a chemical precursor of more complex molecules. It was thus hypothesized that cometary collisions may have represented an important source of volatiles for the primitive Earth, which may have been the precursors for the nonbiological synthesis of biochemical compounds that preceded the first organisms (Oró, 1961).

Based on Urey's (1957) collisional probability, and assuming that cometary nuclei had densities in the range of 0.01–0.5 g cm^{-3} and diameters in the range of 1–10 km, Oró (1961) calculated that during its first 2×10^9 years the Earth had accreted up to 2×10^{18} g of cometary material, which would lead to localized

TABLE 1.1. Cometary Matter Trapped by Solar System Bodies.

	Trapped cometary matter (grams)	Time-span	Reference
Venus	4.0×10^{20}	2×10^9 years	Lewis (1974)
Moon	2.0×10^{20}	Late-accretion period	Wetherill (1975)
Earth	$2.0 \times 10^{14-18}$	2×10^9 years	Oró (1961)
	$1.0 \times 10^{25-26}$	Late-accretion period	Whipple (1976)
	3.5×10^{21}	Late-accretion period	Sill and Wilkening (1978)
	7.0×10^{23}	4.5×10^9 years	Chang (1979)
	2.0×10^{22}	4.5×10^9 years	Pollack and Yung (1980)
	1.0×10^{23}	2.0×10^9 years	Oró et al. (1980)
	$1.0 \times 10^{24-25}$	1.0×10^9 years	Delsemme (1984, 1991)
	$6.0 \times 10^{24-25}$	1.0×10^9 years	Ip and Fernández (1988)
	$1.0 \times 10^{23-26}$	4.5×10^9 years	Chyba et al. (1990)*

*See also Chyba (1991); Chyba and Sagan, this volume.

primitive environments in which the high concentrations of chemical precursors faciliated prebiotic synthesis. Other estimates of cometary material trapped by the Earth have been made for different periods of time and under widely different assumptions (Table 1.1). For instance, Whipple (1976) calculated that during the late accretion period our planet captured 10^{25-26} g of cometary material from a short-lived (10^8 years) cometary nebula of mass 10^{29} g, located within Jupiter's orbit (Table 1.1). According to these figures, the secondary atmospheres of Mars, Venus and the Earth may have accumulated entirely from comets that replenished the gases and volatile compounds lost during the accretion period of these planets (Whipple, 1976). More conservative estimates have been made for this same period of time for our planet, including calculations by Sill and Wilkening (1978), who suggested that during the heavy-bombardment phase of the solar system, the Earth acquired $\sim 10^{21}$ g of cometary material rich in clathrates and other volatiles (Table 1.1).

An attempt to update Oró's (1961) original calculations by assuming a cometary reservior of 10^9 random parabolic comets (Everhart, 1969), each of which followed the absolute magnitude–mass relationship (Allen, 1973), led to an estimate that during its first 2×10^9 years the Earth captured 10^{21} g of cometary material. The use of Wetherill's (1975) collisional probability tables to account for the higher impact rate of the early solar system increased this figure a hundred-fold (Table 1.1). In recent years these simple assumptions have been superceded by more complex and detailed models involving a wide range of dynamical parameters, different masses for the Oort cloud, and lunar cratering rates adjusted for the different surface area and gravitational attraction of the planets (Chyba et al., 1990; Chyba and Sagan, this volume). Independent calculations also suggest that carbonaceous chondrites and cometary-derived interplanetary dust particles may have been even more important sources of exogenous volatiles than collisions themselves (Anders and Owen, 1977; Anders, 1989; Zahnle and Grinspoon, 1990; Chyba and Sagan, 1992).

Although it is possible that massive comets with high collisional velocities could have eroded as much volatiles as they delivered to the terrestrial planets (Walker, 1986; McKinnon, 1989; Melosh and Vickery, 1989; Hunten, 1993), this phenomenon would not have affected severely the larger worlds of the inner solar system (Chyba, 1990). It is generally accepted that impactors were delivering volatiles and perhaps even some complex organic molecules to the primitive Earth, as a complement of endogenous prebiotic synthesis. The different contributions that comets may have done to the origins of life have been summarized elsewhere (Oró et al., 1995). As argued by Joshua Ledeberg (1992), "... I share the idea that cometary fragment infall might account for a substantial part of terrestrial organic matter. This would leave a wider range of possibilities about the primordial chemistry than if we are confined to the early Earth's atmosphere" (Ledeberg, 1992). Current calculations support this eclectic view. As shown in Table 1.1, even if only 10% of the bodies colliding with the primitive Earth were comets, they would still account for the bulk of oceanic waters, and for more than ten times the amount of carbon incorporated into the biosphere (Whipple, 1976; Lazcano-Araujo and Oró, 1981; Chyba, 1987; Chyba et al., 1990; Delsemme, 1991, 1992; Owen et al., 1992).

Thus, comets appear to be the best candidates for bridging cosmic phenomena with the origins of life. In fact, comets may be *too* good. As argued by Miller (1991a, b) and Zhao and Bada (1991), data summarized in Table 1.1 may represent gross overestimates of the cometary influx to the primitive Earth. The idea that comets and meteorites made major contributions of organic material and volatiles to the primitive Earth has been challenged by Miller (1991a, b), who has argued that incoming extraterrestrial organic compounds deposited on the planet would be removed as the hydrosphere underwent periodic $\approx 350°C$ passages through hydrothermal submarine vents every 10^7 years. If the rate of volatile influx is increased to compensate for this destructive process, then we are faced with an overabundance of carbon. This has led Miller (1991a, b) to conclude that the amount of extraterrestrial organic material added by comets and meteorites to the primitive Earth was small compared to terrestrial-based synthesis.

This criticism is shared by Zhao and Bada (1991) who, based on the concentration of the extraterrestrial amino acids α-amino isobutyric acid (AIB) and racemic isovaline associated with the K/T boundary clays (Zhao and Bada, 1989), have estimated the efficiency of impact delivery of extraterrestrial organics to the Earth. Their results are not very encouraging: according to their calculations, the estimated concentration of AIB in the primitive oceans would be ~ 2 nM, which suggests that exogenous delivery was an inefficient process when compared to abiotic synthesis occurring on the Earth (Zhao and Bada, 1991). However, these calculations assume that the prebiotic hydrosphere was as massive as the extant one, and do not take into account the simultaneous arrival of both water and organic compounds in a volatile-rich impactor.

The efficiency of endogenous prebiotic synthesis is strongly dependent on the reducing state of the primitive atmosphere (Miller and Orgel, 1974). In a CH_4-rich primitive atmosphere the abiotic production of organic matter would have occurred more readily (Miller and Orgel, 1974; Stribling and Miller, 1987), and would have been substantially larger than the exogenous cometary contribution (Miller, 1991a, b; Zhao and Bada, 1991). However, under a CO_2-rich atmosphere the amount of accreted exogenous sources of organic compounds would be comparable to the products of Earth-based, endogenous synthesis (Chyba and Sagan, this volume). Since there is no direct evidence of the composition of the primitive atmosphere, it is difficult to choose between these two opposing points of view. Current model photochemical calculations favor short-lived amounts of prebiotic CH_4, but the actual levels of CO_2 in the primitive atmosphere are still speculative (Kasting, 1993).

The issue of the relative contribution of endogenous prebiotic sources of organic matter as compared to the cometary contribution is far from solved, and assigning values may be misleading. "Exogenous delivery" refers to the volatiles and organic compounds acquired during the late accretion period, i.e., during the final stages of planetary formation. On the other hand, "endogenous" refers to organic compounds synthesized in the primitive Earth from outgassed volatile precursors injected into the terrestrial paleoatmosphere from the interior, but with an ultimate extraterrestrial origin. It is both clear that cometary collisions (Oró, 1961) and impact shock-waves (Bar-Nun et al., 1970) were more frequent in the primitive Earth, and that electric discharges and other free-energy sources also played a major role in the endogenous synthesis of organic compounds (Miller and Urey, 1959). The primitive Earth may be aptly described by two lines from the Roman poet Virgil: "at no other time did more thunderbolts fall in a clear sky, nor so often did dread comets blaze"—but additional information on both the thunderbolts and the blazing comets is required, in order to constrain their relative prebiotic significance.

1.6   Comets and Prebiotic Synthesis

Despite claims to the contrary, there is no compelling evidence that life may have appeared in comets, or to assume that the emergence of the biosphere was triggered by some unknown substance present only in comets (Bar-Nun et al., 1981; Chyba and Sagan, 1987; Marcus and Olsen, 1991; Oró et al. 1992a, b). However, since the paleontogical evidence suggests that the late accretion period was coeval with the origin and early evolution of life (Figure 1.1), it is reasonable to ask whether comets may have contributed directly or indirectly to prebiotic synthesis of biochemical

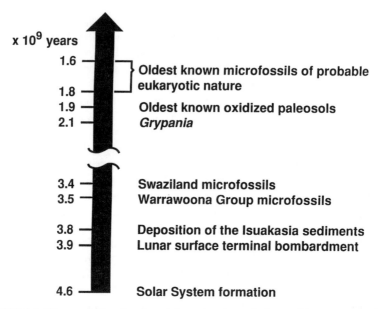

x 10^9 years

1.6 — Oldest known microfossils of probable eukaryotic nature
1.8 —
1.9 — Oldest known oxidized paleosols
2.1 — *Grypania*

3.4 — Swaziland microfossils
3.5 — Warrawoona Group microfossils

3.8 — Deposition of the Isuakasia sediments
3.9 — Lunar surface terminal bombardment

4.6 — Solar System formation

FIGURE 1.1. The possibility that the origin and early evolution of life were coeval with the late accretion period is clearly indicated by the small time space separating the oldest known Archean microfossils and the lunar late terminal heavy bombardment. The 2.1×10^9 years old *Grypania*-like organism is the oldest known fossil eukaryote (Han and Runnegar, 1992), and corresponds to a megascopic algae whose presence suggests the existence of at least 1% of the present level of atmospheric oxygen (modified from Lazcano et al., 1992).

compounds.

It has been argued that in some rare cases the soft landing of intact comets or fragments of comets may have occurred (Clark, 1988; Krueger and Kissel, 1989), and that aerodynamic braking and anisotropic distribution of the shock energy associated with the collision of small comets may have accumulated relative intact cometary compounds in the Earth's surface (Chyba et al., 1990). However, it is likely that in most cases the transformation of the impactor's kinetic energy into heat would lead to very high temperatures shattering the incoming nuclei and vaporizing and destroying all the organic compounds on board (Oró et al., 1980; Chyba et al., 1990; Thomas and Brookshaw, this volume).

Of course, the large array of amino acids, nitrogen bases, and other organic compounds present in carbonaceous chondrites suggest that a certain amount of the organic molecules present in the nonvolatile fraction of the cometary nuclei could survive collision with the Earth. This conclusion is supported by the discovery of isovaline and of α-amino isobutyric acid, two extraterrestrial amino acids associated with the Danish 65 million years old K/T boundary layer clays (Zhao and Bada, 1989), which shows that incoming organic compounds of biochemical significance can survive a violent impact with the Earth, although the possibility that these amino acids are derived from cometary dust swept up by the Earth prior to and after the collision with a comet has also been suggested (Zahnle and Grinspoon, 1990). It is easier to envisage the survival of simpler compounds: the presence of H_2O, OH, NO, N_2, NH, C_2, CN, CO, and other small carbon-bearing species in the surface of the Sun, and of the C_3 and C_5 molecules in hot, circumstellar envelopes (Hinkle et al., 1988; Bernath et al., 1989), shows that these are resilient compounds with strong possibilities of surviving a cometary collision and eventually becoming the starting point for further secondary abiotic synthesis (Oró et al., 1992b).

Indeed, it has been suggested that the collisions would produce highly reducing transient atmospheric environments rich in reactive chemical species. Upon quenching to low temperatures due to the subsequent expansion and cooling of the resulting gas ball, chemical recombination would lead to the synthesis of a great variety of organic compounds (Lazcano et al., 1983). This possibility has been questioned, since theoretical modeling predicts the rapid chemical destruction of atmospheric CH_4, NH_3, and H_2S and other reduced molecules due to the rapid influx of cometary water (Levine et al., 1980), as well as an oxidation of the cometary carbon due to mixing of gases from a CO_2-rich atmosphere with the vapor plumes of impacting comets (Kasting, 1993; Chyba and Sagan, 1992).

However, the lack of existence of these transient atmospheric environments where abiotic synthesis may have taken place is far from proven. As shown by Barak and Bar-Nun (1975), amino acid abiotic synthesis is feasible in shock-wave tube experiments even in the presence of air, and Mukhin et al. (1989) have demonstrated that a number of organic synthesis precursors can be formed using laser-pulse heating of terrestrial rocks and carbonaceous chondrites. These results reinforce the idea that recombination among chemical species formed after the vaporization of the volatile-rich impactors would lead to compounds of biochemical significance.

These experiments are supported by theoretical studies using a thermochemical model, which predicts efficient synthesis of reduced carbon-bearing compounds during shocking of reactants with the cometary components (McKay et al., 1989). Using measurements of physical parameters and chemical abundances of Halley's comet, Oberbeck et al. (1989) and Oberbeck and Aggarwal (1992) have argued that during the expansive, cooling stages of such collisional gaseous environments created by the impact of cometary nuclei larger than 150 m, the excited species would recombine and would give rise to compounds of biological significance. Cometary collisions may have provided the Earth not only with volatiles, but also with an important free-energy source.

1.7 Cometary Collisions and Biological Evolution

The timing of the appearance of life is not documented in the geological record. However, the existance of well-preserved microfossil assemblages in South Africa and in Northwestern Australia indicates that life began more than 3.5×10^9 years ago, and perhaps even much earlier, in order to account for morphological diversity and ecological complexity of the oldest known fossils (Knoll and Barghoorn, 1977; Schopf, 1983, 1994; Schopf and Packer, 1987). The existence of an extended period of evolutionary history prior to the Warrawoona microfossil is supported by molecular phylogenetic analysis suggesting that prior to the divergence of the two major bacterial lineages, simpler prokaryotes endowed with low-specificity biochemical processes existed (Lazcano, 1994a, b). Thus, both paleontological studies and the molecular phylogenies suggest that the late accretion period was coeval with the origin and early evolution of life (Figure 1.1).

Throughout this chapter we have reviewed and summarized the evidence suggesting that comets have contributed to the origins of life. However, it has also been suggested that the intense flux of cometary impactors and other related minor bodies on the primitive Earth may also have created major environmental upheavals that could have delayed or even prevented an earlier appearance of the biosphere. According to this hypothesis, the impact frustration of life was caused by collisions that raised the Earth's surface temperature and evaporated the oceans (Maher and Stevenson, 1988; Oberbeck and Fogelman, 1989a, b; Sleep et al., 1989; Zahnle, this volume). Using different assumptions on the lunar cratering record and the size of impactors, Maher and Stevenson (1988) have argued that life emerged and became extinct several times, until primordial thermophilic forms of life, that appeared 3.8–4.2×10^9 years ago at hydrothermal vents, finally became established and eventually gave rise to all extant forms of life.

The arguments presented by Maher and Stevenson (1988) implied the almost absolute unavoidability of life in a planet like the Earth. This is an attractive idea, but far from proven, and fraught with pitfalls. No evidence of such multiple origins is found among extant organisms. Despite some claims to the contrary (Kandler, 1994), the very basic biochemical and genetic unity of life strongly supports the hypothesis of a monophyletic origin for all known living beings. Was this common ancestor a heat-loving microbe whose preadaptation to the harsh environmental conditions of deep-sea hydrothermal vents allowed its survival in an early Earth warmed up by cometary impacts to boiling temperatures? The hypothesis of totally annihilating collisions is ridden with major problems, which include the lack of detailed calculations of the energy associated with the lunar and terrestrial cratering record, the difficulties involved in computations of the size of the lunar impactors, and the absence of detailed models of the early Earth (Sleep et al., 1989; Chyba, 1990; Zahnle, this volume).

Since the timescale for the origin and early evolution of life is not known, it is quite possible that prior to such annihilating impacts life had already appeared and had extended itself to many different environments, including submarine hot springs and other oceanic niches. The inhabitants of submerged places would have survived the numerous impacts which may have obliterated surface microbes. In the simmering terrestrial environment, only the resilient offsprings of earlier organisms already adapted to live in hot environments such as hydrothermal vents, would have survived. This implies that although the starting point for biological evolution was not a hyperthermophile, nonetheless heat-loving prokaryotes should be located in the stern of universal phylogenetic trees.

The geological record holds no evidence of such a catastrophic sterilizing event, but it has been suggested that the memory of such intense heating is found in the phylogenetic distribution of hyperthermophily among extant organisms. Recent attempts to isolate thermophilic organisms have been extremely successful, and have led to the characterization of prokaryotes that can live and reproduce in abyssal environments with temperatures as high as 105–1100°C (Huber et al., 1989; Gottschal and Prins, 1991; Daniel, 1992). These discoveries have important evolutionary implications. As reviewed by Kandler (1992), until recently the antiquity of hyperthermophily was considered extremely unlikely, since it was believed that temperature-resistant enzymes had a very restricted biological distribution, and that no correlation between antiquity and hyperthermophily had been observed.

There is considerable evidence suggesting that the last common ancestor of the two extant prokaryotic lineages was a hypertherophilic bacteria (Woese, 1987; Daniel, 1992; Stetter, 1994). However, the putative causal connection between hyperthermophily and a primitive environment heated by collisions is weak. That hyperthermophily is a primitive, ancestral character is firmly established, but despite some claims to the contrary (Pace, 1991; Holm, 1992), its phylogenetic distribution does not necessarily imply that the first life forms originated in hot, hydrothermal vents (Miller and Bada, 1988). A hypertherophilic lifestyle requires a variety of adaptations that may well be of secondary origin (Forterre, 1995), a possibility that is strongly supported by the evidence that the last common ancestor of archaebacteria and eubacteria was preceded by a long period of biological evolution characterized by simpler organisms (Lazcano, 1993, 1994a, b).

As the solar system matured, leftovers from the accretion period became less abundant, and the likelihood of giant collisions negligible. Massive collisions may have delayed the origin of life, but their sterilizing effects were not repeated in later times. As shown by the continuity of the fossil record since Warrawoona times, after 3.5×10^9 years ago life has never disappeared from our planet. Once the Precambrian environment became oxidizing (Holland, 1994), a collision would probably lead to an extended destruction of the reduced chemical species present in the cometary nuclei. Howver, the conversion of the impactor's kinetic energy into heat would cause major upheavals in the terrestrial environment, and could be the explanation underlying several major Phanerozoic extinctions (Steel, this volume).

The possibility that impacts of extraterrestrial bodies with the Earth may influence biological evolution and lead to the extinction of different taxa has been discussed in the scientific literature for some time. As summarized by Raup (1986), in 1970 Dewey M. McLaren suggested that the 365 million year old mass extinction that marks the end of the Frasnian stage of the Devonian period could have been caused by a giant meteorite. A few years later, a comparison of tektite ages and geological periods led Urey (1973) to argue that the different major extinctions which have taken place in the past 50 million years could have been caused by cometary impacts, but this bold suggestion went unnoticed. It was not until the discovery of iridium anomalies in different parts of the world when Alvarez et al.

(1980) hypothesized that a large asteroid or comet collided with our planet and caused the extinctions that marked the end of the dinosaurs 65 million years ago. According to this idea, the end of the Cretaceous period was due to the impact of an extraterrestrial object that threw large amounts of dust into the upper atmosphere for several months, therefore initiating a chain reaction that began with a darkened world and subsequently led to the cessation of photosynthesis, subfreezing temperatures, and soon to the disappearance of many taxa. The possibility of massive poisoning due to the introduction of cometary HCN to the Cretaceous hydrosphere has also been suggested (Hsü, 1980), but dismissed as somewhat unlikely (Thierstein, 1980).

By the end of the Cretaceous period several major taxa were in decline, but the evidence supporting the impact origin of the K/T boundary extinctions has continued to accumulate. It now includes the presence of extraterrestrial amino acids of possible cometary origin associated with the 65 million year old clays (Zhao and Bada, 1989; Zahnle and Grinspoon, 1990). The hypothesis that the disappearance of a Cretaceous dinosaur-dominated biota was caused by a massive impact is still prevalent and has gained many adherents. A number of biologists weary of the standard description of gradual evolution based on panselectionist orthodoxy have greeted with enthusiasm the idea that an unexpected event like a collision with an asteroid or a comet could lead to major extinctions (Gould, 1983). Chance may have played a larger role in biological evolution than is usually acknowledged, and comets and meteorites may have been one of its instruments.

However, the hypothesis of extraterrestrial-induced unpredictable die-outs took a somewhat different turn when the statistical analysis of a large record of marine extinctions over the past 250 million years led Raup and Sepkoski (1984) to the suggestion that mass extinctions appear to be regularly spaced, and occur with a 26 million year period. Different astronomical explanations have been developed to account for such periodicity, including the possibility that periodic comet showers were triggered by the Sun's oscillation about the galactic plane (Schwartz and James, 1984; Rampino and Stothers, 1984), or by Nemesis, a hypothetical unseen distant solar companion star (Davis et al., 1984; Alvarez and Muller, 1984; Whitmire and Jackson, 1984; Muller, 1985).

No evidence has been found for Nemesis, and the periodical impact extinction hypothesis has been challenged (Kerr, 1985). The issue is still a controversial one (Raup, 1986, 1988), but the fascination in the role that comets may have played in the origin and evolution of life has not diminished. These traveling chemical fossils of the early solar system come from distant times and faraway places, but they have found a place in contemporary evolutionary theory as the chance agents that may have caused major extinctions. Comets and related minor bodies may have altered the course of subsequent biological events, opening the way for the development of other species, including our own (Gould, 1983). As discussed throughout this book, comets may have been a mixed blessing for the biosphere or, as the Spanish-born Roman philosopher Seneca wrote, "when this rare and strangely shaped fire appears everyone wants to know what it is and, forgetting everything else, seeks information about the strange visitor, uncertain whether it is to be marveled at or feared" (Barrett, 1978).

Acknowledgments: We are indebted to Dr. Gail R. Fleischaker for calling our attention to the work of Chamberlin and Chamberlin (1908) and for providing us with a readable copy of their paper. We thank Professor Stanley L. Miller for several useful conversations on the work of T.C. Chamberlin and on the role of comets in prebiotic evolution, and Drs. C.F. Chyba, P. Forterre and C.R. O'Dell for copies of their work prior to publication. Support for work reported here has been provided by UNAM.IN 105289 (A. Lazcano) and NASA Grant NAGW-2788 (J. Oró)

1.8   References

Alden, W.C. (1929), Thomas Chrowder Chamberlin's contributions to glacial geology. *Jour. Geol.*, **37**, 293–319.

Allen, C.S. (1973), *Astrophysical Quantities* (The Athlone Press, London).

Alvarez, W. and Muller, R.A. (1984), Evidence from crater ages for periodic impacts on the Earth. *Nature*, **308**, 718–720.

Alvarez, L.W., Alvarez, W., Asaro, F., and Michel, H.V. (1980), Extraterrestrial cause for the Cretaceous-Tertiary extinction. *Science*, **208**, 1095–1108.

Anders, E. (1989), Pre-biotic organic matter from comets and asteroids. *Nature*, **342**, 255–257.

Anders, E. and Owen, T. (1977), Mars and Earth: Origin and abundance of volatiles. *Science*, **198**, 453–465.

Aumann, H.H., Gillett, F.C. Beichmann, C.A., de Jong, T., Houck, J., R. Low, F., Neugebauer, G., Walker, R.G. and Wesselius, P. (1984), Discovery of a shell around Alpha Lyrae. *Astrophys. Jour. Lett.*, **278**, L23–L27.

Bailey, M.E., Clube, S.V.M., and Napier, W.M. (1990), *The Origin of Comets* (Pergamon Press, Oxford) p. 452.

Barak, I. and Bar-Nun, A. (1975), The mechanism of amino acid synthesis by high temperature shock waves. *Origins Life* **6**, 483–506.

Bar-Nun, A., Bar-Nun, N., Bauer, S.H., and Sagan C. (1970), Shock synthesis of amino acids in simulated primitive environments. *Science*, **168**, 470–473.

Bar-Nun, A., Lazcano-Araujo, A., and Oró, J. (1981), Could life have originated in cometary nuclei? *Origins Life*, **11**, 387–394.

Barrett, A.A. (1978), *J. Roy. Soc. Can.*, **72**, 81.

Benz, W., Slattery, W.L., and Cameron, A.G.W. (1986), The origin of the Moon and the single impact hypothesis. I. *Icarus*, **66**, 515–535.

Benz, W., Slattery, W.L., and Cameron, A.G.W. (1987), The origin of the Moon and the single impact hypothesis. II. *Icarus*, **71**, 30–45.

Bernath, P.F., Hinkle, K.H., and Keady, J.J. (1989), Detection of C_5 in the circumstellar shell of ICR+10216. *Science*, **244**, 562–564.

Berzelius, J.J. (1834), Über Meteorsteine, 4. Meteorstein von Alais. *Ann. Phys. Chem.*, **33**, 113–123.

Beust, H., Lagrange-Henri, A.M., Vidal-Majdar, A., and Ferlet, R. (1990), The β Pictoris circumstellar disk X. Numerical simulations of infalling evaporating bodies. *Astron. Astrophys.*, **236**, 202–216.

Briggs, R., Ertem, G., Ferris, J.P., Greenberg, J.M., McCain, P.J., Mendoza-Gómez, X.C., and Schutte, W. (1992), Comet Halley as an aggregate of interstellar dust and further evidence for the photochemical formation of organics in the interstellar medium. *Origins Life*, **22**, 287–307.

Brown, H. (1952), Rare gases and the formation of the Earth's atmosphere. In G.H. Kuiper (ed.), *The Atmospheres of the Earth and Planets* (Chicago University Press, Chicago), pp. 258–266.

Brown, J.C. and Hughes, D.W. (1977), Tunguska's comet and non-thermal ^{14}C production in the atmosphere. *Nature*, **268**, 512–514.

Butlerow, A. (1861), Formation sintetique d'une substance sucreé. *Compt. Rend. Acad. Sci.*, **53**, 145–147.

Cameron, A.G.W. (1980), A new table of abundances of the elements in the solar system. In L.A. Ahrens (ed.), *Origin and Distribution of the Elements* (Pergamon Press, New York), pp. 125–143.

Cameron, A.G.W. (1988), Origin of the solar system. *Annu. Rev. Astron. Astrophys.*, **26**, 441–472.

Cameron, A.G.W. and Benz, W. (1989), Possible scenarios resulting from the giant impact. *Proc. Lunar Planet. Sci. Conf. XX*, 715.

Chamberlin, T.C. (1893), The diversity of the glacial period. *Am. Jour. Sci.*, **45**, 171–200.

Chamberlin, T.C. (1894), Proposed genetic classification of Pleistocene glacial formations. *Jour. Geol.*, **2**, 517–538.

Chamberlin, T.C. (1896), Nomenclature of glacial formations. *Jour. Geol.*, **4**, 872–876.

Chamberlin, T.C. (1904), Fundamental problems of geology. Carnegie Institution of Washington Yearbook No. 2: 261–270.

Chamberlin, T.C. (1911), The seeding of planets. *Jour. Geol.* **19**, 175–178.

Chamberlin, T.C. and Chamberlin, R.T. (1908), Early terrestrial conditions that may have favored organic synthesis. *Science*, **28**, 897–910.

Chang, S. (1979), Comets: Cosmic connections with carbonaceous meteorites, interstellar molecules and the origin of life. In M. Neugebauer, D.K. Yeomans, J.C. Brandt and R.W. Hobbs (eds.), *Space Missions to Comets* (NASA CP 2089, Washington, DC), pp. 59–111.

Chyba, C.F. (1987), The cometary contribution to the oceans of the primitive Earth. *Nature*, **330**, 632–635.

Chyba, C.F. (1990), Impact delivery and erosion of planetary oceans in the early inner solar system. *Nature*, **343**, 129–133.

Chyba, C.F. (1991), Terrestrial mantle siderophiles and the lunar impact record. *Icarus*, **92**, 217–233.

Chyba, C.F. (1993), Comets in other planetary systems? *Adv. Space Res.* (in press).

Chyba, C.F. and Sagan, C. (1987), Cometary organics but no evidence for bacteria. *Nature*, **329**, 208.

Chyba, C.F. and Sagan, C. (1992), Endogenous production, exgenous delivery and impact-shock synthesis of organic molecules; an inventory for the origins of life. *Nature*, **355**, 125–132.

Chyba, C.F., Thomas, P.J., Brookshaw, L., and Sagan, C. (1990), Cometary delivery of organic molecules to the early Earth. *Science*, **249**, 366–373.

Chyba, C.F. Thomas, P.J., and Zahnle, K.J. (1993), The 1908 Tunguska explosion: Atmospheric disruption of a stony asteroid. *Nature*, **361**, 40–44.

Clark, B.C. (1988), Primeval procreative comet pond. *Origins Life*, **18**, 209–238.

Daniel, R.M. (1992), Modern life at high temperatures. *Origins Life*, **22**, 33–42.

Davis, M., Hut, P., and Muller, R.A. (1984), Extinction of species by periodic comet showers. *Nature*, **308**, 715–717.

Delsemme, A.H. (1984), The cometary connection with periodic chemistry. *Origins Life*, **14**, 51–60.

Delsemme, A.H. (1991), Nature and history of the organic compounds in comets: An astrophysical view. In R.L. Newburn, M. Neugebauer, and J. Rahe (eds.), *Comets in the Post-Halley Era, Vols. I-II* (Dordrecht, Boston), pp. 377–427.

Delsemme, A.H. (1992), Cometary origin of carbon, nitrogen and water on the Earth. *Origins Life*, **21**, 279–298.

Donn, B.D. (1976), The study of Comets (NASA SP-393, Washington, DC).

Eberhardt, P., Krankowski, D., Schutte, W., Dolder, U, Lämmerzahl, P., Berthelier, J.J., Woweries, J., Stubbermann, U., Hodges, R. R., Hoffman, J.H., and Illiano, J.M. (1987), The CO and NH_2 abundance in comet P/Halley. *Astron. Astrophys.*, **187**, 481–487.

Encrenaz, T. and Knacke, R. (1991), Carbonaceous Compounds in Comets. In R.L. Newburn, M. Neugebauer and J. Rahe (eds), *Comets in the Post-Halley Era, Vols. I-II* (Dordrecht, Boston), pp. 107–137.

Everhart, E. (1969), Close encounters of comets and planets. *Astrophys. Jour.*, **74**, 735–739.

Farley, J. (1977), *The Spontaneous Generation Controversy: From Descartes to Oparin* (John Hopkins University Press, Baltimore).

Fenton, C.L. and Fenton, M.A. (1952), Giants of Geology (Doubleday, New York).

Forterre, P. (1995), Thermoreduction, a hypothesis for the origin of prokaryotes. C.R. Acad. Sci. Paris, **318**, 1–8.

Gottschal, J.C. and Prins, R.A. (1991), Thermophiles: A life at elevated temperatures. *Trends in Ecol. and Evol.*, **6**, 157–161.

Gould, S.J. (1983), *Hen's Teeth and Horse's Toes: Further Reflections in Natural History* (W.W. Norton, New York).

Greenberg, M.J. (1983), Chemical evolution of interstellar dust – a source of prebiotic material? In C. Ponnamperuma (ed.), *Comets and the Origin of Life* (Reidel, Dordrecht), pp. 111–127.

Greenberg, M.J. and Grim, R. (1986), The origin and evolution cometary nuclei and comet Halley results. In B. Battrick, E.J. Rolfe and R. Reinhard (eds.), *20th ESLAB Symposium on the Exploration of Halley's Comet* (ESA Report SP-250), pp. 255–263.

Grieve, R.A.F. and Robertson, P.B. (1979), The terrestrial cratering record I. Current status of observations. *Icarus*, **38**, 212–219.

Grün, E., Bar-Nun, A., Benkhoff, J., Bischoff, A., Düren, H., Hellmann, H., Hesselbarth, P., Hsiung, P., Keller, H.U., Klinger, J., Knölker, J., Kochan, H., Kohl, H., Kölzer, G., Krankowsky, D., Lämmerzahl, P., Mauersberger, K., Neukum, G., Oehler, A., Ratke, L., Roessler, K., Schewm, G., Spohn, G., Stöffler, D. and Thiel, K. (1991), Laboratory simulation of cometary processes: Results from first KOSI experiments. In R.L. Newburn, M. Neugebauer, and J. Rahe (eds.), *Comets in the Post-Halley Era, Vols. I-II* (Dordrecht, Boston), pp. 277–297.

Hän, T.-M. and Runnegar, B. (1992) Megascopic eukaryotic algae from the 2.1-billion-year-old Negaunee Iron-formation, Michigan. *Science*, **257**, 232–235.

Hayashi, C., Nakasawa, K. and Nakasawa, Y. (1985), Formation of the solar system. In D.C. Black and M.S. Matthews (eds.), *Protostars and Planets II* (University of Arizona Press, Tucson), pp. 1100–1153.

Hinkle, K.H., Keady, J.J., and Bernath, P.F. (1988), Detection of C_3 in the interstellar shell of IRC+10216. *Science*, **241**, 1319–1320.

Hobbs, L.M., Vidal-Majdar, A., Ferlet, R., Albert, C.E. and Gry, C. (1985), The gaseous component of the disk around Beta Pictoris. *Astrophys. Jour. Lett.*, **293**, L29–L33.

Holland, H.D. (1994), Early Proterozoic atmospheric change. In S. Bengtson (ed.), *Early Life on Earth. Nobel Symposium No. 84*, Columbia University Press, New York, pp. 237–244.

Hollis, J.M., Snyder, L.E., Suenram, R.D. and Lovas, F.J. (1980), A search for the lowest energy conformer of interstellar glycine. *Astrophys Jour.*, **241**, 1001–1006.

Holm, N.G. (1992), Marine hydrothermal systems and the origin of life. *Origins Life*, **22**. Special issue.

Hong, J. H. and Becker, R. S. (1979), Hydrogen atom initiated chemistry. *J. Mol. Evol.*, **13**, 15–26.

Hoyle, F. and Wickramasinghe, C. (1984), *From Grains to Bacteria* (University College Cardiff Press, Bristol).

Hsü, K.J. (1980), Terrestrial catastrophe caused by cometary impact at the end of Cretaceous. *Nature*, **285**, 201–203.

Huber, R., Kurr, M., Jannasch, H.W. and Stetter, K.O. (1989), A novel group of abyssal methanogenic archaebacteria (*Methanopyrus*) growing at 110° C. *Nature*, **342**, 833–834.

Huebner, W.F. (1987), First polymer in space identified in comet Halley. *Science*, **237**, 628–630.

Hunten, D.M. (1993), Atmospheric evolution of the terrestrial planets. *Science*, **259**, 915–920.

Ibandov, K.I., Rahmonov, A.A. and Bjasso, A.S. (1991). Laboratory simulation of cometary structures. In R.L. Newburn, M. Neugebauer and J. Rahe (eds.), *Comets in the Post-Halley Era, Vols. I-II* (Dordrecht, Boston), pp. 299–311.

Ip, W.H. and Fernández, J.A. (1988), Exchange of condensed matter among the outer and terrestrial protoplanets and the effect on surface impact and atmospheric accretion. *Icarus*, **74**, 47–61.

Irvine, W.M., Leschine, S.N. and Schloerb, F.P. (1980), Thermal history, chemical composition and relationship of comets to the origin of life. *Nature*, **283**, 748–749.

Joss, P.C. (1974), Are stellar surface heavy-elements abundances systematically enhanced? *Astrophys. Jour.*, **191**, 771–774.

Kamminga, H. (1988), Historical perspective: the problem of the origin of life in the context of developments in biology. *Origins Life*, **18**, 1–11.

Kandler, O. (1992), Where next with the archaebacteria? *Biochem. Soc. Symp.* **58**, 195–207.

Kandler, O. (1994), The early diversification of life. In S. Bengston (ed.), *Early Life on Earth. Nobel Symposium No. 84*. (Columbia University Press, New York), pp. 152–160.

Kasting, J.F. (1990), Bolide impacts and the oxidation state of carbon in the Earth's earliest atmosphere. *Origins Life*, **20**, 199–231.

Kasting. J.F. (1993), Earth's earliest atmosphere. *Science*, **259**, 920–926.

Kerr, R.A. (1985), Periodic extinctions and impacts challenged. *Science*, **227**, 1451–1453.

Khare, B.N., Sagan, C. Thompson, W.R., Arakawa, E.T., Suits, F., Callcott, T.A., Williams, M.W., Shrader, S., Ogina, H., Willingham, T.O., and Nagy, B. (1984), The Organic aerosols of Titan. *Adv. Space Res.*, **4**, (12) 59–68.

Kissel, J. and Krueger, F.R. (1987), The organic component in dust from comet Halley as measured by the PUMA mass spectrometer on board Vega 1. *Nature* , **326**, 755–760.

Knoll, A.H. and Barghoorn, E.S. (1977), Archean microfossils showing cell division from the Swaziland system of South Africa. *Science*, **198**, 396–398.

Kondo, Y. and Bruhweiler, F.C. (1985), IUE observations of Beta Pictoris: an IRAS candidate for a proto-planetary system. *Astrophys. Jour. Lett.*, **391**, L1–L5.

Korth, A., Marconi, M.L., Mendis, D.A., Krueger, F.R., Richter, K.A., Lin, R.P., Mitchell, O.L., Andersen, K.A., Carlson, C.W., Réme, H., Savaud, J.A., and d'Uston, C. (1989), Probable detection of organic-dust-borne aromatic $C_3H_3^+$ ions in the coma of comet Halley. *Nature*, **337**, 53–55.

Kresák, L. (1978), The Tunguska object: A fragment of comet Encke? *Bull. Astron. Inst. Czechosl.*, **29**, 129–134.

Krueger, F.R. and Kissel, J. (1989), Biogenesis by cometary origin: Thermodynamical aspects of self-organization. *Origins Life*, **19**, 87–93.

Lagrange, A.M., Ferlet, R., and Vidal-Majdar, A. (1987), The Beta Pictoris circumstellar disk IV. Redshifted UV lines. *Astron. Astrophys.*, **173**, 289–292.

Lagrange-Henri, A.M., Vidal-Majdar, A., and Ferlet, R. (1988), The β Pictoris circumstellar disk VI. Evidence for material falling on to the star. *Astron. Astrophys.*, **190**, 275–282.

Langevin, Y., Kissel, J., Berhaus, J.L., and Chassefiere, E. (1987), First statistical analysis of 5000 mass spectra of cometary grains obtained by PUMA (Vega 1) and PIA (Giotto) impact ionization mass spectrometers in the compressed modes. *Astron. Astrophys.*, **187**, 761–766.

Lazcano, A. (1992a), Origins of life: The historical development of recent theories. In L. Margulis and L. Olendzenski (eds.), *Environmental Evolution: Effects of the Origin and Evolution of Life on Planet Earth* (MIT Press, Cambridge), pp. 57–59.

Lazcano, A. (1992b), *La Chispa de la Vida* (Pangea, México).

Lazcano, A. (1993), The significance of ancient paralogous genes in the study of the early stages of microbial evolution. In R. Guerrero and C. Pedrós-Aliós (eds.). *Proceedings of the 6th International Symposium of Microbial Ecology* (Soc. Catalana de Biologìa, Barcelona), pp. 559–562.

Lazcano, A. (1994a), The transition from non-living to living. In S. Bengtson (ed.), *Early Life on Earth*. Nobel Symposium No. 84 (Columbia University Press, New York), pp. 60–69.

Lazcano, A. (1994b), The RNA world, its predecessors and descendants. In S. Bengtson (ed.), *Early Life on Earth. Nobel Symposium No. 84* (Columbia University Press, New York), pp. 70–80.

Lazcano, A., Oró, J., and Miller, S.L. (1983), Primitive Earth environments: Organic synthesis and the origin and early evolution of life. *Precambrian Res.*, **20**, 259–282.

Lazcano, A., Fox, G.E., and Oró, J. (1992), Life before DNA: the origin and evolution of Early Archean cells. In R.P. Mortlock (ed.) *The Evolution of Metabolic Function* (CRC Press, Boca Raton), pp. 237–295.

Lazcano-Araujo, A. and Oró, J. (1981), Cometary material and the origins of life on Earth. In C. Ponnamperuma (ed.) *Comets and the Origins of Life* (Reidel, Dordrecht), pp. 191–225.

Lederberg, J. (1992), Foreword to L. Margulis *Symbiosis in Cell Evolution: Microbial communities in the Archean and Proterozoic Eons* (Freeman, New York), pp. xv–xvi.

Lerner, N.R., Peterson, E., and Chang, S. (1991), Meteoritic amino acids from cometary/interstellar precursors. *Comets and the Origins and Evolution of Life. Abstracts of a Meeting in Eau Claire, Wisconsin*, September 30–October 2, 1991, p. 19.

Levine, J.S., Augustsson, T.R., Boughner, R.E., Natajaran, M., and Sacks, L.J. (1980), Comets and the photochemistry of the paleoatmosphere. In C. Ponnamperuma (ed.) *Comets and the Origin of Life* (Reidel, Dordrecht), pp. 161–190.

Lewis, J.S. (1974), Volatile element influx on Venus from cometary impacts. *Earth Planet. Sci. Lett.*, **22**, 239–244.

Löb, W. (1913), Über das Verhalten des Formamids unter der Wirkung der stillen Entladung. Ein Beilrag zur Frage der Stickstoff-Assimilation. *Berichte der Deutschen Chem. Gessellschaft*, **46**, 684–697.

MacMillan, W.D. (1929), The field of cosmogony. *Jour. Geol.* **37**, 341–356.

Maher, K.A. and Stevenson, D.J. (1988), Impact frustration of the origin of life. *Nature*, **331**, 612–614.

Marcus, J.N. and Olsen, M.A. (1991), Biological implications of organic compounds in comets. In R.L. Newburn, M. Neugebauer, and J. Rahe (eds.), *Comets in the Post-Halley Era, Vols. I–II* (Dordrecht, Boston), pp. 439–462.

Matthews, C.N. and Ludicky, R. (1986), The dark nucleus of comet Halley: Hydrogen cyanide polymers. In B. Battrick, E.J. Rolfe, and R. Reinhard (eds), *20th ESLAB Symposium on the Exploration of Halley's Comet* (ESA Report SP-250), pp. 273–277.

McKay, C.P., Boruki, W.R., Kujiro, D.R., and Church, F. (1989), Shock production of organics during cometary impacts. *Lunar Planet. Sci. Conf. XX*, 671–672.

McKinnon, W.B. (1989), Impacts giveth and impacts taketh away. *Nature*, **338**, 465–466.

Melosh, J. and Vickery, A. (1989), Impact erosion of the primordial Martian atmosphere. *Nature*, **338**, 487–489.

Miller, S.L. (1957), The mechanism of synthesis of amino acids by electric discharges. *Biochem. Biophys. Acta.*, **23**, 480–487.

Miller, S.L. (1974), The first laboratory synthesis of organic compounds under primitive Earth conditions. In J. Neyman (ed.), *The Heritage of Copernicus: Theories "Pleasing to the Mind"* (MIT Press, Cambridge), pp. 228–242.

Miller, S.L. (1991a), The relative importance of prebiotic synthesis on the Earth and input from comets and meteorites. In R.A. Wharton, D.T. Andersen, Sara E. Bzik, and J.D. Rummel (eds.). *Fourth Symposium on Chemical Evolution and the Origin and Evolution of Life* NASA Conference Publication No. 3129 (Washington DC), p. 105.

Miller, S.L. (1991b), Comets and meteorites were not a significant source of organic compounds on the primitive Earth. *Comets and the Origins and Evolution of Life. Abstracts of a Meeting in Eau Claire, Wisconsin*, September 30–October 2, 1991, pp. 22–23.

Miller, S.L. and Bada, J.L. (1988), Submarine hot springs and the origin of life. *Nature*, **334**, 609–611.

Miller, S.L. and Orgel, L.E. (1974), *The Origins of Life on Earth* (Prentice Hall, Englewood Cliffs, NJ).

Miller, S.L. and Urey, H.C. (1959), Organic compound synthesis on the primitive Earth. *Science*, **130**, 245–252.

Mitchell, D.L., Lin, R.P, Anderson, K.A., Carlson, C.W., Curtis, D.W., Korth, A., Réme, H., Sauvard, J.A., d'Uston, C., and Mendis, D.A. (1987), Evidence for chain molecules enriched in carbon, hydrogen and oxygen in comet Halley. *Science*, **237**, 626–628.

Moulton, F.R. and Chamberlin, T.C. (1900), Certain attempts to test the nebular hypothesis. *Science* **11**, 311–312.

Mukhin, L.M., Gerasimov, M.V., and Safonova, E.N. (1989), Origin of precursors of organic molecules during evaporation of meteorites and rocks. *Adv. Space Res.*, **9**, 95–97.

Muller, R.A. (1985), Evidence for a solar companion star. In M.D. Papagiannis (ed), *The Search for Extraterrestrial Life: Recent Developments* (Reidel, Dordrecht), pp. 233–243.

Navarro-González, R., Castillio-Rojas, S., and Negrón-Mendoza, A. (1991), Experimental and computational study of the radiation-induced decomposition of formaldehyde.

Implications to cometary nuclei. *Origins Life*, **21**, 39–49.

Negrón-Mendoza, A., Chacón, E., Navarro-González, R., Draganic, Z.D., and Draganic, I.G. (1992), Radiation-induced syntheses in cometary simulated models. *Adv. Space Res.* **12**: 63–66.

Oberbeck, V.R. and Aggarwal, H. (1992), Comet impacts and chemical evolution of the bombarded Earth. *Origins Life*, **21**, 317–338.

Oberbeck, V.R. and Fogelman, G. (1989a), Impacts and the origin of life. *Nature*, **339**, 434.

Oberbeck, V.R. and Fogelman, G. (1989b), Estimates of the maximum time require to originate life. *Origins Life*, **19**, 549–560.

Oberbeck, V.R., McKay, C.P., Scattergood, T.W., Carle, G.C., and Valentin, J.R. (1989), The role of cometary particle coalescence in chemical evolution. *Origins Life*, **19**, 35–55.

O'Dell, C.R., Wen, Z., and Hu, X. (1993), Discovery of new objects in the Orion Nebula on HST images: shocks, compact sources and protoplanetary disks. *Astrophys. Jour.* (in press).

Oparin, A.I. (1924), Proiskhozhdenie Zhizni (Moskovskii Rabochii, Moscow). Translated and published as an Appendix in J.D. Bernal (1967). The Orgin of Life (Weidenfeld and Nicolson, London).

Oparin, A.I. (1938), The Origin of Life (Macmillan, New York).

Oró, J. (1960), Synthesis of adenine from ammonium cyanide. *Biochem. Biophys. Res. Comm.*, **2**, 407–412.

Oró, J. (1961), Comets and the formation of biochemical compounds on the primitive Earth. *Nature*, **190**, 389–390.

Oró, J. (1963), Synthesis of organic compounds by high-energy electrons. *Nature*, **197**, 971–974.

Oró, J. and Mills, T. (1989), Chemical evolution of primitive solar system bodies. *Adv. Space Res.*, **9**, 105–120.

Oró, J., Kimball, A., Fritz, R., and Master, F. (1959), Amino acid synthesis from formaldehyde and hydroxylamine. *Arch. Biochem. Biophys.*, **85**, 115–130.

Oró, J., Holzer, G., and Lazcano-Araujo, A. (1980), The contribution of cometary volatiles to the primitive Earth. *Life Sciences and Space Research XVIII*, pp. 67–82.

Oró, J., Miller, S.L., and Lazcano, A. (1990), The origin and early evolution of life on Earth. *Annu. Rev. Earth Planet. Sci.*, **18**, 317–356.

Oró, J., Mills, T., and Lazcano, A. (1992a), The cometary contribution to prebiotic chemistry. *Adv. Space Res.*, **12**, 33–41.

Oró, J., Mills, T., and Lazcano, A. (1992b), Comets and the formation of biochemical compounds–a review. *Origins Life*, **21** 267–277.

Oró, J., Mills, T., and Lazcano, A. (1995), Comets and life in the universe. *Adv. Space Res.*, **15**, 81–90.

Owen, T., Bar-Nun, A., and Kleinfeld, I. (1992), Possible cometary origin of heavy noble gases in the atmospheres of Venus, Earth and Mars. *Nature*, **358**, 43–46.

Pace, N.R. (1991), Origin of life—Facing up to the physical environment. *Cell*, **65**, 531–533.

Pollack, J.P. and Yung, Y.L. (1980), Origin and evolution of planetary atmospheres. *Ann. Rev. Earth Planet. Sci.*, **8**, 425–487.

Rampino, M.R. and Stothers, R.B. (1984), Terrestrial mass extinctions, cometary impacts and the Sun's motion perpendicular to the galactic plane. *Nature*, **308**, 709–712.

Raup, D.M. (1986), *The Nemesis Affair: A Story of the Death of the Dinosaurs and the Ways of Science* (W.W. Norton, New York).

Raup, D.M. (1988), Extinction in the geological past. In D.E. Osterbrock and P.H. Raven (eds.), *Origins and Extinctions* (Yale University Press, New Haven), pp. 109–119.

Raup, D.M. and Sepkoski, J. Jr. (1984), Periodicity of extinctions in the geological past. *Proc. Natl. Acad. Sci. USA*, **81**, 801–805.

Sagan, C., Thompson, W.R., and Khare B.N. (1992), A laboratory for prebiological organic chemistry. *Accounts of Chemical Research* **25**, 286–292.

Schopf, W.J. ed (1983), The Earth's Earliest Biosphere: its origin and evolution (Princeton University Press, Princeton, NJ).

Schopf, W.J. (1994), The oldest known records of life: Early Archean stromatolites, micro-fossils, and organic matter. In S Bengtson (ed.), Early Life on Earth. Nobel Symposium No. 84. Columbia University Press, New York, pp. 193–206.

Schopf, W.J. and Packer, B.M. (1987), Early Archean (3.3 billion to 3.5 billion years-old) microfossils: New evidence of ancient microbes. *Science*, **237**, 70–73.

Schutte, W.A., Allamandola, L.J., and Sandford, S.A. (1992), Laboratory simulation of the photoprocessing and warm-up of cometary and pre-cometary ices: production and analysis of complex organic molecules. *Adv. Space Res.*, **12**, 47–51.

Schwartz, R.D. and James, P.B. (1984), Periodic mass extinctions and the Sun's oscillation about the galactic plane. *Nature*, **308**, 712–713.

Sill, G.T. and Wilkening, L.L. (1978), Ice clathrate as a possible source of the atmospheres of the terrestrial planets. *Icarus*, **33**, 13–22.

Sleep, N.H., Zanhle, K.J. Kasting, J.F., and Morowitz, H.J. (1989), Annihilation of ecosystems by large asteroid impacts on the early Earth. *Nature*, **342**, 139–142.

Slettebak, A. (1975), Some interesting bright southern stars of early type. *Astrophys. Jour.*, **197**, 137–138.

Smith, B.A. and Terrile, R.J. (1984), A circumstellar disk around β Pictoris. *Science*, **226**, 1421–1424.

Stetter, K.O. (1994), The lesson of Archaebacteria. In S. Bengtson (ed.), Early Life on Earth. Nobel Symposium No. 84. Columbia University Press. New York, pp. 143–151.

Strazzulla, G. and Johnson, R.E. (1991), Irradiation effects on comets and cometary debris. In R.L. Newburn, M. Neugebauer, and J. Rahe (eds.), Comets in the Post-Halley Era, Vols. I–II (Dordrecht, Boston), 243–275.

Stribling, R. and Miller, S.L. (1987), Energy yields for hydrogen cyanide and formaldehyde synthesis: The HCN and amino acid concentrations in the primitive oceans. *Origins Life*, **17**, 261–273.

Strom, K., Strom, S.E., Edwards, S., Cabrit, S., and Skrutskie, M.F. (1989), Circumstellar material associated with stellar-type pre-main sequence stars: a possible constraint on the timescale for planet building. *Astron. J.*, **97**, 1451–1470.

Suess, H. and Urey, H.C. (1956), Abundances of the elements. *Rev. Mod. Phys.*, **28**, 53–62.

Theirstein, H.R. (1980), Cretaceous oceanic catastrophism. *Paleobiology*, **6**, 244–247.

Thomas, P.J. (ed.) (1992), *Comets and the Origin and Evolution of Life. Origins Life*, **21**. Special issue.

Urey, H.C. (1957), The origin of tektites. *Nature*, **179**, 556–557.

Urey, H.C. (1973), Cometary collisions and geological periods. *Nature*, **242**, 32–33.

Vidal-Majdar, A., Hobbs, L.M., Ferlet, R., Gry, C., and Albert, C.E. (1986), The circumstellar gas cloud around Beta Pictoris. II. *Astron. Astrophys.*, **167**, 325–332.

von Helmholtz, H. (1871), The Origin of the Planetary System. In *Selected writings of Hermann von Helmholtz* (Wesleyan University Press, 1971, p. 284). Quotation and reference are from J. Farley (1977). *The Spontaneous Generation Controversy: From Descartes to Oparin* (Johns Hopkins University Press, Baltimore), p. 142.

Walker, J.C.G. (1986), Impact erosion of planetary atmospheres. *Icarus*, **68**, 87–89.

Wetherill, G.W. (1975), Late heavy bombardment of the moon and terrestrial planets. In

Proceedings of the 6th Lunar Science Conference (Lunar and Planetary Institute, Houston), pp. 1539–1561.

Wetherill, G.W. (1990), Formation of the Earth. *Annu. Rev. Earth Planet. Sci.*, **18**, 205–256.

Whipple, F.L. (1976), A speculation about comets and the Earth. *Mem. Soc. Royale Sci. Liege*, **9**, 101–111.

Whitmire, D.P. and Jackson, A.A. (1984), Are periodic mass extinctions driven by a distant solar companion? *Nature*, **308**, 713–715.

Woese, C.R. (1987), Bacterial evolution. *Microbiol. Rev.*, **51**, 221–271.

Wöhler, M.F. (1858), Über die Bestandteile des Meteorsteines von Kaba in Ungarn. Sitzber. *Akad. Wiss. Wien, Math-Naturwiss. Kl.*, **33**, 205–209.

Wöhler, M.F. and Hörnes, M. (1859), Die organische Substanz im Meteorsteine von Kaba. *Sitzber. Akad. Wiss. Wein, Math- Naturwiss. Kl.*, **34**, 7–8.

Zahnle, K. and Dones, L. (1992), Impact origin of Titan's atmosphere in Proceedings Symposium on Titan, Toulouse, France. (ESA SP-338), 14–25.

Zahnle, K. and Grinspoon, D. (1990), Comet dust as a source of amino acids at the Cretaceous/Tertiary boundary. *Nature*, **348**, 157–159.

Zhao, M. and Bada, J.L. (1989), Extraterrestrial amino acids in Cretaceous/Tertiary boundary sediments at Steuns Klint, Denmark. *Nature*, **339**, 463–465.

Zhao, M. and Bada, J.L. (1991), Limitations on impact delivery of organics to the Earth based on extraterrestrial amino acids in K/T boundary sediments. Comets and the Origins and Evolution of Life. Abstracts of a Meeting in Eau Claire, Wisconsin, September 30–October 2, 1991, 41.

2
The Origin of the Atmosphere and of the Oceans

A. Delsemme

ABSTRACT The atmosphere of the Earth, its oceans as well as most carbon contained in its carbonates and in organic matter, seem to have been brought by a large bombardment of comets that lasted almost one billion years before diminishing drastically to its present-day value.

2.1 Introduction

The origin of our atmosphere and oceans has never been properly elucidated. The reason is that there is no geological record for the first billion years of the Earth's evolution. "During that lost interval, all the volatiles of the Earth could have been derived and recycled many times while the evidence for the exact mechanism of supply was obliterated completely" (Turekian, 1972).

2.2 Hypothesis of the Volcanic Origin

The view that air and water were outgassed by volcanic activity has been the prevailing paradigm for about a century. This view was first proposed by the Swedish geologist G. Hogbom in 1894. It was substantiated more recently, in particular by Rubey's (1951–1955) extensive work. The chemical composition of volcanic gases is highly variable; however, a crude average can be established from Anderson's (1975) comprehensive review. It yields some 60% water steam, 30% CO_2 while the residual 10% are made by a mixture of SO_2, H_2, CO, and N_2. The fact that volcanic gases do not contain any substantial amount of oxygen is not inconsistent with the composition of an early atmosphere, since it is known that oxygen is the byproduct of the photosynthetic activity of vegetation and, earlier, of blue–green algae (Berkner and Marshall, 1965).

However, it has also become clear that a large fraction of the volcanic gases is not juvenile, but is constantly recycled by known geologic processes. Water comes mostly from the present oceans, CO_2 from the sedimentary carbonates, and sulfur from dissolved sulfates, that are transported by plate tectonics into subduction zones. The subduction zones appear where plates collide, which heats carbonates and water enough to produce new volcanic activity. The recycled fraction seems to be extremely large, but if there is a contribution of juvenile gases, they must have been stored inside the Earth for a long time. Are they the end of a phenomenon that has produced the bulk of the atmosphere and of the oceans very early, during that "lost interval" of the first billion years? There are no very good ways to know. One of the least ambiguous ways to study the problem is to rely on noble gases. They are chemically inert, hence their fractionation must be interpreted by physical processes only.

To do this, it is first necessary to distinguish between primordial noble gases (those dating back from the accretion of the Earth), and radiogenic noble gases (those steadily formed by the decay of radioactive elements, like ^{4}He from U or Th, or ^{40}Ar from ^{40}K, for instance). A less important but not negligible source is the spallation reaction coming from cosmic rays. Primordial ^{3}He, as well as radiogenic ^{4}He, ^{40}Ar, and ^{129}Xe, have been unambiguously detected in mantle-derived material (Eberhart, 1981) but no unambiguous answer on the degassing of the Earth has been obtained. It is interesting to note that submarine basalts indicate a *uniform* ^{3}He/^{4}He ratio of 1.4×10^{-5} (ten times smaller than the primordial ration of 1.4×10^{-4}, Reynolds et al., 1975); the uniformity of the ratio suggests a well-mixed region in the upper mantle, in spite of the fact that the helium abundance is fifty times as large in the Atlantic as in the Pacific basalts. The presence of CO_2 and H_2O in the upper mantle is also indicated by stable isotope data (Eberhart, 1981).

The lack of conclusive answers on the early degassing of the Earth clearly comes from the large number of factors controlling the noble gas abundance patterns in deep-sea basalts (Dymond and Hogan, 1978): in particular, the exchange with atmospheric gases, the diffusion in the magma source during transit through the crust, the partition coefficients during partial melting and crystallization, bubble formation in erupting lava, hydrated phases that absorb large amounts of noble gases, etc. Attempts to deduce the outgassing history from noble gas isotopes are discussed in Alexander and Ozima (1978). The key questions left unanswered are summarized by Eberhardt (1981) as being:

(a) To what extent is the Earth degassed?
(b) Was the degassing uniform with time?
(c) Did an early catastrophic degassing occur?
(d) Is the degassing within the mantle uniform?

Standing in contrast, the elemental and isotopic patterns of the noble gases left in the atmosphere itself are much easier to understand, and their interpretation comes as a surprise: there is no trace left of a *primary* atmosphere.

2.2.1   The Missing Primary Atmosphere

During its accretion, as soon as the Earth was large enough to develop a sufficient gravity, it could have captured a sizeable atmosphere from the gases existing in the primeval solar nebula. These gases are the same as those now present in the Sun, therefore the composition of this so-called "primary" atmosphere has been dubbed "solar." Signer and Suess (1963) have shown that the noble gases came from several reservoirs with different patterns in elemental, as well as in isotopic, abundances that have never been well mixed and can therefore be easily distinguished. The first reservoir of noble gases is clearly "solar" because it shows the undisturbed solar abundance ratios.

They dubbed the second reservoir "planetary" not because its primeval source had been identified, but only because the atmosphere of the terrestrial planets showed a more or less similar depletion pattern for their noble gases, like the contents of the carbonaceous chondrites. As Pepin (1989) mentions, the terminology was unfortunate since it led too many people to believe that the source of the "planetary" component could be identified with the carbonaceous chondrites.

In the "planetary" component, the pattern of progressively smaller depletion with increasing atomic mass for ^{20}Ne, ^{36}Ar, and ^{84}Kr suggests a mass-dependent fractionation of "solar" gases, whereas the very heavy ^{130}Xe remains about "solar" in the atmospheres of Venus, the Earth, and Mars. Standing in contrast, in carbonaceous chondrites, ^{130}Xe reaches ten times the "solar" value, which suggests that chondrites are not the unique source of the "planetary" component.

However, the best proof that the primeval reservoir of the "planetary" component is neither the solar component properly fractionated, nor the carbonaceous component only, is found in the characteristic patterns of the isotopic ratios of the six isotopes of krypton of masses 78, 80, 82, 83, 84, and 86 as well as those of the seven isotopes of xenon with masses 124, 126, 128, 130, 132, 134, and 136. It is clear that here, mechanisms capable of fractionating the isotopes have been at work, but the fractionation of xenon was not coupled to that of krypton (Pepin, 1989), and the possible mechanisms are not unambiguously understood (Pepin, 1991).

There is however one single inescapable conclusion. The "primary" atmosphere, with its solar composition, is completely at variance with the elemental and isotopic signatures of the noble gases in the present atmospheres of the terrestrial planets. Either this "primary" atmosphere was lost very early, or it was never captured to begin with.

Finally, since geophysics has never been able to explain in detail the origin of the "planetary" component in the noble gases of our atmosphere, the traditional explanation of the volcanic origin of the volatiles is not based on any empirical observations. The hypothesis is based only on a blind extrapolation of present mechanisms, back through the one billion year "lost interval" that hides the formation and very early history of the Earth. In order to understand what happened during that "lost interval," we must leave geophysics and look at the astronomical evidence about the origin of the Solar System.

2.2.2   The Origin of the Solar System

Laplace (1796) is the first who tried to solve the problem of the origin of the Solar System by using observational facts. In the last pages of his "Exposition du Systéme de Monde," he lists first all the quasi-regularities of the Solar System, then proposes that originally, a "fluid of immense extent enveloped the Sun as an atmosphere," whose primeval rotation enforced all those quasi-regularities. For a long time however, nineteenth century astronomers believed that Laplace's model was wrong, because the Sun turned much too slowly to result from the central agglomeration of this "Solar Nebula" (dubbed this way by a false analogy with what we now call spiral galaxies).

The problem of the present angular momentum of the Sun disappeared when the solar wind was discovered. Its spiral motion away from the Sun produces a magnetic brake that slows the solar rotation continuously. The apparent contradiction with a very fast primeval rotation had disappeared and Laplace's hypothesis came back in favor some 50 years ago. However, it still was a speculative scenario difficult to analyze numerically. For instance, von Weiszächer (1944) proposed to arrange the gas eddies in epicycles in order to explain the distances of the planets (Bode's law). Soon, his analysis was shown to be in quantitive disagreement with fluid mechanics, but it attracted the attention on a possible mechanism that had been neglected so far: viscous turbulence can dissipate energy and redistribute momentum in the primitive nebula.

To make a long story short, the basic theory of the viscous accretion disk has been given by Lynden-Bell and Pringle (1974). Numerical models have been developed, in particular by Cameron (1985), Lin and Papaloisou (1985), Wood and Morfill (1988), Morfill (1988), and Morfill and Wood (1989). There are still some difficulties with the model. For instance, the origin of the rather high viscosity needed to dissipate the angular momentum has not been properly clarified, although its most likely source is a strong convection driven by gas turbulence. Gas turbulence in the disk is itself maintained by the gravitational collapse of the interstellar cloud that feeds the disk. However, there are other possible mechanisms. A gravitational torque (Larson, 1984) could also dissipate angular momentum. Such a gravitational torque could be introduced by a spiral structure in the disk, or in general by mass distributions that are not axisymmetrical. If the gas is ionized enough, magnetic forces could also play an important role, as the action of the solar wind on the present Sun demonstrates. After all, the solar wind could still be the subsiding remnant of the mechanism involved in the shedding of the disk five billion years ago.

2.3   Existence of Accretion Disks

Observational evidence has established recently the ubiquity of large disks of dust around very young stars. The crucial observations came first from the infrared (typically from 10 to 100 micrometers), a large number of very young stars were much brighter than expected from their visual magnitudes (Rowan-Robinson, 1985). This infrared excess was interpreted as coming from the radiation of a large circumstellar disk of cold dust, and most of these stars were shown to be very young T-Tauri stars (Bertout, 1989). Some, like FU Orionis, were still probably accreting mass (Hartmann and Kenyon, 1985). Finally, Optical pictures in the visual, obtained by hiding the central star in the field of the telescope, have detected and resolved dusty disks of sizes 500–1000 Astronomical Units (AU) among the nearest candidates, like in β Pictoris (Smith and Terrile, 1984).

The existence of numerous accretion disks has therefore been substantiated recently and has become the accepted explanation on the way Nature succeeds in making single stars: namely, by shedding the angular momentum in excess to the expanding margin of the disk during the buildup of the central mass. The first consequence of the explanation is that many single stars are likely to make a planetary system by the evolution of their accretion disk. Of course, the accretion disks are only visible around very young stars because at that time, dust is fine enough to radiate a large amount of infrared in spite of its small total mass. As soon as it coalesces and accretes into larger objects (like planetesimals or eventually planets), their total cross-section becomes much smaller and they become much more difficult to detect from afar.

Accretion disks have also become popular in a different context, namely, to explain the mechanism which produces the energy released in X-ray stars and quasars. The gravitational potential well which surrounds very dense compact objects is deep enough for a particle spiraling inward to transform a reasonable fraction of its rest mass to radiation. The deeper the gravitational well, the hotter the accretion disk, but this is not the type of interest here; we have mentioned them now not so much to avoid any confusion later, as to emphasize the generality of the accretion disk mechanism to extract energy and angular momentum from a central spinning mass.

For our concern, observational evidence revealing the ubiquity of large disks of dust around very young stars has given a sudden respectability to the theory that describes Laplace's "Solar Nebula" by a viscous accretion disk, even if the cause of the viscosity has not been properly elucidated.

2.4   Numerical Models for a Protosolar Accretion Disk

We will limit our discussion to the viscous accretion disks because their study has been developed more than that of other possible mechanisms. Following rather general assumptions that we will not review here, in particular on the viscosity behavior of the disk, there remains two parameters, the collapse time and the viscosity coefficient, that can be combined into one.

Larson (1984) has established the order of magnitude of the collapse time, by the following considerations. The center of the dense interstellar nodule collapses fast and its outer parts falls down more slowly; in order to accrete one solar mass, the outskirts of the nodule will reach the accretion disk some 10^5 years later. The actual rate will vary because of density fluctuations, but it is useful to know that the average rate is about 10^{-5} solar masses per year. Cameron (1985) proposes accretion rates a few times larger to explain the high luminosities observed in very young (T-Tauri) stars.

The viscosity coefficient is the second parameter. It sets the dissipation rate of the inner angular momentum, hence the lifetime of the disk evolution. By combining the uncertain rate of collapse and the uncertain coefficient of viscosity in one single variable, this variable can be adjusted to empirical data. Such an adjustment changes the temperature everywhere in the disk, but not its radial gradient. It is fortunate that the temperature gradient in the mid-plane of the disk, as a consequence of the virial theorem, reflects the shape of the gravitational potential well made by the protoSuns's mass.

Morfill (1988) can be used as an example of a rather evolved model. Its midplane temperature varies with $r^{-0.9}$ (r being the heliocentric distance) except at temperature plateaus due to the latent heat of condensation of the two major constituents, namely silicates within 0.4 AU, and water from 4 to 8 AU. Figure 2.1 shows Morfill's model. The two unknown parameters have been adjusted to the aggregation temperatures for the different terrestrial planets and two satellites of the giant planets, derived by Lewis (1974) from their empirical mean densities.

The success of the model adjustment (solid line) comes from the approximate temperature gradient in r^{-1} deduced by Lewis from the empirical condensation temperatures of the planets from a gas of solar composition. The two dotted lines represent the model for an accretion rate $\dot{M}$ larger (or smaller) by a factor of ten. A change in the viscosity coefficient would produce the same type of shift.

Lewis's (1972a, b) model assumed not only thermochemical equilibrium between gas and dust in the solar nebula, but also that planets accreted from dust captured only at the exact heliocentric distance of the planet. In spite of the gross oversimplification, the model predicted rather well the uncompressed densities of most of the planets. Lewis' model could not accurately predict the range of condensation temperatures for each planet, because it also depends on the location of the selected abiabat, hence on the model used for the solar nebula. However, its major virtue was to demonstrate that, at the time when dust sedimented from gas, the temperature gradient predicted by theory was empirically confirmed (Fig. 2.1).

2.5   The Chondrites as Clues on Planetary Formation

Those primitive meteorites call chondrites provide empirical evidence on the way planetary bodies were formed. The chondrites come from primitive bodies because, with the exception of a few very volatile elements, most of their elements have remained accurately in the same abundance ratios as in the Sun. This establishes not only that they derived from the same primeval reservoir as the Sun's, but also that they have never been through any process of differentiation, such as those that have separated the cores from the mantles of the different planets.

The chondritic meteorites come from the asteroid belt, that is, roughly from 2 AU to 4 AU. This was established from accurate triangulation of three orbits of chondrite meteorites observed as meteors during their entry in the atmosphere and recovered on the ground later. The results were entirely confirmed by the orbits of about 30 bright meteors identified as chondritic but unrecovered later (Wetherill and Chapman, 1988). Chondrites are assumed to be the fragments of asteroidal collisions. Their parent bodies had a radius of the order of 100 km.

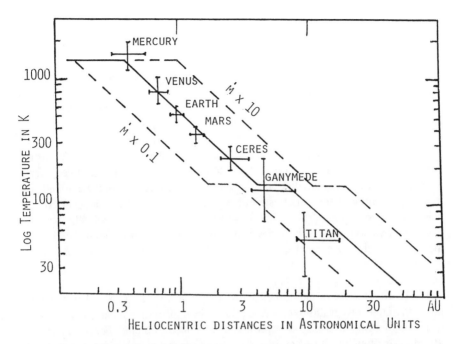

FIGURE 2.1. Mid-plane temperature of the accretion disk, as a function of the distance to the Sun, at the time of dust sedimentation. The solid line is the adjustment of the disk theoretical model (Morfill, 1988). For a given viscosity coefficient, the two dotted lines correspond to an accretion rate $\dot{M}$ ten times smaller or ten times larger. The crosses are Lewis' (1974) aggregation temperatures of planets and satellites derived from their densities. The crosses are not error bars: they show the width of the zones from which planetesimals were presumably collected, and their corresponding temperature ranges. The two horizontal pleateaus in the profile are produced by the condensation latent heat of the silicates (near 0.4 AU) and of water ice (from 4 AU to 7 AU). The success of the adjustment comes from the fact that the empirical gradient derived by Lewis from the planet's densities is also predicted by theory. The actual temperatures of condensation for the planets remain however somewhat uncertain because they depend on the pressure in the nebula, hence on the adiabat chosen by Lewis.

This size is implied by their concordant radiogenic ages of four billion years or more, implying a rather fast cooling. This small size explains why they had no differentiation induced by gravitation, since their gravity was never larger than 10^{-2} g.

Chondrites are stony meteorites (made mostly of silicates) classified as carbonaceous, ordinary and enstatite chondrites according to their diminishing degree of oxidation. The enstatite chondrites are completely reduced, and the carbonaceous chondrites are completely (CI, CM) or almost completely (CO, CV) oxidized. The most oxidized carbonaceous chondrites are those that contain the most volatile elements and, in particular, very large amounts of organic compounds (there is typically 6% carbon in the CI type). The chondrite classes seem to sample different regions of the accretion disk across the asteroid belt. Although the evidence is indirect, the infrared spectra of asteroids seem to imply that the dark C asteroids have carbonaceous chondritic surfaces, whereas the five light asteroids look more like ordinary chondrites. See, however, the controversy about the identification of the five asteroids in Wetherill and Chapman (1988). Another important clue comes from the fact that the C asteroids begin to outnumber the five asteroids at distances beyond 2.6 AU (Morrison, 1977).

The silicate matrix of chondrites shows that it was made by a moderate compression of fine dust grains of different origins. In spite of their close contact, these often submicrometer-sized grains are chemically unequilibrated. For instance, oxidized grains touch reduced grains, some have been altered by liquid water and some have not, some refractory grains are in close contact with volatile grains. The matrix also imprisons larger objects, such as millimeter-sized chondrules or CAI (calcium–aluminum inclusions). The chondrules show signs of transient (and often partial) melting (1500–1600 K). The CAI are refractory grains probably made at temperatures higher than 1600 K.

This heterogeneous composition seems to imply a process entirely comparable to a sedimentation. In our rivers, when water turbulence subsides, sand sediments to the bottom, bringing together grains of quartz, feldspar, mica, calcite, or silicates of widely different origins. Since chondrites have never felt the gravity of a large planet, the sedimentation must have taken place in the solar accretion disk.

This is exactly what the models of the viscous accretion disks predict. During the gravitational collapse of the interstellar nodule into the disk, the gravitational energy of the infall kept a high rate of trubulence in the gas. This turbulence kept the dust in suspension perhaps for some 10^5 years. However, when the collapse rate subsided, before completely stopping, there was a time when the dust was not supported any more by turbulent eddies in the gas. In the quasi-quiescent disk, it fell down to the mid-plane: this is a sedimentation that is going to make thin equatorial rings of dust around the Sun, much wider than but otherwise entirely comparable to Saturn's rings.

2.6   From Dust to Planets

After sedimentation in the quiescent disk, the dust grains move in practically circular orbits (their different thermal histories come from the previous turbulence in the gas, and from the different heights of their infall). A few years ago, it was thought by Goldreich and Ward (1973) that gravitational instablilties in the dust rings would suffice to make numberless 10 km planetesimals in a few years only. Now, Weidenschilling (1988) has given solid arguments, showing that some turbulence is still fostered in the dust ring by the small differential velocity existing between gas and dust: this low degree of turbulence is sufficient to prevent Goldreich and Ward's process. Weidenschilling concludes that planetesimals grew more slowly, by coagulation of grain aggregates that had numberless soft collisions due to their different settling rates and to the drag-induced decay of the orbits. His mechanism can form meter-sized bodies in a few thousand years; this is therefore the characteristic time needed to remove dust from its thermal equilibrium with gas in the mid-plane of the nebula. Further growth from meter- to kilometer-sized bodies takes again a few thousand years. The following growth of the planetesimals goes on indefinitely through accumulation by soft low-speed collisions from nearby quasi-circular orbits.

However, the nature of the process changes slowly as the planetesimals grow in size. The reason is that the gravitational influence of the growing bodies can be less and less neglected. The orbital perturbations due to close encounters are of the same order of magnitude as the escape velocity, which is itself proportional to the planetesimal size. A large body of work on this accumulation process has been summarized by Wetherill (1980). Without trying to repeat all his conclusions here, it is important to note for our purpose that the size distribution of the bodies widens considerably, so that larger and larger planetary embryos will appear in a time scale in the range of 10–100 million years. The largest of these embryos will eventually be hit by the smaller objects still present in their distance zones, until these zones are clear of solid matter. Most of the material accumulating to form the present Earth has therefore been collected from a zone extending radially from 0.8 AU to 1.3 AU, by radical diffusion of the orbits due to gravitational encounters. Numerical models have shown that such a radial diffusion occurs with approximately the same timescale as the accumulation of the terrestrial planets (Wetherill, 1980).

2.7   Temperature History of the Earth's Material

In order to find the temperature of the dust when it sedimented out of the nebular gas, we must find a cosmothermometer working at the right time and place. We know that igneous differentiation processes have erased any fossil trace of this temperature on the Earth and, for that matter, on all the planets. Chondrites are the only undifferentiated objects known in our neighborhood. Hence we have no choice and must use chondrites as cosmothermometers.

We mentioned earlier that chondrites come from the asteroid belt. From the reflection spectra of asteroids in the visible and in the infrared, there is consensus among astronomers that (dark) carbonaceous chondrites come from (dark) C asteroids, and (light) ordinary chondrites come from (light) S asteroids (Gaffey and McCord, 1979). But asteroids C outnumber asteroids S only beyond 2.6 AU (Morrison, 1977). It is true that the distribution of S asteroids extends beyond 2.6 AU, but their fraction falls off with a scale length of 0.5 AU (Zellner 1979). The present distribution can be interpreted as resulting from the secular orbital diffusion through the original 2.6 AU limit, which was sharper 4.5 billion years ago.

To know the accretion temperature of chondrites, the best cosmothermometer probably is the fractionation and loss of their volatile metals Pb, Bi, Tl, and In. Using these latter elements, Larimer (1967) and Anders (1971) find, for a nebular pressure of 10^{-4} bars, a mean accretion temperature of 510 K for ordinary chondrites and of 450 K for the C3 type of carbonaceous chondrites (the other types of carbonaceous chondrites give slightly lower temperatures). If we take into account lower pressures used in recent models of the nebula, these values would become 480 K and 430 K. Several other assessments are consistent with adopting a 450 K temperature to separate the two classes.

We can deduce that, at the epoch of dust sedimentation from the nebular gas, the temperature at 2.6 AU, in the central plane of the nebula, was 450 K. Since the temperature gradient predicted by theory for the epoch has been empirically confirmed, we conclude that the accretion temperatures of the planetesimals that were going to form the Earth were between 900 K (at 1.3 AU) and 1400 K (at 0.8 AU), bodies, that is, to remove them from thermochemical equilibrium with the nebular gas. Figure 2.2 summarizes the situation that we have just described.

2.8   Thermochemical Equilibrium in Solar Nebula

In the first place, we must consider whether the kinetics of the chemical reactions are sufficiently rapid for the dust to reach chemical equilibrium with the gas of the nebula. The chemical kinetics of a gas of solar composition has been very carefully and thoroughly discussed by Lewis et al. (1979) and Lewis and Prinn (1980).

The largest time constants that turn out to be significant are in the range of one century near 1000 K since, after sedimentation to the mid-plane, the grains need 10^3–10^4 years to agglomerate into large objects (Weidenschilling, 1988), there is no doubt that thermochemical equilibrium will be reached in the Earth's zone before solids are completely separated from the gas phase. This means that all dust (which is originally in submicrometer-sized grains) has been completely dehydrated, degassed, and reduced (except a fraction of the silicate grains) before its incorporation into larger and larger planetesimals and their later integration into planetary bodies.

Lewis et al. (1979) have, in particular, considered the carbon problem, and they have rightly been puzzled by the retention of carbon in the terrestrial planets.

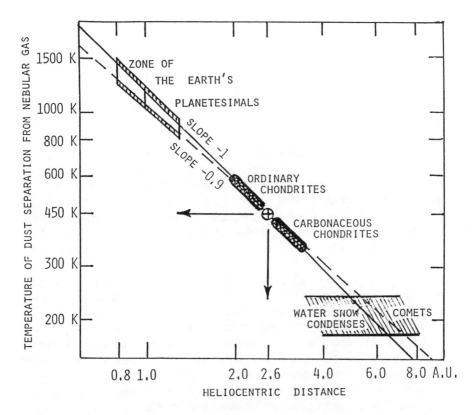

FIGURE 2.2. The same diagram as in Figure 2.1 is enlarged here in the vicinity of the Earth, with a view to improve our knowledge of the temperature in the Earth's zone, when dust sedimented from the nebular gas. Of course, all clues on this temperature have been erased by the igneous differentiation of our planet. The best cosmothermometer for this epoch lies in bodies that have never differentiated in the asteroid belt. It tells us that at the time of sedimentation, all ordinary chondrites were at a temperature higher than 450 K, and all carbonaceous chondrites at a temperature lower than that. Reflection spectra and colors of asteroids tell us that 2.6 AU. represents the separation distance. Since the temperature gradient was in the vicinity of −1, possibly −0.9, but definitely larger (in absolute value) than −0.75, it is concluded that the temperature in the Earth's zone was such that the dust did not contain any volatiles. All water remains as steam in the nebular gas, carbon was in CO and nitrogen in gaseous N_2. Hence the Earth accreted from an outgassed material, completely depleted of volatiles.

Searching for a mechanism of carbon retention, they correctly deduce that the only way to imprison carbon in the solid phase is to put it in a solid solution inside reduced (metallic) iron grains. In spite of their efforts, they reach an amount two or three times smaller than the observed amount on the Earth (carbonates) or on Venus (CO_2); they do not either address the question on how to extract the carbon from iron and bring it to the surface of the Earth or Venus.

However, in their efforts to reach an amount of carbon retention in the grains as large as possible, their choice of the adiabat in the solar nebula has put the Earth's zone near the peak of graphite activity (see Figure 2.3). For any adiabat, graphite activity goes through a maximum in the vicinity of the line separating the domains of CH_4 and CO in Figure 2.3. Bringing the Earth's zone here removes the 2.6 AU zone from the same location, whatever the model. There, Lewis et al.'s (1979) quite reasonable efforts to bring enough carbon on the Earth have removed the possibility of explaining the observed separation of the C asteroids (carbonaceous chondrites) from the S asteroids (ordinary chondrites) at the distance of 2.6 AU.

Using Lewis et al.'s (1979) own data, if a proper adiabat (between C and CD on Figure 2.3) is used to get the 2.6 AU heliocentric distance close to the line separating CH_4 and CO on Figure 2.3, the amount of carbon available in solid solution in the iron grains becomes in the Earth's zone several hundred times smaller than the observed amounts in the carbonates of the Earth, and the situation is even much worse for the CO_2 present on Venus.

Hence if we accept the accretion temperatures of chondrites as the best available cosmothermometer at the epoch of dust separation from gas, we have also to accept that the bulk of the observed carbon in the terrestrial plants has an exogenous origin.

The same conclusion can be reached for water, because at those temperatures no hydrated silicates exist, hence the total amount of water is in steam in the nebula; in the same fashion, all nitrogen is in gaseous N_2. The presence of the atmosphere and of the oceans on the Earth would therefore be quite mysterious, if the formation of the giant planets had not created the unavoidable mechanism that was going to put a veneer of volatile material on the terrestrial planets.

2.9   Discussion: Was the Earth Outgassed?

Is it possible to escape from the conclusion that the planetesimals that made the bulk of the Earth were completely outgassed and devoid of any volatiles? Let us consider the chain of arguments in detail.

(a) The cosmothermometer at 2.6 AU is valid. The heterogeneous microscopic composition of the chondrites, combined with their total lack of igneous differentiation, leaves little doubt that the variable fractionation and loss of their volatile metals dates back from their sedimentation as fine dust and is due to a temperature gradient. The temperature separating the two classes is confirmed by several independent somothermometers: the fractionation of the volatile metals (Pb, Bi, Tl, and In) (Larimer, 1967); oxidation state (olivine/pyroxene ratio) (Larimer, 1968); the presence of FeS and the absence of magnetite in carbonaceous chondrites confirms

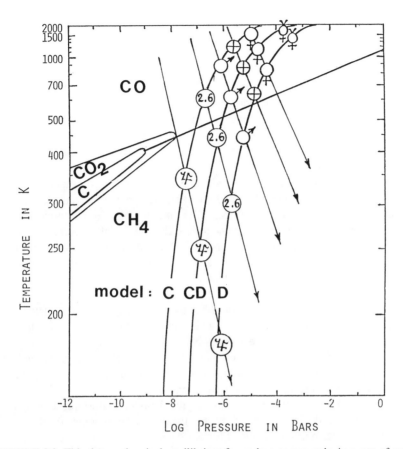

FIGURE 2.3. This thermochemical equilibrium for carbon compounds, in a gas of solar composition, is used to understand the carbon chemistry in the solar nebula at the time of dust sedimentation. The quasi-vertical curves represent the adiabats of Cameron's (1985) models C and D, whereas curve CD is an interpolation of the two models C and D that brings all temperatures in agreement with those of Figure 2.2. The different steady-state models C, CD, and D can be interpreted as a slow evolutionary sequence (shown by the arrows) corresponding to the time when the inflow of nebular gas onto the disk is slowly subsiding. This allows gas turbulence to diminish, eventually letting the dust separate from the gas and the sediment to the mid-plane of the disk. Identification with Figure 2 shows that this happens when adiabat CD is reached. Then, dust settles out and is removed from its chemical equilibrium with gas. At the Earth distance (circle with inside cross) this happens near 900 K; chemical kinetics are fast and dust grains are at thermochemical equilibrium with gas before their sedimentation. Beyond 2.6 AU, chemical kinetics prevails at temperatures lower than 450 K; there, metastable carbonaceous compounds separate from the nebular gas and are imprisoned in carbonaceous chondrites. The other lines on the diagram that separate the zones here CO, CH_4, CO_2, or C (graphite) are each the major constituent. If the conditions of adiabat CD are applicable as implied by Figure 2.2, then all carbon was in gaseous CO are applicable as implied by Figure 2.2, then all carbon that was in gaseous CO in the Earth zone; carbon on Earth must be exogenous and must have come later.

the temperature range (Anders, 1971); even if the Fischer–Tropsch Type (FTT) reactions are no more considered as unambiguous (Kerridge 1991), their possible presence in carbonaceous chondrites would only bring a further confirmation of the 450 K temperature separating the carbonaceous from the ordinary chondrites (Anders, 1971). Finally, there is not much doubt that the identification of the S (stony) asteroids with the ordinary chondrites and the C (carbonaceous) asteroids with the carbonaceous chondrites. If the present 2.6 AU distance has changed somewhat because of mass loss within the solar nebula, the 1 AU distance of the Earth would have shifted in the same proportion and the net result of our computation would be the same.

(b) The temperature gradient at sedimentation of dust is not much in doubt, because the theoretical prediction of Morfill's (1988) model is confirmed by the empirical gradient deduced by Lewis (1974); both are close to -1, which is also derived from a simple-minded argument based on the virial theorem. The virial theorem implies that, in the absence of any latent heat (due, for instance, to the condensation of silicates or of water) the temperature gradient is the same everywhere as that of the gravitation potential. This is close to -1 as soon as the mass is concentrated in the Sun.

The virial theorem can be used only if the angular acceleration of the disk is negligible. This is always the case when sedimentation takes place. Besides, the virial theorem tells only how much heat is available from the gravitational collapse; at steady state, the heat reradiated by the disk photosphere depends on its temperature, hence on the radial distance. This explains the corrected gradient of -0.9 introduced in Morfill's approximation.

Wood and Morfill (1988) also discuss simplified models of the three major stages of the accretion disk. Their stage 1 describes the period during which the disk mass is steadily increasing; it is irrelevant for our concern. Their stage 2 describes the epoch when the rate of inflow equals the mass rate fed to the Sun. During this steady state, the temperature gradient in the mid-plane of the disk is -1.5. Stage 3 is reached when the infall rate diminishes enough to stop the gas turbulence; the disk still feeds the Sun until its mass diminishes enough to make it transparent. At that time, the temperature gradient in the mid-plane has fallen down to -0.75. Since the sedimentation of dust is triggered by the disappearance of gas turbulence, the temperature gradient in the mid-plane falls down slowly from -1.5 to -0.75. This is consistent with the empirical value close to -1 for the epoch of separation of dust from the gas. Also see Lin and Papaloizou (1985).

Cassen et al. (1985) have also compared the temperature distribution at different stages. They find that the photospheric temperature has a radial gradient of -0.75 regardless of the details of the viscous mechanism. However, when the disk is optically thick, its opacity implies that the mid-plane temperature remains closer to the results of the virial theorem, hence this remains consistent with the Morfill's (1988) temperature gradient of -0.9 in the mid plane. We conclude that the consensus between theory and empirical data is satisfactory.

(c) At sedimentation, the mean temperature of earthy dust is high enough to degass it. After sedimentation, earthy dust was on circular orbits located between

0.8 AU and 1.3 AU. Dust grains started to stick together and their sizes grew slowly. After a few thousand years, they had become large enough to be called planetesimals, and the eccentricities of their orbits started to grow steadily, in step with their growing sizes (Wetherill, 1980). As seen on Figure 2.2, their accretion temperatures were at least 800 K, even if the least steep gradient of -0.9 is used. The limiting gradient of -0.75 (which would yield 700 K for the coolest grains), is ruled out before the grains are imprisoned in meter-sized bodies; and even 700 K would outgas the grains.

Of course we deal with a stochastic process; this means that, in the very final steps of accumulation, some of the major collisions with lunar-sized planetary embryos could come from objects whose initial heliocentric distance of accumulation were outside our definition of the Earth's zone (0.8–1.3 AU). An object that accumulated in a zone nearer to the Sun would be even more outgassed than the protoearth; but if a large embryo were coming from the inner asteroid belt (2.0–2.6 AU) it could bring ordinary chondritic material to Earth. Such a possibility was ruled out until recently, because it was believed that no object ever became larger than Ceres in the asteroid belt. However, Wetherill (1991) has shown by numerical Monte-Carlo experiments that it is conceivable that larger runaway embryos could have formed in the asteroid belt, then ejected later by Jupiter secular resonances. The self-clearing of the inner belt is assisted by collisions with terrestrial planets. However, the process induces higher eccentricities and inclinations than those observed for the terrestrial planets. Such a collision, that would not greatly change the orbit of the Earth, is a *grazing* collision, which seems more proper to explain the origin of the Moon (Cameron, 1988) without depositing volatile material deep into the Earth's mantle.

(d) The chemistry of the solar nebula is reasonably well understood, including its chemical kinetics (Prinn and Fegley, 1989). The fact that FTT reactions are likely beyond 2.6 AU, and there is kinetic inhibition in the vapor-phase hydration of the silicates as well as in the chemical reduction for CO and N_2 and all water was in steam. No water, no nitrogen, no organic carbon to speak of, were present in the protoEarth.

2.10   Formation of the Giant Planets

To understand the origin of the volatile material present on the terrestrial planets and more specifically on the Earth, the formation of the giant planets must be discussed first. The study of their interiors (Hubbard, 1984) shows that the giant planets must have a large dense core, hence they must have developed first a solid embryo before being able to accumulate a large atmosphere.

Three different arguments concur in suggesting that a solid embryo accreted early for Jupiter. In the first place, this embryo had to exist early enough to be the cause of the gross mass depletion of the asteroid belt and of the small mass of Mars (Wetherill, 1980). In the second place, the embryo had to be massive enough to capture the gaseous atmosphere of Jupiter before the dissipation of the nebular gas. The third argument comes form the empirical data on the interior of Jupiter; they imply that it has a dense solid core of about 30 Earth masses. Such a large core is more easily predicted by a fast runaway growth scenario (Safronov, 1991).

The first argument implies that Jupiter's embryo was already massive enough (ten Earth masses) early enough to stop the accumulation of asteroids into a single planet. In less than 10^7 years, it had to enlarge the eccentricities of the residual planetesimals of the Jupiter's zone, to make their orbits penetrate the zone of the asteroids and somewhat that of Mars. Their interactions with the asteroids scattered their orbits, enlarged their relative velocities, and transformed the accretion process in the belt into a fragmentation process, ejecting a large amount of that mass that was then in the 1.4–2.8 AU zone. The orderly accumulation of objects much larger than (1000 km) Ceres could not have been stopped after a few times 10^7 years, and these lunar-to-Mars-sized bodies could not have all been destroyed later.

The second argument comes from the early dissipation of the nebular gas. Although this dissipation is not yet thoroughly understood, it has been linked by Horedt (1978) to the existence of the T-Tauri wind. T-Tauri stars are very young stars that are still in their contracting stage before reaching the main sequence in the Hertzprung–Russell diagram. Such a diagram drawn for 50 T-Tauri stars in the Taurus cloud shows the spread of their ages to be 10^7 years. Most of them are first surrounded by an accretion disk. The disk mass and its inflow rate seem to decrease steadily with their increasing age, but there is also an outflow (comparable to the solar wind but much larger). This is what is called the T-Tauri wind. The T-Tauri wind seems to halt the inflow. The mass outflow is very large first ($10^{-6} M_\odot$/yr) but it falls down quickly to 10^{-8} or $10^{-9} M_\odot$/yr. The accretion disk disappears and the wind stops just before the stars reach the main sequence. The observed luminosity of 3 $L_\odot$ for a T-Tauri star of one solar mass, assumed to come from the gravitational energy of its contraction, confirms a lifetime of 3×10^7 years. Hence it seems that, to capture enough nebular gas, Jupiter's embryo must be massive enough in no more than 10^7 years. Mizuno (1980) has established that, to produce this gas collapse, the embryo must reach about ten Earth masses. This result is universal, in the sense that it does not depend on the location in the solar nebula. This explains why there was no gravitational collapse of the nebular gas on the terrestrial planets.

We are left with the problem of explaining how a solid planetary embryo can reach a mass of ten terrestrial masses in ten million years, whereas the classical mechanism of accretion requires one hundred million years for the Earth in a gas-free medium, and probably forty million years in a gaseous medium (Wetherill, 1980). Such a time scale for the accumulation of the terrestrial planets is also confirmed by Safronov (1991) from other considerations.

However, two possible scenarios for the accumulation of the growing proto-planets are not mutually exclusive. They are:

(a) an orderly growth, without separation of the largest body from the general mass distribution of the smaller bodies; and

(b) a faster runaway growth of the largest body, which grows much beyond the general mass distribution of the smaller bodies.

Safronov (1991) suggests that a runaway growth took place in the zone of Jupiter, but that there was an orderly growth in the asteroid belt as well as in the terrestrial planet zone. The two processes start in the same fashion by mechanism (a). However, at about the time Jupiter's embryo is close to the Earth's mass, mechanism (b) starts until it reaches 30 Earth masses. Safronov claims that a core of 30 Earth masses is needed instead of 10 Earth masses. This is linked to the need for the final accumulation of five or six runaway embryos before their final accretion into a single body. The reason is the narrow width of the feeding zones, but this is irrelevant to our discussion. The essential point is that the runaway growth mechanism produces a 30 earth-mass embryo, hence the gaseous collapse into a full-size Jupiter in ten million years. The observational data that support the existence of a large core in Jupiter (Hubbard, 1984) (see Table 2.1) can therefore be interpreted as a third argument for a ten million-year runaway growth.

Why have the giant planets been able to accumulate solid cores much larger than one Earth mass? This may be connected to the fact that, as the radial temperature of the accretion disk diminishes with the heliocentric distance, the nature of the solid grains changes. Dust was formed in the zone of the terrestrial planets, from anhydrous silicates and reduced iron grains. When we move to the future asteroid belt, the volatile metal abundances grow in the grains, until they progressively become chondritic near 2.6 AU. Beyond that distance, the diminishing temperature allows silicates to keep more and more organic matter as well as hydration water. This explains why bodies become darker beyond 2.6 AU (Figure 2.2). Up to about 5 AU we call the material present in these bodies "carbonaceous chondrites."

It is a matter of semantics to decide where comets begin. We choose to call "comets" those planetesimals that contain water, not only as hydration water in a mineral, but also as free water ice. It is remarkable that water ice condenses near 5.2 AU in the nebula with the temperature gradient adopted on Figure 2. It is probably not a coincidence that it is the place where Jupiter's solid embryo was formed. Using elemental abundances (Anders and Grevesse, 1989), it is easy to verify that the total solid fraction available at 5.2 AU (dust plus CHON ice) is 4.6 times as massive as the dust (silicates plus reduced iron) available in the Earth zone. Finally, the dust grains loaded by frost condense any extra water vapor that reaches them. The diffusive redistribution of water vapor in the solar nebula (Stevenson and Lunine, 1988) makes it a runaway phenomenon.

The previous discussion has made it clear that the solid cores of the giant planets have accreted from comets and comets only; the extra mass of solid ices available in comets is the major reason why the solid cores were so large. The fate of the numerous planetesimals (= comets) that did not accrete on the giant planets must now be discussed in detail.

2.11   Orbital Diffusion of Comets

Safronov (1972) has shown how the total mass of the perturbed comets is linked to the total mass of the planetary embryos. When the giant planets' embryos became large enough, they were able to eject comets out of the solar system, as well as to store a small fraction of them in a sphere of some 50,000 AU radius. Stellar perturbations removed most the perihelia of the latter comets from the zones of the planets, hence their orbits became secularly stable and formed the primitive Oort cloud whose source was what I will call the Safronov comets.

Beyond Neptune and up to several hundred AU, the planetesimals left over by the accretion disk did not form very large objects, and are now called the Kuiper belt of comets. In the Kuiper belt, resonances with the giant planets also work but with a much longer time scale; up to now, they have been removing comets from the Kuiper belt, either by ejecting them out of the solar system, or by throwing them into the Oort cloud. I will call them the Kuiper comets. Because of the longer time scales involved, they do not play an active role in the origin of the planets, although they are the source of the present short-period comets. Besides, as repeated perturbations from stars, from galactic tides, and from giant molecular clouds have considerably depleted the original Oort cloud, most of its present comets are now Kuiper comets.

The scattering of the cometary orbits present as planetesimals in the zones of the giant planets grows in step with the sizes of the embryos. Since this gravitational scattering is a random process, a fraction of the perturbed comets (including those that are ejected out of the solar system) is sent first through the zone of the terrestrial planets. Using Safronov's (1972) method, I have computed the total mass of those comets that pass through a sphere of 1 AU centered on the Sun. I have multiplied this mass by the collision probability of each individual comet with the Earth. In my early estimates (Delsemme 1991a, b and 1992), I had not changed the masses selected for the giant planets' embryos by Safronov (1972). They were only 9 terrestrial masses for Jupiter's embryo, 12 for Saturn, 14 for Uranus, and 17 for Neptune. Since then, I found it easier to compute it than to convince people that it is negligible in the present approximation.

Recently, I recomputed new masses for the embryos of the giant planets, based on Hubbard's (1984) empirical data on the density distribution in their interiors, and assuming that the embryos have the same composition as the one I found for comet Halley (Delsemme, 1991a) namely 23% rocks, 41% water, and 36% CHON. The masses I found for the embryos are given in Table 2.1.

The new mass for Jupiter's embryo seems to confirm Safronov's (1991) recent arguments that, since a runaway accretion is implied to make Jupiter's embryo with a time scale of at most 10 million years, then the feeding zone remains several times narrower than the planet's zone even up to an embryo larger than 20 masses (this is due to the lower relative velocities of the bodies). This implies that there will be first a transition to several independent embryos and that the gaseous collapse will not start before 30 Earth masses, a value close to that of Table 2.1 deduced from Hubbard's (1984) data.

In Delsemme (1993) I have revised my former assessments by using the new masses given in Table 2.1 for the embryos's masses. The new results are higher than those published earlier. They are submitted in Table 2.2. The missing mass in the present asteroid belt is assumed to be ten Earth masses. In order to translate

TABLE 2.1. Giant Planet Embryo Mass.

Planet	Total mass*	Embryo's mass*
Jupiter	317 m	29 m
Saturn	95 m	19 m
Uranus	14.6 m	13 m
Neptune	17.3 m	15.4 m

*m represents the mass of the Earth.

the masses into a uniform layer covering the Earth, the density of silicates was assumed to be 3.0 and that of carbonaceous compounds was assumed to be 2.0. The distribution of silicates, water, organics, and gases is based on the comet Halley's data previously mentioned. The contribution of the asteroid belt is based on a chondritic average with more weight on ordinary chondrites. Finally, the amplification factor due to resonances, far encounter effects and the focusing due to terrestrial gravity turns out to be a global factor of 3.5.

Because of the numerous uncertainties associated with such a model, the results must of course be understood as orders of magnitude only. However, it is encouraging to see that there is a general consensus on the amount of matter brought down to the Earth by this bombardment. Matsui and Abé (1986) find that comets brought down to Earth four times as much water as the mass of the oceans. Fernández and Ip (1981, 1983) had revised Safronov's evaluations upward. Ip and Fernández (1988) now find that ten times the present mass of our oceans has been brought down to Earth by comets. From the visible craters on the Moon, Chyba (1991) finds that the total mass of the oceans corresponds to a bombardment of only 10% of the lunar craters.

The model of Table 2.2 brings one order of magnitude more water and three orders of magnitude more gases than what remains now in the oceans and in the atmosphere. This is not only explainable, but probably required to justify the considerable losses of volatiles due to the numerous impacts predicted in the last stages; I mean not only the cometary impacts, but also those of larger planetesimals that have accreted in the zones of the giant planets before being ejected by them, as well as lunar-to-Mars-sized bodies predicted by theory, mostly from the Earth zone (Wetherill, 1980), but even possibly some from the asteroid zone (Wetherill, 1991). At least one of these objects could be responsible for the grazing collision that would have created the Moon (Cameron, 1985). Since all these events are stochastic, it is of course impossible to introduce them in a precise chronology.

TABLE 2.2. Thickness of Uniform Layer Covering the Earth.

Origin of terrestrial bombardment	Chondritic silicates	Water	Carbon compounds	Atmosphere
Chondrites from asteroids	2.0 km	0.02 km	0.01 km	—
Comets from Jupiter's zone	9.6 km	35 km	13.0 km	1900 bars
Comets from Saturn's zone	1.6 km	5.2 km	2.0 km	140 bars
Comets from Uranus' zone	0.15 km	0.5 km	0.2 km	23 bars
Comets from Neptune's zone	0.06 km	0.2 km	0.08 km	10 bars
Totals	13 km	41 km	15 km	2100 bars
Mass in 10^{25} g	1.9	2.0	1.5	0.2

2.12   Chronology

A possible chronology model is described in Table 2.3. It summarizes the different assessments coming from the previous speculations. These order-of-magnitude estimates ignore the considerable difficulties that are still present everywhere.

The chronology model of Table 2.3 assumes that the sedimentation of solid particles to the mid-plane of the disk occurred almost simultaneously everywhere, so the small planetesimals also appeared simultaneously. We assumed that a bifurcation appeared after the first million years: there was a fast runaway growth in the region of the giant planets, whereas there was a much slower orderly growth for the terrestrial planets. The two regimes of growth can coexist (Wetherill, 1989) although the details that can trigger one or the other have not been completely quantified (for instance, it could be the difference in the sticking factors of snows and silicates). The model explains the absence of a planet in the asteroid zone and the small size of Mars by the early appearance of Jupiter. It also assumes that, if Saturn has captured gravitationally a smaller amount of gas from the solar nebula,

TABLE 2.3. Time Scales for the Early Solar System.

Chronology for the Beginning of the Planetary System

Sedimentation of solids to the mid-plane	0 yrs
10 km planetesimals on circular orbits	10^4 yrs
10^2–10^3 km planetesimals	10^5 yrs
Planetary embryos (Moon to Mars size)	1 M yrs
Runaway growth of Jupiter embryo	10 M yrs
Runaway growth of Saturn embryo	20 M yrs
Dissipation of nebular gas	30 M yrs
Earth accumulation: 99% finished	40 M yrs
Runaway growth of Uranus embryo	50 M yrs
Runaway growth of Neptune embryo	80 M yrs

Time Scales for Orbital Diffusion of Comets

Chondrites from zone of asteroids	28 M yrs
Comets from zone of Jupiter	70 M yrs
Comets from zone of Saturn	175 M yrs
Comets from zone of Uranus	410 M yrs
Comets from zone of Neptune	820 M yrs

that is because the dissipation of the nebula had already begun. Finally, the nebular gas had completely disappeared when Uranus and Neptune's embryos reached their final sizes, explaining the minimal amount of hydrogen and helium in their atmospheres.

The missing primary atmosphere of the Earth, testified by the patterns in the abundances of the noble gas isotopes in our present atmosphere, is explained by the slow orderly growth of the protoEarth which reached its final size after the dissipation of the nebular gas. The present atmospheres of the terrestrial planets and the oceans of the Earth are explained by the late accreting veneer brought by the orbital diffusion of the large number of comets that were deflected by the building up of the giant planets' solid cores. Comets were the building blocks of these cores. All of those that missed a direct hit on the giant planets' embryos were ejected faraway by the strong gravitational attraction of the large embryos. Most of them were lost from the solar system, but some of them were stored in the Oort cloud, and some of them hit the terrestrial planets when their ejection orbits crossed the inner solar system.

The time scales for the orbital diffusion of comets have been normalized in such a way that they grow in proportion to the revolution periods in the zone where the diffusion comes from (Everhart, 1977). Of course the diffusions begin at different times, namely, the time when the accretion of the relevant embryo has become large enough; this means about at the end of the runaway growth of the embryo. The times scales are exponential time scales; they express when the Earth bombardment had diminished by a factor $1/e$.

2.13   Chronology Discussion

The chronology model leads to an accretion rate of comets on the Earth that is steadily diminishing with time. This rate is displayed on Fig. 2.4 for the first two and a half billion years after "age zero" (the epoch of dust sedimentation from the nebular gas). The contribution of the asteroid belt and of Jupiter's comets comes early; it is half finished during the first two hundred million years. Since the different zones decay at different rates, the total accretion rate, represented by the dotted line, subsides drastically after six hundred million years. The contribution of the zone of Jupiter prevails for the first 400 million years, then Saturn takes over up to one billion years, then Uranus up to 1.4 billion years, and finally Neptune. Neptune's contribution contains the 2:1 resonance with Neptune in the Kuiper Belt: after 4.6 billion years (beyond the limit of Figure 2.4) it still contributes 5×10^{18} g/M yr, or 50 comet collisions per million years: coming from the short-period comets that we still observe nowadays. This is a fair estimate for such a simplistic model. The difference between the times needed by the (short) runaway growth of Jupiter and the (longer) orderly growth of the Earth is the decisive factor which marshals the fraction of volatiles imprisoned in the Earth mantle. If our chronology model is correct, when the chondritic and cometary bombardment induced by Jupiter started, the Earth embryo had probably reached some 90% of the present mass of the Earth, and numerous collisions with 10^{22}–10^{23} g bodies were still to come.

We presume that the accumulation of the Earth was essentially finished in forty million years, assuming that it was somewhat increased at the beginning by the presence of nebular gas. It is not implausible either that it has ended by a small period of runaway growth due to a large value of the gravitational focusing factor for the largest two or three bodies involved. However, if the nebular gas had dissipated much earlier and if there was no runaway growth, the standard orderly growth would yield one hundred million years for the accumulation of the Earth, in which case one-half of the chondritic contribution of the asteroid belt and one-fourth of the comets coming from Jupiter's zone would be buried deep into the Earth's mantle. This represents a total of somewhat less than 10^{24} g of volatile material buried in the mantle, as opposed to a veneer of 6×10^{25} g brought after the accumulation of the Earth is essentially finished. Of course, this 1% of the total mass of exogenous bodies brought early onto the Earth does not seem very significant, until it is realized that its volatile material may explain a non-negligible amount that would be recycled again and again in the mantle.

The chronology model of Table 2.3 implies the more likely view of a forty million year accumulation for the Earth. In this case, the contribution to the lower mantle is minimal, and the early contribution to the upper mantle becomes undistinguishable from the late major contribution of the next six hundred million years. This comes

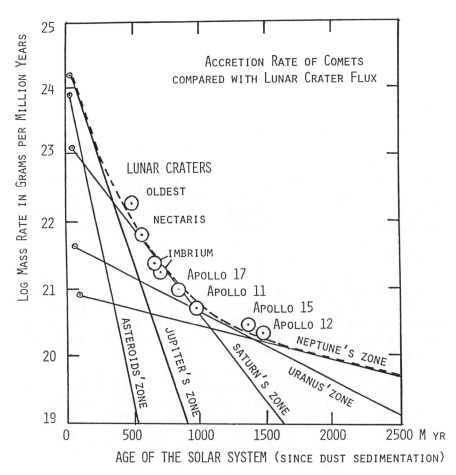

FIGURE 2.4. Accretion rate of comets on the Earth, in grams per million years, as a function of time since dust sedimentation in the mid-plane of the accretion disk 4.56 million years ago. The contributions of the zones of the giant planets start at different times and decay at a different rate. The mass of each contribution is deduced from Table 2.2 and the exponential time scales are from Table 2.3. The exponential time scales for orbital diffusion have been normalized by using the Everhart (1977) results of random walk numerical experiments for the orbital diffusion of comets by the giant planets, and the exponential lifetimes of decay were chosen in proportion to the orbital periods of the giant planets. Each of the contributions of the different planet zones is a straight (solid) line on this logarithmic diagram. The sum of all contributions is the dashed line with an exponential rate varying with time, to be compared with the cratering record on the Moon (Carr et al., 1984), whose data have been normalized to our units. The agreement is surprisingly good, not only for the mass rates, but also for the subsiding rate with time.

from the "gardening" of the crust and upper mantle by large impacts and because of the subsequent convection of the upper mantle.

2.14   Observational Confirmations

2.14.1   Cratering Record

Observational data on impact craters now exist for all terrestrial planets, the Moon, and the satellite systems of Mars, Jupiter, Saturn, and Uranus. See a review by Weissman (1989). The similarity of the cratering density distribution versus size on the Moon, Mars, and Mercury (Carr et al., 1984) suggests that the massive bombardment concerned the whole inner solar system and was simultaneous on all three bodies. Wetherill (1975) has also clearly shown that the mass of the impactors was much too large to be only the subsiding tail of the remnant bodies in the terrestrial planets' zones.

The only accurate chronology has been obtained from lunar data. The lunar rocks brought back from the Apollo and Luna missions have been dated by those radioactive isotope techniques that establish the solidification age of the relevant rock. The younger solidification ages were systematically found where the number of craters per square kilometer was smaller. Assuming that the missing craters had been erased from the record by the latest tectonic events that had melted the rocks, it was possible to establish an absolute chronology of the residual crater density. Carr et al. (1984) data have been normalized to the units of Figure 2.4 and are represented by circles on the diagram. To do this, we assumed that dust sedimentation from gas (our "time zero") occurred 4.56 billion years ago. The comparison is encouraging, in particular, because it seems to explain the subsiding rate observed in the lunar craters, by the combination of the different time scales derived from the orbital diffusion of comets by the different giant planets' zones and of the different masses coming from each zone. The mass rates are equally well predicted.

My model uses different exponential time rates of diffusion, chosen in proportion to the orbital periods in the asteroid belt and of the four giant planets. These lifetimes are normalized by using the results of Everhart (1977) random walk numerical experiments for the orbital diffusion of comets by the giant planets. I have also compared with Kazimirchak–Polonskaya (1972), which confirms the order of magnitude of my estimate.

Chyba (1987) has used the visible craters of the Moon to assess the mass flux of the impacts on the Earth for the same period. He finds that the total mass of our oceans could be explained by a bombardment of comets corresponding to 10% of the lunar craters. His assessment is in complete agreement with the present model if the bulk of the lunar craters comes mostly from comets.

2.14.2   Geochemistry

As judged from comet Halley, elemental abundances in comets are almost chondritic. C, N, and O are even closer to solar than to chondritic abundances, as can be seen in Figure 2.5 (Delsemme, 1991a). In particular, the siderophile elements are in solar (or chondritic) proportions in the cometary silicates, and their contribution to the outer layers of the Earth (crust and upper mantle) is of the order of 2×10^{25} g (see Table 2.2). This is quite enough to solve the apparent paradox of the "siderophile excess," which is a puzzle that has not been easy to explain in geochemistry.

The term siderophile dates back to V.M. Goldschmidt who suggested in 1922 a geochemical classification of the elements, based on their tendency to concentrate in one of the three principal solid phases that appear in a cooling magma. The elements that follow metallic iron are called siderophiles, as opposed to those that follow sulfur (chalcophile) and those that follow silicates (lithophile). Because of the formation of an iron core at the center of the Earth, liquid iron should have scavenged the siderophile elements like Ni, Co, Ir, Au, Os, and Pd. However, they are still found in chondritic proportions in some samples of the crust and the mantle (Morgan et al., 1981).

Rama Murthy (1991) suggests a possible explanation, based on a purely geophysical model, for this "excess" of siderophiles. In his explanation, the "excess" of volatiles remains unexplained. The "excess" volatiles are those elements that are found in the biosphere and cannot be derived by weathering of igneous rocks (Rubey, 1955). The exogenous origin of both the "excess volatiles" and the "excess siderophiles" on the terrestrial planets is a natural explanation that does not require any "ad hoc" hypothesis, because it derives of necessity from the formation mechanism of the planets by planetesimals, and of the growth of the giant planets. This growth has deflected chondritic and cometary bodies whose bombardment of the Earth brought the needed amounts of "excess" volatiles and of "excess" siderophiles.

Chyba (1991) has also shown that the mass flux of the impacts corresponding to the visible craters on the Moon, when properly extrapolated by the gravitation focusing action of the Earth, explains the right order of magnitude for the "excess" siderophiles.

2.14.3   Geochemical Model

Chemical models for the interior of the Earth are still essentially indeterminate. We know the existence of a dense core containing about 31% of the Earth's mass, only because the global density of the Earth and its moment of inertia concur with seismologic data to reveal its density and size. The results are consistent with a core of metallic iron whose density is slightly diminished by a minor fraction of lighter element(s). The bulk of the Earth is the lower mantle, which is probably its least understood part. For instance, its iron content is not certain, and its redox state has been disputed. Is the ratio Fe_2O_3/FeO high or very low? Is it partially

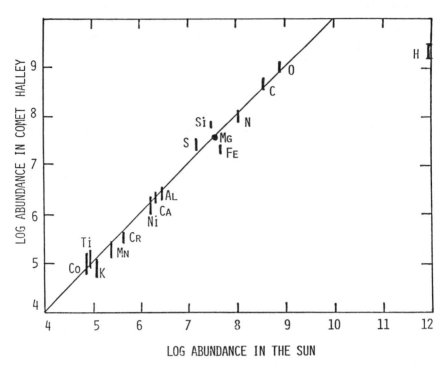

FIGURE 2.5. The abundance of known elements in comet Halley is compared with the same elements in the Sun. Except hydrogen (and presumably the noble gases) all elements seem to be in solar proportions in comet Halley. A curious exception is the Fe/Si ratio which is only 25% of that in the Sun. The silicon seems to be overabundant by a factor of 2, and the iron underabundant by the same factor, in comet Halley. This diagram is shown to establish that most siderophile elements are in solar proportions in comet Halley and, assumedly in all comets, explaining the surprising "siderophile excess" of the Earth crust, as well as their depletion in the upper mantle.

equilibrated with the Fe core? Is the mantle homogenized by mixing throughout its 2900 km depth? Are mantle plumes (responsible for volcanic activity at dozens of hot spots around the world, including Iceland and Hawaii) coming only from the upper mantle, 670 km thick, or do they come from much deeper?

Our accretion model predicts a 99% fraction of outgassed and partially reduced grains in planetesimals, equilibrated with the nebular gas in the temperature range of 800 K (at 1.3 AU) to 1300 K (at 0.8 AU). The last 1% is a veneer of cometary material that contributes to the much larger fraction of volatiles and oxidized material present in the outer mantle and the crust of the Earth. The model predicts a small enrichment of refractory elements and a large depletion of volatile elements that should already become apparent in the upper mantle.

The chemical composition of the upper mantle is known incompletely, and mostly through mantle-derived rocks brought to the surface by volcanic eruptions. Wänke et al. (1984) have compared the abundances of metals in such samples with their abundances in chondrites. They conclude that the refractory elements are indeed slightly enriched (by a factor of 1.30 ± 0.15), whereas the moderately volatile elements are depleted (factors of 0.1 to 0.2) and the very volatile elements are strongly depleted (factors from 10^{-2} to 10^{-4}). The highly siderophile elements are also strongly depleted, but the reason of this depletion is assumedly different, since they have been scavenged by the formation of the iron core.

Ringwood (1977) and Wänke (1981) have also proposed a two-component model for the formation of the terrestrial planets. Ringwood used 90% ordinary chondrites and 10% carbonaceous chondrites. Wänke used 15% of C1 carbonaceous chondrite for his component B; his residual 85% does not correspond to any class of chondrite: it is highly reduced and free of all elements with a volatility at least equal to that of Na. The two models, in particular that of Wänke, seem to go in the right direction, but the two components are arbitrary and the mixture is adjusted empirically. None of the models try to justify the amount of the two components by a numerical argument based on the history of the solar system.

Standing in contrast, the model described in this paper is based on a quantitative mechanism which is an inescapable consequence of: (a) the dust sedimentation in the accretion disk for the Earth zone; and (b) the existence of the giant planets.

2.14.4   Noble Gases

In the section "The Missing Primary Atmosphere," I mentioned that the isotopic pattern shown by the abundances of krypton and xenon on Earth remains a mystery. The source for the so-called "planetary" component of the noble gases is not chondritic at least for Kr and Xe. Without looking at the isotopic patterns, the relative abundance of ^{130}Xe reaches ten times the "solar" value in carbonaceous chondrites, whereas ^{130}Xe remains about "solar" in the atmospheres of Venus, the Earth, and Mars.

In our present model, volatiles are brought about by comets. Could comets solve the mystery of the anomalous abundances of xenon and of the surprising fractionation of the krypton and xenon isotopes? Low-temperature trapping of gases on ices is a possible source for rare-gas enrichment. An extreme form of gas trapping is the possible existence of gas clathrates. The existence of clathrates of gas has been proposed a long time ago (Delsemme and Swings, 1952) to explain some peculiarities in the vaporization rates of comets. Gas adsorption in snows was studied in the laboratory (Delsemme and Wenger, 1970) and clathrates mentioned as adsorption limits (Delsemme and Miller, 1970). The enrichment factors of Kr/Ar and Xe/Ar obtained by trapping the gases on ice was studied at temperatures varying from 30 K to 75 K by Bar-Nun et al. (1988) and by Bar-Nun and Kleinfeld (1989) corresponding to the very low temperatures of comet formation in the outskirts of the solar accretion disk. Since large enrichment factors are found, stochastic processes involving only one or two collisions with large comets may explain variations in planetary atmospheres. In particular, Owen et al. (1991) think that a plausible reservoir for the "planetary" component with the anomalous abundances of noble gases and anomalous ratios on Venus is the first clear indication of the presence of a cometary component in a planetary atmosphere.

Of course, the abundances of no noble gases have ever been measured in comets, and since there might be large variations from one comet to another, only a rather large sampling could yield a final solution to the problem. In the meantime, it is encouraging that studies in the laboratory point to a possible solution to the puzzling "planetary" abundances of the noble gases.

2.14.5 Deuterium

Standing in contrast to noble gases, the deuterium-to-hydrogen ratio has been measured in comet Halley (from the HDO/H_2O ratio) by the mass spectrometer of the Giotto space probe (Eberhardt et al., 1987). Unfortunately, the error bar is large (5×10^{-5} to 5×10^{-4}) but it is centered on the same value as seawater, which is 1.5×10^{-4}. The results are therefore consistent with our deduction that the total amount of water in our oceans comes from comets. The average value of this ratio in diffuse interstellar matter is ten times smaller (Vidal-Madjar, 1983). The fact that D/H is close to that in comet Halley for Titan, for Uranus, and for Neptune is consistent with our scenario that these bodies were accumulated primarily from comets. The fact that Jupiter and Saturn's D/H ratios have values much closer to that of diffuse interstellar matter seems to imply that the nebular gas ratio was close to the interstellar value, hence that the condensation of snows in the mid-plane of the accretion disk has enhanced this value by a factor of ten for comet Halley.

This enhancement is predicted to be temperature-dependent. If it comes from the deuterium-exchange reaction between water vapor and molecular hydrogen, then the tenfold enhancement is reached at about 200 K (Geiss and Reeves, 1981). This implies that the water of comet Halley as well as that of our oceans had a D/H ration quenched near 200 K at separation from the nebular gas. Because of kinetic limitations (Grinspoon and Lewis, 1987) this may derive from a much lower condensation temperature, as that suggested by trapping noble gases on ice (Bar-Nun, et al., 1988).

TABLE 2.4. The Volatile Fraction of Comet Halley.

78.5% H_2O	2.6% N_2	1.5% C_2H_2	0.1% H_2S
4.5% $HCO \cdot OH$	1.0% HCH	0.5% CH_4	0.05% CS_2
4.0% H_2CO	0.8% NH_3	0.2% C_3H_2	0.05% S_2
3.5% CO_2	0.8% N_2H_4		
1.5% CO	0.4% $C_4H_4N_2$		

92.0% with O	5.6% with N	2.2% hydrocarbons	0.2% with S

2.15   Nature of the Early Atmosphere

The possibility of building a consistent model of the volatile fraction escaping from the cometary nucleus has been reached for the first time with comet Halley (Delsemme, 1987, 1991a). It was also shown that the apparent diversity of different comets is not inconsistent with the bulk homogeneity in the cosmic material that has accreted into comets. In particular, the abundance ratios of the light elements in all the bright comets that have been observed recently (Delsemme, 1987) seem to be the same. This bulk homogeneity is not inconsistent with the large diversity in the microscopic composition of dust grains discovered in comet Halley (Jessberger et al., 1988), which suggests that individual interstellar grains coming from different stellar environments have been preserved in comets (Delsemme, 1991c).

The model of the volatile fraction of comet Halley, given in Table 2.4, is not final and must be used in a heuristic way only, but it is not in disagreement with an early atmosphere which would have been altered, processed and recycled many times in order to provide the bulk of oceanic water, the nitrogen still present in our atmosphere, the carbon of the carbonates and of the biosphere, and the residual gases recycled by present-day volcanoes.

Table 2.4 is in quantitive agreement with the mean elemental ratios of comet Halley and of several other recent bright comets, and in semiquantitive agreement with spectroscopic molecular data; however, the error bars are large. I have used $C_4H_4N_2$ as a symbol of heterocyclics with N, like pyrimidine, which is almost as volatile as water (boiling point 123° C at atmospheric pressure). Several heterocyclics including pyrimidine have been identified in the CHON grains of comet Halley by Krueger and Kissel (1987). At the present time, we must wait for better data coming from other comets, before discussing the possible evolution of the early atmosphere, which is clearly out of the scope of this chapter.

2.16   Prebiotic Organic Syntheses

The origin of the atmosphere and of the oceans that has just been described has direct connections with the origins of life on Earth. In particular, the carbonaceous chondrites that were brought onto the Earth (Table 2.2) from the asteroid belt, mostly during the first two hundred million years (Fig. 2.4), contained numerous amino acids. As an example, 74 different amino acids were extracted from the Murchison meteorite, a CM carbonaceous chondrite which fell in 1969 near Murchison, Australia (Cronin et al., 1988). Among these amino acids, eight of the protein amino acids (glycine, alanine, valine, leucine, isoleucine, proline, aspartic acid, and glutamic acid) have been identified along with eleven less common amino acids used by life. The total concentration of amino acids in the Murchison meteorite is about 60 parts per million (ppm). Assuming from the statistics (Sears and Dodd, 1988) that 3% of the meteorites are carbonaceous chondrites, this is still 6×10^{19} g (6×10^7 million tons!) of amino acids that would have reached the Earth. Assuming that everything would be destroyed by heat during the first 130 million years, 6×10^5 million tons would still reach the atmosphere later, and 6×10^3 million tons after the first 260 million years.

The Murchison meteorite also contains a complex mixture of aliphatic and aromatic hydrocarbons, of carboxylic acids and of nitrogen heterocycles. The latter are of particular interest because of the use of purines and pyrimidines as coding elements of the biological DNA and RNA. One pyrimidine (uracil) and four purines (xanthine, hypoxanthine, guanine, and adenine) used in biological DNA and RNA have been found at the ppm level in the Murchison meteorite (Cronin et al., 1988).

Comets have also brought to the Earth much more organic compounds than the carbonaceous chondrites, and on the average much later. Their flux to the Earth has drastically subsided, but in a way it is not yet finished (Figure 2.4). Their chemistry is still very poorly known, and the information available comes almost entirely from comet Halley. Its dust particles revealed a large amount of fine grains coated with mostly unsaturated organic material (Krueger and Kissel, 1987). The presence of purines and pyrimidines were inferred from the mass spectra, but amino acids were not detected; if present, they were at least a factor of 30 less abundant than the purines and pyrimidines.

The argument that the heat of the impacts with the Earth is going to destroy all this organic material, is brought to rest when one realizes that a large fraction of it is brought to Earth by cometary dust. This dust, visible in the beautiful dust trails of comets, is braked by the upper atmosphere and gently brought to the ground. Prebiotic organic compounds came from space, and were brought again and again for aeons until they found the right "little pond" to get life started.

2.17   Summary

We have first established that the volcanic origin of our atmosphere and our oceans is an assumption that has never been demonstrated by any empirical data. It was

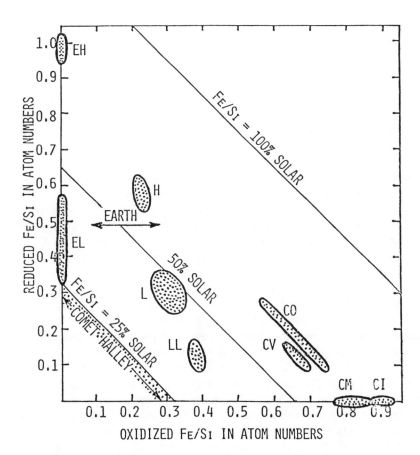

FIGURE 2.6. This plot of the reduced Fe/Si ratio versus the oxidized Fe/Si ratio for chondritic meteorites is usually referred to as a Urey–Craig diagram. The position of the comet Halley grains is indicated, as well as the place that would represent the Earth if it were homogenized. The arrows indicate the limits of our ignorance as far as a reduced phase of iron in the mantle is concerned. One understands that the outgassed and partially reduced fraction, collected near 1000 K, would correspond rather well to the enstatite chondrites of group H. An admixture of a small volatile fraction coming from comets would easily bring the mixture onto the Earth's position. Such a model should not be overinterpreted, because we do not know the zone of origin of the H chondrites.

based only on a blind extrapolation of the present recycling of volcanic gases back to a period that cannot be explored by geophysical means.

The only way to handle the problem is to consider the formation of the Earth in the more general paradigm of the origin of the solar system and the formation of the planets. This paradigm has received considerable support from recent observational evidence. It explains that Nature has found a way to make single stars by getting rid of the excess angular momentum through an accretion disk. Most of the mass falls first onto the accretion disk, before being fed into the central star. When the mass stops falling from afar, the turbulence stops in the disk and its dust sediments to the mid-plane of the disk, making big flat rings of solid particles that are the possible source of a planetary system. In the case of the Sun, these particles were mainly silicate and metallic iron dust for the terrestrial planets.

This sedimentation is convincingly documented by meteorites that came from the asteroid belt: the undifferentiated chondrites. Chondrites tell us that the gas dust separation occurred at a temperature close to 450 K at a distance of 2.6 AU from the Sun. Theory and observations concur to a value for the temperature gradient in the nebula, which implies that the dust particles that were going to agglomerate to form the Earth, were removed from the nebular gas in a range of temperatures going from 800 K to 1200 K (Figure 2.6).

These temperatures imply that the silicate and metallic iron particles were outgassed before their agglomeration into larger objects. So water was in nebular steam, carbon in gaseous CO, and nitrogen in gaseous N_2. The first 99% of the protoEarth were therefore completely devoid of volatiles. Isotopic abundances of the noble gases on the Earth confirm that there is no trace left of an early atmosphere. Volatile metals are depleted as shown by upper mantle samples.

The giant planets formed at a location where the temperature was much lower; hence their early building blocks were not stony, but icy, with a large fraction of volatile materials: the comets. The mass of the giant planets became large enough to deflect comets to the inner solar system; the final 1% of the Earth mass was a veneer of volatile material brought about by a late cometary bombardment due to the growth of the giant planets. Of this veneer, gases were depleted by a factor of 2000, and water by a factor of 100, by the numerous collisions with larger and larger objects predicted for the final stages of accretion. Some of the collisions with lunar- or Mars-sized bodies, as the one which has been proposed for the formation of the Moon, are likely to have completely vaporized and lost all the volatiles to space, which implies that the buildup of the atmosphere and the oceans started anew from the last of these collisions, whose epoch is difficult to specify since it was a stochastic event.

A small fraction of the volatile veneer, at most of the order of 0.1%, has been brought onto the Earth before the major accretion was complete, hence was buried in the mantle. The residual 0.9% has been brought mostly during the first half billion years of the Earth's existence, as testified by the ages of the lunar craters which were the relics of the same bombardment. In a general way, this bombardment has considerably rarefied but is not entirely finished, since comets must eventually hit the Earth from time to time. In particular, short-period comets still come from the Kuiper belt by the same orbital diffusion process that started some four and a half billion years ago, and long-period comets derive from those "new" comets coming from the Oort cloud, where they were also stored by an orbital diffusion that started in the same way.

2.17.1 Verified Predictions of the Model

- The cratering record for the Moon, Mars, and Mercury confirms an early massive bombardment larger that the possible "tail" of the early bodies from the terrestrial planets' zones.

- The lunar record predicts a similar bombardment of the Earth, of the right order of magnitude, and covering mostly the first half billion years.

- The diffusion rate of the cometary orbits predicts the subsiding mass rate of the flux reaching the Earth, by the combination of four exponential time constants produced by the four giant planets. The fact that time constants in proportion to the periods of the four giant planets explain satisfactorily the decline in time of the observed cratering flux on the Moon is particularly striking.

- The 1% volatile fraction coming from comets explains the presence in the right proportions of the many siderophile metals still present in the Earth's crust. They were not scavenged by the formation of the iron core of the Earth, because they were brought later as a veneer when the Earth was cooler. It also explains the origin of the oceans' water and of the atmosphere. The large excesses of water and gases brought by comets were needed to compensate for the large volatile depletions coming from the bombardment of large bodies during the last stages of the accretion. The "siderophile excess" and the "volatile excess" still considered as a puzzle by geologists are explained.

- The fact that refractory elements are slightly enriched in samples of the upper mantle, whereas the volatile elements are depleted, and the very volatile elements are strongly depleted, is explained by the accretion temperature of the planetesimals in the zone of the Earth, implying that the bulk of the Earth in not chondritic in the usual sense, but was formed by reduced iron grains with silicate grains depleted in most volatile elements.

- The "planetary" component of the noble gases on the Earth seem to be derived from anomalous enrichment factors by low-temperature trapping in the cometary snows.

- The ten-fold enhancement of the deuterium-to-hydrogen ratio was quenched at 200 K by the condensation of water snow from the nebular gas, which was eventually stored in comets before being brought into our oceans.

2.17.2   Unverified Predictions of the Model

The prediction that the lower mantle is completely outgassed cannot be verified yet. This is still a vigorously debated controversy in geophysics, whether the mantle plumes responsible for volcanic activity, in particular in the middle of the oceans like in Hawaii, come only from the upper mantle or throughout the 2900 km depth of the whole mantle. The present model takes sides and claims that no lower mantle source is likely; however, it is quite clear that the "gardening" of the Earth during the late phase of impacts, followed by vigorous convection in the upper mantle, may have buried volatiles down to 600 km or more, that could become the source of hot plumes.

The mean chemical composition of comets is still poorly known. For this reason, the composition of the early atmosphere and its evolution is still difficult to predict, in particular, because of its temperature history during the first half billion years which corresponds to the final phases of accretion for the volatiles.

2.18   Conclusion

The idea that the primary organic syntheses started in space is not new. Without even going back to Chamberlin and Chamberlin (1908), the possibility of an organic cosmochemistry was presumed by Oró (1961) and Bernal (1968). Whipple (1979), Chang (1979), and Delsemme (1979) concurred (in the same meeting) that comets may have provided most of the volatiles present nowadays on the Earth. Later, Delsemme (1981) took one step further, showing that the new paradigm on the origin of the solar system implies that the Earth first accreted from outgassed planetesimals than received a veneer of comet material as the result of the growth of the giant planets. This chapter has quantified this idea and shown that it connects a series of apparently unrelated problems, like the age record of the lunar craters, the reason why siderphile elements of the Earth crust were not scavenged by the formation of the Earth's iron core, the reason why the "siderophile excess" and the "volatile excess" are linked, the source of the "planetary" component of the noble gas abundances on the Earth, the origin of the deuterium in the oceanic water, as well as the depletion of the volatile and siderophile metals in the upper mantle. In order to understand better our early atmosphere, we must now study comets.

2.19   References

Alexander, E.C., Ozima, M. (eds.) (1978), Terrestrial Rare Gases,*Adv. Earth Planet. Sci.*, **3** (Japan Sci. Soc. Press, Tokyo), 229.

Anders, E. (1971), Meteorites and the early solar system. *Ann. Rev. Astronom. Astrophys.*, **9**, 1–34.

Anders, E., Grevesse, N. (1989), Abundances of the elements: meteoritic and solar. *Geochim. Cosmochim. Acta*, **53**, 197–214.

Anderson, A.T. (1975), Some basaltic and andesitic gases. *Rev. Geophys. Space Phys.*, **13**,

37–55.

Bar-Nun, A. Kleinfeld, I., Kochavi, E. (1988), Trapping of gas mixtures by amorphous water ice. *Phys. Rev B*, **38**, 7749–7754.

Bar-Nun, A., Kleinfeld, I. (1989), On the temperature and gas composition in the region of comet formation. *Icarus*, **80**, 243–253.

Berkner, L.V. Marshall, L.C. (1965), On the origin and rise of oxygen concentration in the earth's atmosphere. *J. Atmos. Sci.*, **22**, 225–261.

Bernal, J.D. (1968), *Origins of Prebiotic Systems....* S.W. Fox (ed.) (Academic Press, New York), pp. 65–68.

Bertout, C. (1989), *Annu. Rev. Astronom. Astrophys.*, **27**, 351–395.

Cameron, A.G.W. (1985), Formation and evolution of the primitive solar nebula. In D.C. Black and M.S. Matthews (eds.), *Protostars & Planets II* (University of Arizona Press, Tucson), pp. 1073–1099.

Cameron, A.G.W. (1988), Origin of the solar system. *Annu. Rev. Astron. Astrophys.*, **26**, 441–472.

Carr, M.H., Saunders, R.W., Strom, R.G., Wilhelms, D.E. (1984), *Geology of the Terrestrial Planets* (NASA SP-469, Washington DC)

Chamberlin, T.C. and Chamberlin, R.T. (1908), Early terrestrial conditions that may have favored organic synthesis. *Science*, **28**, 897–910.

Cassen P., Shu F.H., Tereby S. (1985). In D.C. Black and M.S. Matthews (eds.), *Protostars and Planets II* (Univ. of Arizona Press, Tucson), pp. 448–483.

Chang, S. (1979), Comets: Cosmic connections with carbonaceous meteorites, interstellar molecules and the origin of life. In M. Neugebauer, D.K. Yeomans, J.C. Brandt and R.W. Hobbs (eds.), *Space Missions to Comets* (NASA SP-2089, Washington, DC), pp. 59–111.

Chyba, C.F. (1987), The cometary contribution to the oceans of primitive Earth. *Nature*, **330**, 632–635.

Cronin, J.R., Pizzarello, S., Cruikshank, D.P. (1988), Organic matter in carbonaceous chondrites, planetary satellites, asteroids and comets. In J.F. Kerridge & M.S. Matthews (eds.), *Meteorites and the Early Solar System* (Univ Arizona, Tucson), pp. 819–857.

Delsemme, A.H. (1979), Scientific returns from a program of space missions to comets,. In Neugebauer et al. (eds.) *Space Missions to Comets*, (NASA SP-2089, Washington DC), pp. 139–178.

Delsemme, A.H. (1981), Nature and origin of organic molecues in comets. In C. Ponnamperuma (ed.), *Comets and the Origin of Life* (D. Reidel Publishing Company, Dordrecht, Holland), pp. 33–42.

Delsemme, A.H. (1981), In C. Ponnamperuma (ed.), *Comets and the Origin of Life* (D. Reidel Publishing Company, Dordrecht, Holland), pp. 141–159.

Delsemme, A.H. (1987). In *Diversity and Similarity of comet* (European Space Agency, ESA-SP-278, Paris), pp. 19–30.

Delsemme, A.H. (1991), Nature and history of the organic compounds in comets: An astrophysical view. In R.L. Newburn, M. Neugebauer, and J. Rahe (eds.), *Comets in the Post-Halley Era, Vols. I-II* (Dordrecht, Boston), pp. 377–427.

Delsemme, A.H. (1991b), Origin of the biosphere of the Earth. In J. Heidmann & M.J. Klein (eds.), *Lecture Notes in Physics 390: Bioastronomy* (Springer-Verlag, New York), pp. 117–123.

Delsemme, A.H. (1991c), International Halley Watch. In Z. Sekanina (ed.), *The Comet Halley Archives* Summary Volume (NASA-JPL, Pasadena), pp. 317–330.

Delsemme, A.H. (1992), Cometary origin of carbon, nitrogen and water on the Earth.

Origins Life Evol. Biosphere, **21**, 279–298.

Delsemme, A.H. (1993), Cometary origin of the biosphere: A progress report. *Adv. Space Res.*, **15**, 49–57.

Delsemme, A.H. and Miller, D.C. (1970), Physico-chemical phenomena in comets -II: Gas adsorption in the snows of the nucleus. *Planet. Space Sci*, **18**, 717–730.

Delsemme, A.H. and Wenger, A. (1970), Physico-chemical phenomena in comets -I: Experimental study of snows in a cometary environment. *Planet Space Sci.*, **18**, 709–716.

Delsemme, A.H. and Swings, P. (1952), Gas hydrates in cometary nuclei and interstellar grains. *Ann. Astrophys.*, **15**, 1–6.

Dreibus, G. and H. Wänke (1989), Supply and loss of volatile constituents during the accretion of terrestrial planets. In S.K. Atreya, J.B. Pollack, and M.S. Matthews (eds.), *Origin and Evolution of Planetary and Satellite Atmospheres* (Univ. Arizona Press, Tucson), pp. 268–288.

Dymond, J. and Hogan, L. (1978), Factors controlling the noble gas abundance patterns of deep-sea basalts. *Earth planet. Sci. Lett.*, **38**, 117–128.

Eberhart, P. (1981). In *Basaltic Volcanism Study Project* (Pergamon, New York), pp 1025–1031.

Eberhart, P., Dolder, U. and Schulte, W., Krankowsky, D., Lammerzahl, P., Hoffmann, J.H., Hodges, R.R., Bertheller, J.J, and Illiano, J.M. (1987), The D/H ratio in water from Comet P/Halley, *Astron. and Astrophys.*, **187**, 435–437.

Everhart, E. (1977), The evolution of comet orbits as perturbed by Uranus and Neptune. In A.H. Delsemme (ed.), *Comets, Asteroids, Meteorites* (Univ. Toledo), pp 99–104.

Fernandez, J.A. and Ip, W.H. (1981), Dynamical evolution of a cometary swarm in the outer planetary region. *Icarus*, **47**, 470–479.

Fernandez, J.A. and Ip, W.H. (1983), On the time-evolution of the cometary influx in the region of the terrestrial planets. *Icarus*, **54**, 377–387.

Gaffey, M.J. and Mc Cord, T. B (1979), Mineralogical and petrological characteristics of asteroid surface materials. In T. Gehrels (ed.), *Asteroids* (Univ. Arizona), pp. 688–723.

Geiss, J. and Reeves, H. (1981), Deuterium in the solar system. *Astron. Astrophys.*, **93**, 189–199.

Goldreich, P. and Ward, W.R. (1973), The formation of planetesimals, *Astrophys. J.*, **183**, 1051–1061.

Grinspoon, D.H. and Lewis, J.S. (1987), Deuterium fractionation in the presolar nebula: kinetic limitations on surface catalysis. *Icarus*, **72**, 430–436.

Hartmann, L.W., Kenyon, S.J. (1990), Optical veiling disk accretion, and the evolution of T Tauri Stars. *Astrophys. J.*, **349**, 190–196.

Horedt, G.P. (1978), Blow-off of the protoplanetary cloud by a T Tauri like solar wind. *Astron. Astrophys.*, **64**, 173–178.

Hubbard, W.B. (1984), *Planetary Interiors* (Van Nostrand-Reinhold, New York).

Ip, W.H. and Fernandez, J.A. (1988), Exchange of condensed matter among the outer and terrestrial protoplanets and the effect on surface impact and atmospheric accretion. *Icarus*, **74**, 47–61.

Jessberger, E.K., Christoforidis, A and Kissel, J. (1988), Aspects of the major element composition of Halley's dust. *Nature*, **332**, 691–695.

Kazimirchak-Polonskaya, E.I. (1972), The major planets as powerful transformers of cometary orbits. In G.A Chebotarev, E.I. Kazimirchak-Polonskaya, B.G. Marsden (eds.), *The Motions, Evolution of Orbits and Origins of Comets* (D. Reidel Publishing Co., Dordrecht, Holland), pp. 373–397.

Kerridge, J.F. (1991) (personal communication). See also Anders, E. and Kerridge, J.F.

(1988), Future directions in meteorite research. In J.F. Kerridge and M.S. Matthews (eds.), *Meteorites and the Early Solar System* (Univ. of Arizona, Tucson), pp. 1155–1186.

Krueger, F.R. and Kissel, J. (1987), The chemical composition of the dust of Comet P/Halley as measured by "PUMA" on board VEGA-1. *Naturwiss.*, **74**, 312–316.

Laplace, P.S. (1796), *Exposition du Systeme du Monde* (Vve Courcier, Paris), pp. 431 in the 4th edition of 1813.

Larimer, J.W. (1967), Chemical fractionations in meteorites -I: Condensation of the elements. *Geochim. Cosmochim. Acta*, **31**, 1215–1238.

Larimer, J.W. (1968), An experimental investigation of oldhamite, CaS; and the petrologic significance of oldhamite in meteorites. *Geochim. Cosmochim. Acta*, **32**, 965–982, and Experimental studies on the system Fe-MgO-SiO2-O2 and their bearing on the petrology of chondritic meteorites, 1187–1207.

Larson, R.B. (1984), Gravitational torques and star formation. *Mon. Not. Royal Astron. Soc.*, **206**, 197–207.

Lewis, J.S. (1972 a), Low temperature condensation from the solar nebula. *Icarus*, **16**, 241–252.

Lewis, J.S. (1972 b), Metal/silicate fractionation in the solar system. *Earth Planet. Sci. Lett.*, **15**, 286–290.

Lewis, J.S. (1974), The temperature gradient in the solar nebula. *Science*, **186**, 440–443.

Lewis, J., Barshay, S.S. and Noyes, B. (1979), Primordial retention of carbon by the terrestrial planets. *Icarus*, **37**, 190–206.

Lewis, J.S. and Prinn, R.G. (1980), Kinetic inhibition of CO and N2 reduction in the solar nebula. *Astrophys. J.*, **238**, 357–364.

Lin, D.N.C. and Papaloizou, J. (1985), On the dynamical origin of the solar system. In D.C. Black and M.S. Matthews (eds.), *Protostars and Planets II)* (Univ. of Arizona, Tucson), pp. 981–1072, in particular Fig 7.

Lynden-Bell, D. and Pringle, J.E. (1974), The evolution of viscous discs and the origin of the nebular variables. *Mon. Not. Royal Astron. Soc.*, **168**, 603–637.

Matsui, T. and Abe, Y. (1986), Impact induced atmospheres and oceans on Earth and Venus. *Nature*, **322**, 526–528.

Mizuno, H. (1980), Formation of the giant planets. *Progr. Theoret. Phys.*, **64**, 544–557.

Morfill, G.E. (1988), Protoplanetary accretion disks with coagulation and evaporation. *Icarus*, **75**, 371–379.

Morfill, G.E. and Wood, J.A. (1989), Protoplanetary accretion disc models: the effects of several meteoritic, astronomical, and physical constraints. *Icarus*, **82**, 225–243.

Morgan, J.W., Wandless, G.A., Petrie, R.K., and Irving, A.J. (1981), Composition of the earth's upper mantle. I- siderophile trace elements in ultramafic nodules. *Tectonophys.*, **75**, 47–67.

Morrison, D. (1977). In A.H. Delsemme (ed.), *Comets, Asteroids, Meteorites* (Univ. of Toledo, Ohio), pp. 177–184.

Oró, J. (1961), Comets and the formation of biochemical compounds on the primitive earth. *Nature*, **190**, 389–390.

Owen, T., Bar-Nun, A., and Kleinfeld, I. (1991). In Newburn et al. (eds.), *Comets in the post-Halley Era, vol. I* (Kluwer Publ., Netherlands), pp. 429–437.

Pepin, R.O. (1989), Atmospheric compositions: key similarities and differences. In Atreya et al. (eds.), *Origin and Evolution of Planetary and Satellite Atmospheres* (Univ. of Arizona, Tucson), pp. 291–305.

Pepin, R.O. (1991), On the origin and early evolution of terrestrial planet atmospheres and meteoritic volatiles. *Icarus*, **92**, 2–79.

Prinn, R.G. and Fegley, B. (1989). In Atreya et al. (eds.), *Origin and Evolution of Planetary and Satellite Atmospheres* (Univ. of Arizona, Tucson), pp. 78–137.

Rama Murthy, V. (1991), Early differentiation of the Earth and the problem of mantle siderophile elements: a new approach. *Science*, **253**, 303–306.

Reynolds, J.H., Frick, U., Neil, J.M., and Phinney, D.L. (1975), Rare-gas-rich separates from carbonaceous chondrites. *Geochim. Cosmoshim. Acta*, **42**, 1775–1797.

Ringwood, A.E. (1977), *Composition and Origin of the Earth*, (School of Physics, Publ. 1299, Australian National Univ., Canberra).

Rowan-Robinson, M. (1985), Infrared observations of interstellar clouds. *Physica Scripta*, **T11**, 68–70.

Rubey, W.W. (1951), Geologic history of sea water: an attempt to state the problem. *Bull. Geol. Soc. Am.*, **62**, 1111–1147.

Rubey, W.W. (1955). In Poldervaart (ed.), *Crust of the Earth*, (Geol. Soc. of America, New York), pp. 630–650.

Safronov, V.S. (1972). In G.A Chebotarev, E.I. Kazimirchak-Polonskaya, B.G. Marsden (eds.), *The Motions, Evolution of Orbits and Origins of Comets* (D. Reidel Publishing Co., Dordrecht, Holland), pp. 329–334.

Safronov, V.S. (1991), Kuiper Prize Lecture: Some problems in the formation of the planets. *Icarus*, **94**, 260–271.

Sears, D.W.G. and Dodd, R.T. (1988), Overview and classification of meteorites. In J.F. Kerridge and M.S.Matthews, (eds.), *Meteorites and the Early Solar System* (Univ. of Arizona, Tucson), pp. 3–31.

Signer, P. and Suess, H.E. (1963). In Geiss and Goldberg (eds.), *Earth Science and Meteoritics* (North Holland, Amsterdam), pp. 241–278.

Smith, B.A. and Terrile, R.J. (1984), A circumstellar disk around β Pictoris. *Science*, **226**, 1421–1424.

Stevenson, D.J. and Lunine, J.I. (1988), Rapid formation of Jupiter by diffusive redistribution of water vapor in the solar nebula. *Icarus*, **75**, 146–155.

Turekian, K.K. (1972). In *Chemistry of the Earth* (Holt, Rinehart & Winston, New York), pp. 102.

Van Hise, N. (1904). In *A Treatise on Metamorphism* (United States Geological Survey, Mon. 40), pp. 970, 973, & 974.

Vidal-Madjar, A. (1983). In Audouze et al. (eds.), *Diffuse Matter in Galaxies* (ASI Series C No 110, Reidel Dordrecht), pp. 57–94.

Von Weiszäcker (1944) quoted by Kuiper in A. Hyneck (ed.), *Astrophysics: a topical symposium* (Univ. of Chicago Press, Chicago).

Wänke, H. (1981), Constitution of terrestrial planets, *Phil. Trans. Roy. Soc. London, Ser.A*, **303**, 287–302.

Wänke, H., Dreibus, G., and Jagouts, E. (1984). In Kroner et al. (eds.), *Archaean Geochemistry* (Springer Verlag, Berlin), pp. 1–24.

Weidenschilling, S.J. (1988), Formation processes and time scales for meteorite parent bodies. In J.F. Kerridge and M.S. Matthews (eds.), *Meteorites and the Early Solar System* (Univ. of Arizona, Tucson), pp. 348–371.

Weissman, P. (1989). In Atreya et al. (eds.), *Origin and Evolution of Planetary and Satellite Atmospheres* (Univ. of Arizona, Tucson), pp. 230–267.

Wetherill, G.W. (1975), Late heavy bombardment of the moon and terrestrial planets. In *Proceedings of the 6th Lunar Science Conference* (Lunar and Planetary Institute, Houston), 1539–1561.

Wetherill, G.W. (1980), Formation of the terrestrial planets. *Annu. Rev. Astron. Astrophys.*, **18**, 77–113.

Wetherill, G.W. (1989). In Binzel et al. (eds.), *Asteroids II* (Univ. Arizona, Tucson), pp. 666–670.

Wetherill, G.W. (1990), Comparison of analytical and physical modeling of planetesimal accumulation. *Icarus*, **88**, 336–354.

Wetherill, G.W. (1991), Occurrence of Earth-like bodies in planetary systems. *Science*, **253**, 535–538.

Wetherill, G.W. and Champman, C.R. (1988), Asteroids and meteorites. In J.F. Kerridge and M.S. Matthews (eds.), *Meteorites and the Early Solar System* (Univ. Arizona, Tucson), pp. 35–67.

Wetherill, G.W. and Cox, L.P. (1985), The range of validity of the two-body approximation in models of terrestrial planet accumulation. *Icarus*, **63**, 290–303.

Wetherill, G.W. and Stewart, G.R. (1989), Accumulation of a swarm of small planetesimals. *Icarus*, **77**, 330–357.

Whipple, F.L. (1979), Scientific need for a cometary mission. In Neugebauer et al. (eds.), *Space Missions to Comets* (NASA SP-2089, Washington DC), pp. 1–32.

Wood, J.A., Morfill, G.E. (1988), A review of solar nebula models. In J. F. Kerridge and M.S. Matthews (eds), *Meteorites and the Early Solar System* (Univ. of Arizona, Tucson), pp. 329–347.

Zellner, B. (1979). In *Asteroids*, T. Gehrels (ed.), University of Arizona Press, Tucson, pp. 783–806.

3
Organic Chemistry in Comets From Remote and In Situ Observations

J. Kissel, F.R. Krueger, and K. Roessler

ABSTRACT Radiation induced chemistry is the major source of organic matter in space. Comets as small bodies, that were kept in the cold parts of our solar system since its formation provide a unique source to study such genuine material.

When the VEGA and GIOTTO spacecrafts flew by comet P/Halley in 1986 the mass-spectrometers Puma and PIA measured the composition of cometary dust particles impacting at speeds of well above 65 km s^{-1}. Ion formation upon impact leads to mostly atomic ions. However, a small fraction of the ions measured could be related to molecules. A sophisticated analysis allowed for the first time to point to the chemical nature of cometary organics based on actual mass spectra.

The next logical step for in situ cometary exploration is a rendezvous-type mission. This had been planned by NASA and the German BMFT, but was unfortunately canceled in the spring of 1992. In the meantime the European Space Agency (ESA) has dedicated its next major mission, Rosetta, to perform a comet rendezvous.

A time-of-flight secondary ion mass spectrometer (CoMA) can provide much higher mass resolution up to molecule masses of some 3000 Da.

3.1   Introduction

Early observations of comets with ground based telescopes and spectrometers have revealed that comets do release organic material as they pass the Sun. The source of this material has long been subject to speculations. When Whipple (1950) proposed his "dirty snowball" model for the comet nucleus, the nature and complexity of the organic material was not known. The formation of molecules in the free interstellar space with low-particle density is largely prohibited by reaction kinematics as it is not possible to satisfy the conservation of momentum and energy for a two-body collision that would lead to the formation of a molecule. Only larger molecules, however (> some 10 non-H atoms), have a vapor pressure low enough to form and remain in solid state even in evolutionary phases, where a slight warm-up may occur due to, for example, star passage. As multiparticle collisions are extremely unlikely in interstellar space, synthesis of larger molecules must occur step by step. That may well happen on the surfaces of solid bodies. The most abundant molecules H_2 and CO are formed that way. The formation of more complex molecules (i.e., consisting of more atoms) has first to overcome the high binding energy of these two molecules. It is again a solid surface that helps. In interstellar space these solid bodies assume a very low surface temperature of a few Kelvin. Should one molecule be dissociated by an energetic event, the reaction products remain in place. While electrons are always mobile, protons (H^+) become mobile at 12 K, and radicals above some 50 K, depending on the substrate. Two ways for the formation of organic products have been proposed and are still widely discussed, as in the following:

- Greenberg, see, e.g., Greenberg (1978), has proposed that on the surfaces of small interstellar grains molecules are accumulated and further processed by UV radiation. As these particles become part of a protoplanetary nebula, they are embedded in cometary nuclei in the outer part of the forming solar system. This means that a great deal of organic chemistry has already taken place before the formation of the cometary nucleus. He and others have performed a large number of laboratory experiments showing that modification of simple ices into complex molecules does indeed take place at low temperatures under the action of UV radiation.

- It has been shown that other types of radiation, especially cosmic rays, can be used to produce complex molecules from simple ices at low temperature. For cosmic rays, however, the penetration depth would be limited. Assuming today's observed cosmic ray flux, Draganic et al. (1984) concludes a processing depth of some 100 m. Again many authors have published the results of laboratory experiments showing the feasibility of the concept.

In both cases very long times (order of 10^9 yr) are needed for the production of large quantities of molecules. If accumulation is achieved prior to comet nucleus formation one may assume that the entire cometary nucleus has the potential of gas emission. The situation would be different, however, should processing take place after nucleus formation. As a comet loses part of its active surface during each perihelion passage in the latter case, emission of volatile organic material would decrease and finally cease with time (i.e., with the comet nucleus age), and an inactive nucleus would be the "leftover." Even then it is not clear, however, whether the supply of volatile material is completely exhausted or simply sealed off by a more refractory layer which would allow reactivation after its penetration. Since we know that over 20 bodies in the outer solar system expose ices on their surface, what is it that makes comets so unique?

Formed in the outer, cold regions of the early solar system cometary nuclei are supposedly unaltered from the time of their formation until today. Some of these small bodies are deflected into the inner part of the solar system, where their surfaces are warmed up by solar radiation. The release of volatile material provides the first insight into the chemistry occurring in the cold stage of formation through observation of the volatile reaction products. Interpretation has to be cautious, however. Once the volatiles are released they react with the solar radiation and the solar wind. The result is that we can observe only so-called "daughter products." Only close to the nucleus we may expect the presence of the original volatiles, the so-called "parent molecules."

Knowing more details about the possible formation processes of those organic volatiles, we might be better able to infer the materials in the comet from what we measure, once they are released. In the next chapter we will give an overview on processes leading to organic molecules in space.

3.2   Radiation and Hot Atom Chemical Processes in Space

The majority of chemical reactions in space, in particular, at low temperatures are directly or indirectly radiation induced, see Duley and Williams (1984), Winnewisser and Armstrong (1989), Huebner (1990), and Strazzulla and Johnson (1991). All ion-molecule interactions start with the radiolytic or photolytic elimination of an electron out of the molecule. There are, however, some reactions which are not radiation induced: those in high temperature spots (hot interstellar clouds, hot planets, and surfaces of comets at or near perihelion). Polymerization of H_2CO or HCN has been reported to proceed at temperatures below 80 K, by tunnel effects even at 5–10 K, see: Rettig et al. (1992) and Schutte et al. (1993). Some of the not yet examined possibilities in icy systems like comets are chemical reactions in thermal equilibrium of evaporating and recrystallizing gases with the inner and outer surfaces, in particular, when minerals of the montmorillonite type are present as catalysts. The new class of suprathermal reactions (outside the thermal equilibrium) by atoms, ions, molecules or clusters with kinetic energies >0.5 eV per atom (McFadden et al., 1987; Goldanski, 1984; Greenberg, 1978, 1984; Heyl and Roessler, 1992) in general is triggered by radiation, even if they can also be induced by shock waves, impact or explosive processes, see: Stöcklin (1969), Roessler (1991, 1992a, b), Heyl and Roessler (1992).

3.2.1   Radiation Sources in Space

Table 3.1 summarizes energies, fluxes, and fluences for energetic ions from solar wind and flares, T-Tauri winds in the early history of the solar system and solar cosmic rays.

Table 3.2 lists solar wind fluxes for bodies in the solar system. Some of the most interesting objects for the study of solar wind effects are the asteroids. Since they do not possess an atmosphere their surface is hit by the ions. In the lifetime of the solar system, the asteroids were subject to a fluence of 4×10^{24} cm^{-2} of keV ions creating some 10^{26} cm^{-2} secondary energetic atoms. Even if the surfaces are turned from time to time by mechanical mixing of grains, the enormous number of hot processes can totally change the pristine matter of carbonaceous asteroids. Other interesting objects are the planets and satellites without or almost without an atmosphere such as Mercury. Cometary surfaces are not touched by solar wind when near to the Sun because of the develoment of a coma.

However, on its orbit far from the Sun ($r > 5$ AU) the bare nucleus of P/Halley

TABLE 3.1. Fluxes and Fluences of Energetic Ions and Photons at 1 AU, from Roessler (1992a).

Source	Major constituents	Flux, $cm^{-2}s^{-1}$	Fluence in 4×10^9 yr, cm^{-2}
Solar wind	95 % 1 keV H^+ 4 % 4 keV He^{2+} 0.1 % 10 keV C^{6+}	$2 - 3 \times 10^8$	4×10^{25}
Solar flares ($1\,a^{-1}$)	95 % < MeV H^+ 5 % < MeV He^{2+}	varies	4×10^{19}
Interstellar medium (relative motion)	50 % 18 eV H 45 % 70 eV He 2 % 250 eV C, N, O	6.5×10^5	8×10^{22}
Low energy cosmic rays	87 % < 1 MeV H^+ 12 % < 1 MeV He^{2+} 0.2 % < 1 MeV C^{6+}, O^{8+}	10^1	10^{18}
Solar UV photons (110 - 300 nm)	$\approx$100 % 4.3 - 7.3 eV (300 - 170 nm) $1^o/_{oo}$ 7.3 - 11.3 eV (170 - 110 nm)	2×10^{15} 3×10^{12}	2×10^{32} 4×10^{29}

obtains 20% of the total radiation dose (fluence) the comet experiences along its whole orbit. The total dose (fluence) of cosmic rays to all small icy objects in space and to the first meters of larger bodies such as comets amounts to 2×10^{11} rad or 400–500 eV per molecule of H_2O ice, cf. Roessler (1991). Table 3.3 presents the possibilties of simulating space radiation in the laboratory. Here the application of doses lower by a factor of ten is meaningful since it considers long-term annealing changes of the damage state. Total dose (fluence) arguments are less valid in space chemistry simulation than those of energy density within one cascade.

3.2.2   Hot Species in Space

Table 3.4 lists the sources and energies of suprathermal (hot) species. These may be ions, atoms, molecules or their fragments, radicals, clusters, whole grains, colliding larger bodies, etc.

Except for the cases of photodissociation and dissociative recombination, most of the species are primarily ionized. In solids a fast neutralization takes place by electron pick-up, and the hot species react as neutrals at the end of their trajectories

TABLE 3.2. Solar Wind Fluxes for Objects in the Solar System, from Roessler (1992a).

Object in space	Distance to sun, AU	Flux of solar wind ions $cm^{-2}s^{-1}$
Mercury	0.387	2×10^9
P/Halley at perihelion	0.587	1.3×10^9
Venus	0.723	10^9
Earth	1.000	$2 - 3 \times 10^8$
Mars	1.524	10^8
Asteroids	3.5	2×10^7
Jupiter + moons	5.202	10^7
Saturn + moons	9.555	3×10^6
Uranus	19.21	7×10^5
Neptune	30.109	$3 - 4 \times 10^5$
P/Halley at aphelion	35	2×10^5

TABLE 3.3. Simulation of Space Radiation in the Laboratory, from Roessler (1992a).

Experiment	Exp. energies eV	Type of space radiation
Atomic beam	$1 - 10$	Shockwaves in interstellar medium
VUV photons	$4.3 - 11.3$	Solar photons
Plasma discharge	$10^2 - 10^3$	Ion torus (Io) discharge lightening
Ion implantation	$10^3 - 3 \times 10^5$	Solar wind
Cyclotron ions (nuclear recoil)	$5 - 30 \times 10^6$	Cosmic rays

TABLE 3.4. Sources for Hot Species in Space, from Roessler (1992b).

Source	Primary energy, eV	Approx. penetration into solids, μm
Photodissociation by UV photons	$1-5$	Surface
Dissociative recombination of photolytically produced ions	$1-5$	Surface
Collision of space craft with rest gas	$3-10$	Surface
Motion relative to the interstellar medium	$10^1 - 10^2$	Surface
Nuclear β-decay	$1 - 10^1$	Surface
Nuclear α-decay	10^5	Several 10^{-2}
Nuclear reactions	$10^2 - 10^8$	$10^{-5} - 10^1$
Shock waves in the interstellar medium	$10^1 - 10^3$	$10^{-6} - 10^{-4}$
Discharges, lightening	$10^2 - 10^4$	$10^{-5} - 10^{-3}$
Solar (stellar) wind	$10^3 - 10^5$	$10^{-4} - 10^{-2}$
Ion torus (Io)	$10^3 - 10^5$	10^{-4} – several 10^{-2}
Pick-up ions	$10^4 - 10^5$	$10^{-3} - 10^{-2}$
T-Tauri winds	10^6	$10^{-1} - 100$
Cosmic rays	$10^3 - 10^{23}$	$10^{-4} - 10^8$

TABLE 3.5. All Secondary Knock-on Atoms with E 2 eV Formed in Collision Cascades of keV Ions in Frozen H_2O and CH_4 at 77 K. MARLOWE Computer Simulation, from Roessler (1991).

Primary	Total number of secondaries per cascade (E 2 eV)	
	H_2O	CH_4
1 keV H	25 H, 10 O	18 H, 15 C
4 keV He	57 H, 42 O	63 H, 30 C
10 keV C	150 H, 113 O	203 H, 63 C

when chemical bonds can be formed. Hot atom chemistry induced by energetic ions depends on the Sn/Se-ratio, i.e., the distribution of energy among nuclear (Sn) and electronic stopping (Se). The first is in particular important for the creation of suprathermal species. The electronic part plays an important role in the case of strong ionization by heavy swift ions via Coulomb repulsion or explosion. Projectiles with a kinetic energy in keV higher than the value for the mass in a.m.u. will transfer their energy to the electronic system of the target atoms (Se) leading to excitation and ionization. Species with energies in keV below the a.m.u. value will preferentially transfer energy by elastic collisions (Sn) creating point defects, replacement events, etc. Sn leads to the formation of hot secondary atoms which can undergo in their turn suprathermal reactions. According to the rule of thumb that Sn $\approx$ Se when the energy of the projectile in keV is similar to its mass in a.m.u., solar wind particles deliver about 50% of their energy to nuclear collisions, whereas MeV ions of cosmic rays transfer most of their energy by inelastic interactions. However, some nuclear collisions always occur which bring about the effects observed by MeV ions in CH_4 (see Roessler (1987), Roessler and Eich (1987), Patnaik et al. (1989, 1990), Roessler et al. (1990), Kaiser and Roessler (1992) and Kaiser et al. (1992), and Kaiser (1993)). Another way of transferring energy deposited to the electronic system of insulators into energetic motions of atoms could be via self-trapped excitons.

The secondary atoms are in general much more numerous than the primaries and are responsible for the majority of hot reactions in space. Table 3.5 shows the number of secondary (and tertiary, etc.) hot H, O, and C atoms with kinetic energies E $\geq$ 2 eV formed by typical primary solar wind ions.

Photodissociation by UV photons gives rise to primaries with energies in the eV range, unable to induce secondaries. Table 3.6 lists some of the processes

TABLE 3.6. Photodissociation Leading to Hot Species, from Roessler (1992a).

Photodissociation	Excess energy, eV
$NO \xrightarrow{191\,nm} N + O$	1.8
$H_2O \xrightarrow{186\,nm} H + OH$	1.9
$HCN \xrightarrow{195\,nm} H + CN$	4.3
$CO_2 \xrightarrow{194\,nm} O + CO$	5.5
$NH_3 \xrightarrow{317\,nm} H + NH_2$	2.0
$CH_4 \xrightarrow{280\,nm} H + CH_3$	5.9
$O_2 \xrightarrow{175\,nm} O + O$	1.3

in simple molecules. The excess energy in that table has to be divided onto the fragments inversely to their mass ratio. Photodissociation gives rise to limited local effects, especially "hot radiation chemistry" by H atoms, scission of bonds, etc. The buildup of large stuctures is not possible. However, molecules like CH_2O, CH_3OH, NH_2OH, R-CN, etc., have been prepared by radical recombination in thin layers of ice mixtures simulating icy layers on interstellar grains or cometary surfaces (Greenberg, 1984). It has to be mentioned that for the solid state the penetration of VUV photons is limited to about 1 μm.

3.2.3 Target Systems in Space

The interstellar matter of our galaxy consists of 99% gases (3% ions and 97% neutrals) and 1% solids. Liquids are encountered in very few cases such as only in aerosols in atmospheres and in some oceans of planets and satellites (predicted ethane ocean on Titan). The gaseous systems are in general very diluted, corresponding to ultra-high vacuum, such as listed in Table 3.7. Only a few of the planetary and satellite atmospheres are comparable to terrestrial conditions.

The temperatures of these systems range from a few 10 K in dense molecular clouds to 730 K in the atmosphere of Venus and some 10^5 K in very dilute interstellar medium. The composition can vary from very simple molecules H_2, CO, H_2O, CH_4, NH_3, CO_2, HCN, C_2H_6, etc., to complicated highly unsaturated long chain molecules such as, for example, cyanoacetylene $HC \equiv C - CN$, cyanopentayne $HC - (C \equiv C)_5 - CN$, or organic ring molecules such as the cyclopropenyl ion $C_3H_3^+$ as also found in P/Halley's coma by Korth et al. (1989) or the cyclopropenylidene radical C_3H_2, most of which are only stable at high dilution.

Table 3.8 lists some selected solid systems in space, the temperatures of which range between 10 K for interstellar grains to some 100 K for planets and asteroids

TABLE 3.7. Density of Some Selected Gaseous Systems in Space, from Roessler (1992b).

System	Number Density, atoms or mols. cm^{-3}	Equivalent Pressure, mbar
Interstellar medium	$10^{-1} - 10^{1}$	several $10^{-18} - 10^{-16}$
Diffuse interstellar clouds	$10^{1} - 10^{3}$	several $10^{-16} - 10^{-14}$
Dark interstellar clouds	$10^{3} - 10^{5}$	several $10^{-14} - 10^{-12}$
Inner coma of comets	$10^{5} - 10^{8}$	several $10^{-12} - 10^{-9}$
Atmosphere of Mercury	$< 10^{8}$	$< 10^{-9}$
Atmosphere of Mars	1.76×10^{17}	6.5
Atmosphere of Titan	4.3×10^{19}	1.3×10^{3}
Atmosphere of Venus	2.4×10^{21}	9×10^{4}

in the inner solar system. It can be seen that almost all systems are mixtures of minerals (silicates, oxides, sulphides, iron), ices (H_2O, NH_3, CH_4, CO, CO_2, etc.), carbonaceous matter (C, SiC, a — C:H) and complex organic substances (aromates, polycyclic aromatic hydrocarbons (PAH), pyrimidine derivatives, and other precursors of biomolecules).

3.2.4   Dualism of Radiation Chemistry and Hot Atom Chemistry

Solar photons produce hot hydrogen atoms in cometary comae by photodissociation of H_2O molecules which on their turn are then attacked by the hydrogens via H abstraction. Both processes lead to the destruction of H_2O. The formation of hot secondary atoms by knock-on is in essence a radiolytic destruction of the substrate. Formation, motion, and reaction of hot species are always accompanied by radiation physics and chemistry (photolysis, radiolysis) either by the hot atom itself or by the accompanying radiation field, which change the target and the products likewise. Since the processes are interwoven and lead sometimes to similar products, a discrimination is only possible by a careful study of dose (flux, fluence) and energy density (linear energy transfer) effects.

Astrophysics has hitherto considered the radiation field in space only from the point of view of radiation physics and chemistry, in particular, the formation of ions and radicals in the gas phase. Solids were considered only in the case of the icy layers on interstellar dust and comets. Radiation processes in solids are in general an open field in space research. For example, it is quite unknown at present how annealing during the long irradiation periods modifies the concentration of defects.

Due to its high reaction rates (see Heyl and Roessler, 1992) and in some cases the formation of specific products (e.g., HCN in the system C/N_2) hot atom chemistry in the gas phase may well compete with radiation-induced chemistry and a separate

TABLE 3.8. Some Selected Solid Systems in Space, from Roessler (1992b).

System	Diameter, m	Approx. composition
Interstellar grains	$10^{-7} - 10^{-6}$	Mineral (silicate) nucleus with layers of frosts and carbonaceous (organic) refractories
Interplanetary dust	$10^{-7} - 10^{-3}$	Mineral (silicate) nucleus with layers of frosts and carbonaceous (organic) refractories
Meteorites	$10^{-5} - 10^{1}$	Minerals, carbon, organics
Meteors	$10^{-4} - 10^{-2}$	Minerals, carbon, organics
Ring material of planets	$10^{-5} - 10^{1}$	Minerals, ices
Comets	several $10^{3} - 10^{4}$	Minerals, ices, organics, carbon
Asteroids	$10^{5} - 10^{6}$	Minerals (S), organics and carbon (C)
Satellites	$3 \times 10^{4} - 5 \times 10^{6}$	Mineral nuclei, often icy surfaces
Planets	$3 \times 10^{6} - 1.4 \times 10^{8}$	Mineral nuclei, often icy surfaces

treatment is justified. Even more in the solid state since hot reactions proceed in most of the cases in a collision cascade where secondary hot atoms, excited or at least activated molecules, and a decrease of lattice binding energy add to create an activated zone. Collective or multicenter reactions are possible which cannot be induced by simple ionizing radiation. Furthermore, it can be expected that hot reactions in the solid state are also fast enough to compete with thermal radiation induced processes, see Roessler (1991, 1992a, b). Many studies which are thought to be pure radiation chemistry, such as a treatment with MeV protons, contain a large amount of hot atom chemistry via the hot secondary atoms formed by knock-on. Even UV photolysis gives rise to (isolated) hot atoms.

3.2.5   Fundamental Suprathermal Reaction Mechanisms

Table 3.9 gives a review of the most important suprathermal reaction mechanisms. Abstraction is typical for hot hydrogen atoms which remove H out of molecules to form H_2. Hot C, N, and O can form CH, NH and OH and successively CH_2, CH_3, and NH_2. Hot oxygen can abstract O out of oxygen containing molecules such as oxides or oxyanions. Replacement occurs in particular when projectile and target atom are of similar mass and an almost front-on collision occurs. Collisional addition can lead to partial fragmentation of the intermediate.

The most important hot process is the insertion into $O — H$, $N — H$, $C — C$, $N — N$ bonds, etc. Table 3.10 presents an overview on the formation of intermediate excited states by hot $C —$, $N —$, and $O —$ atoms in some simple targets and on the final products by H elimination, 1,2-H transfer, and H radical uptake. Similar mechanisms will also apply for hot Si, S, and P atoms. Fragmentation of the individual intermediates into two moieties can also take place. In the gas phase more H elimination and fragmemtation to smaller products is observed whereas in the solid state the 1,2-H transfer and H pickup are important too.

A special experiment using radioactive ^{11}C tracer gave some insight into mechanistic pathways in frozen methane at 77 K. It was irradiated with 20 MeV $^3He^{2+}$ ions from a cyclotron (Patnaik et al., 1990). The $^3He^{2+}$ ions generated energetic ^{12}C secondaries by knock-on processes, but also induced a nuclear reaction $^{12}C(^3He, {}^4He)^{11}C$ leading to ≈ 2 MeV ^{11}C recoils. The products were separated by gas chromatography. The ^{12}C species were measured by a FID detector, the ^{11}C compounds by a radioactivity flow detector. After normalization to 100% the yields of the ^{12}C and ^{11}C products were plotted as a function of dose in eV per carbon atom (target molecule). Fig. 3.1 shows for the three most important products that ^{11}C yields are higher than those of ^{12}C for acetylene, lower for ethane and almost equal for ethene. The situation in the cascade is reported in Fig. 3.2. The high energetic $^3He^{2+}$ ion and the similarly energetic product nucleus $^4He^{2+}$ from nuclear reaction will transfer their kinetic energy preferentially by inelastic collisions creating mostly H radicals. The majority of ^{12}C hot recoils will react in zones with high H radical concentration. The ^{11}C atom at the end of its trajectory will increasingly transfer energy by elastic collisions, finally inserting into a CH_4 molecule. Here the number of H radicals will be much lower. The mechanistic scheme for recoil C in

TABLE 3.9. Review of Some Suprathermal Nonequilibrium Reaction Mechanisms.

	Mechanisms	Typical example
Reactions involving one target molecule	Abstraction	$\vec{H} + H_2O \rightarrow H_2 + OH$
	Collision complex with one molecule	$\vec{C} + H_2O \rightarrow [C - OH_2]$
	Insertion	$\vec{C} + H_2O \rightarrow [HOCH]^*$
	Fragmentation	$[HOCH]^* \rightarrow HO + CH$
	Elimination	$[HOCH]^* \rightarrow CO + 2H$
	Substitution	$\vec{C} + H_2O \rightarrow CH_2 + O$
Reactions involving several target molecules	Collision complex with two molecules	$\vec{C} + 2H_2O \rightarrow CO_2 + 4H$
	With three molecules	$\vec{C} + 3NH_3 \rightarrow C(NH)(NH_2)_2 + 4H$
	With many molecules (multicenter)	$\vec{C} + nCH_4 \rightarrow C_{n+1}H_x$, PAHs
Explosive events	Polymerization or oligomerizations	$3C_2H_2 \rightarrow$ benzene
	Spontaneous H elimination and carbonization	$C_2H_2 \rightarrow$ formation of PAHs, a — C:H, etc.

TABLE 3.10. Insertion Processes of Hot C, N, and O Atoms in Simple Molecules, from Roessler (1992b).

Reaction to intermediate	Successive products		
	H elimination	1,2-H transfer	H uptake
$C+H_2O \rightarrow [HCOH]^*$	CO, COH, HCO	H_2CO	CH_3OH
$C+NH_3 \rightarrow [HCNH_2]^*$	HCN, CN, HNC	CH_2NH	CH_3NH_2
$C+CH_4 \rightarrow [HCCH_3]^*$	C_2H_2, C_2	C_2H_4	C_2H_6
$N+H_2O \rightarrow [HN-OH]^*$	NO, HNO	—	NH_2OH
$N+NH_3 \rightarrow [HN-NH_2]^*$	N_2	—	N_2H_4
$N+CH_4 \rightarrow [HN-CH_3]^*$	HNC, CN, CH_2NH	—	CH_3NH_2
$O+H_2O \rightarrow [HOOH]^*$	O_2, HO_2	—	—
$O+NH_3 \rightarrow [HO-NH_2]^*$	NO, HNO	—	—
$O+CH_4 \rightarrow [HO-CH_3]^*$	CO, HCO, HOC	—	—

frozen CH_4 in Fig. 3.3 shows that the insertion product $[HC-CH_3]$ will pick up H in H rich areas to form C_2H_6 and undergo H elimination to C_2H_2 in H poor areas of the trajectory. The formation of C_2H_4 is due to an internal H transfer within the radical and does not depend on H radical concentration. The above described differentiation between ^{11}C and ^{12}C demands also that the reactions proceed very fast, otherwise H diffusion would lead to similar conditions in all parts of the cascade.

Computer simulation of collision cascades with the program MARLOWE (M.T. Robinson, ORNL) in solid CH_4 by Roessler (1992a) indicated that at the end of the cascades of 1 keV carbon atoms in solid CH_4 about 8 – 15 secondary hot carbon atoms (Fig. 3.5) come to rest in a volume with a radius of about 10 Å. This part of the lattice is damaged by swift hot H atoms which create H_2 and CH_3 radicals. This constitutes the "activated" zone discussed in Section 3.2 containing CH_3, CH_2, H atoms or H_2.

A new mechanistic theory, the multicenter reaction, incorporates all the detailed steps discussed above. It had first been postulated on the base of theoretical considerations and was derived in its final form from experimental observation in solid CH_4 (Kaiser, 1991, 1993; Kaiser and Roessler, 1992; Roessler, 1991, 1992 a, b). Irradiation with MeV cyclotron ions (p, [3]He) at 15 K and 77 K led to the formation of percent amounts of aromates in particular polycyclic aromatic hydrocarbons (PAH) with up to six annulated rings (phenanthrene, dibenzopyrene, coronene) and some partly saturated polycyclic hydrocarbons as shown in Fig. 3.4. These findings are in accordance with the formation of amorphous hydrogenated carbon (a — C:H) upon ion irradiation of frozen organic systems.

Each individual C atom may undergo an abstraction reaction to CH and CH_2 or,

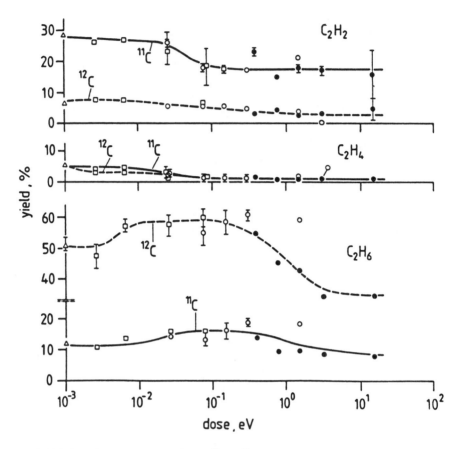

FIGURE 3.1. Yields in percent of total ^{12}C or ^{11}C of the most important products of the ^{12}C(^{3}He, ^{4}He)^{11}C process in frozen CH$_4$ at 77 K with 20 MeV ^{3}He^{2+} from a cyclotron. ^{12}C products are from secondary hot species and radiolysis, ^{11}C by a high energetic (2 MeV) hot secondary only.

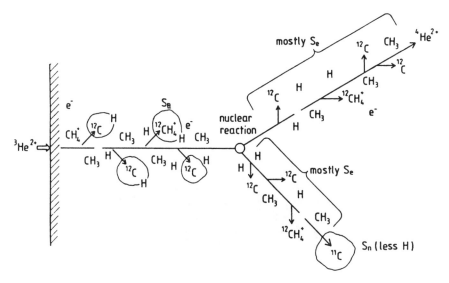

FIGURE 3.2. Schematic situation in a high energy collision cascade in solid CH_4 induced by 20 MeV $^3He^{2+}$ ions including the nuclear reaction $^{12}C(^3He, ^4He)^{11}C$.

more probably, insert into the CH bond of CH_4 to form methylcarbene $[HC — CH_3]^*$ as an excited intermediate.

All these radicals can interact with each other in the activated zone within some $10^{-13} – 10^{-12}$ s and form two- or three-dimensional cyclic or polycyclic structures, a process which is accompanied by strong H elimination. It is questionable whether the mechanism includes the formation of individual methylcarbenes or whether the whole process proceeds as a concerted reaction. Driving forces seem to be thermodynamics of PAH formation and reduction of lattice strain. The best support of the multicenter reaction model came from the observation that the formation of polycyclic structures and of amorphous carbon depends less on the total radiation dose than on the density of energy deposited by the cyclotron beam (Kaiser and Roessler, 1992; Kaiser et al., 1992; Kaiser, 1991, 1993). An interesting proof for a more or less concommitant multicenter reaction to complex organic structures is the fact that $^{12}CH_4$ and $^{13}CH_4$ did not show any differences in product formation at 10 K (Kaiser, 1993). Neither could a ^{13}C enrichment be observed in the dark residues containing PAHs and amorphous carbon (Lecluse, 1993). Multistep reactions would lead to differences in reaction kinetics between ^{12}C and ^{13}C and then to isotopic fractionation.

The above mechanism can be applied to all kinds of solids in space particulary to ice mixtures, and O and N containing compounds. Among the products, polyheterocyclic compounds (PHC) such as shown in Fig. 3.6 can be expected.

Another new but still rather unknown process is the impact of fast clusters or grains with $E >$ some eV/atom. This may lead to collective hot phenomena, i.e., a kind of hot Maxwell–Boltzmann distribution inside the grain and at the surface

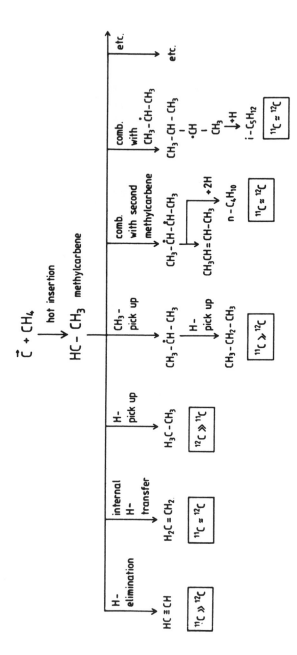

FIGURE 3.3. Mechanistic pathways of hot carbon in frozen CH$_4$.

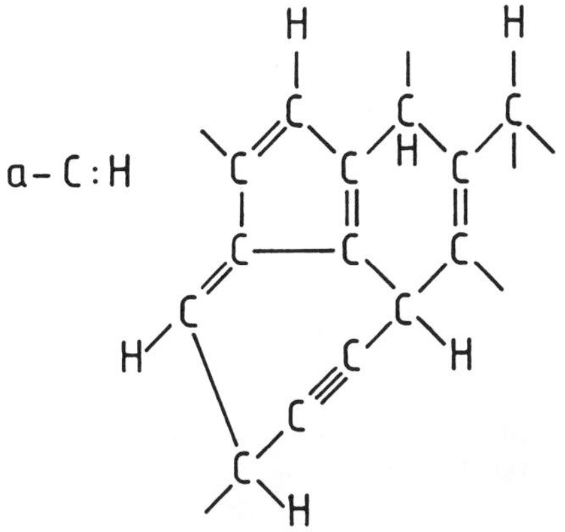

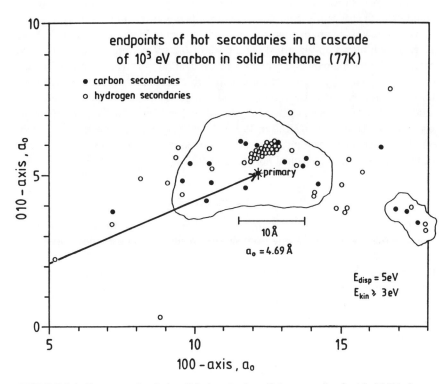

FIGURE 3.4. Example of a polycyclic hydrocarbon.

FIGURE 3.5. Computer simulation (Marlowe) of a collsion cascade of a 1 keV C in frozen CH_4 (77 K). The three dimensional cascade is projected onto the (100) plane. Only the endpoints of the secondary H and C atoms are shown, from Roessler (1992a).

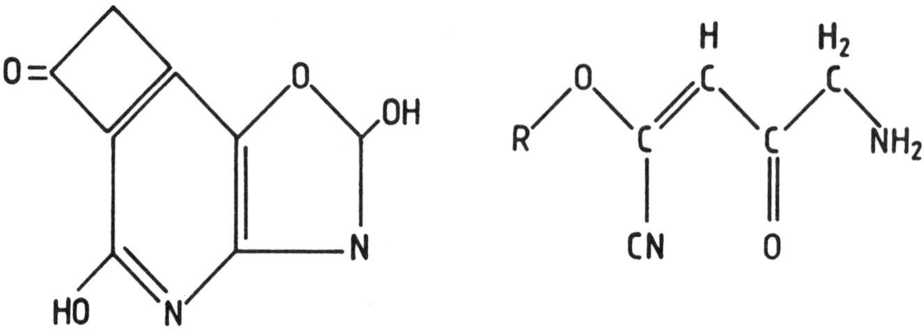

FIGURE 3.6. Typical polyheterocyclic compounds formed by multicenter reactions in C and N containing frosts, from Roessler (1992a).

of a collision partner and wall material. These reactions could likewise induce H elimination and formation of polycyclic structures and modify the chemical composition of the original grain or at least of their upper layer. They seem to be very important for the final state of organic molecules in cometary dust, when colliding with spacecraft materials and the walls of particle impact analyzers.

A very new issue are explosions, such as, for example, those of solid C_2H_2 at 77 K, which blows up vigorously at a dose of 0.2 eV per molecule of 20 MeV protons. The mechanism is a spontaneous cyclization to PAHs and related structures accompanied by sudden H elimination which creates the pressure necessary for the explosion. Likewise solid CH_4 did explode at 10 K when irradiated with a dose of ≈ 120 eV per target atom of 9 MeV He^{2+} ions. Here the mechanism was pure H recombination (Kaiser, 1993). All these explosive processes with their multicenter character belong in essence to nonequilibrium chemistry and should be treated together with typical hot atom reactions.

3.2.6   Complex Organic Matter in Space

The most important issue for space chemistry and simulation is the build-up of complex organic matter starting from simple precursors and the modification and carbonization of already complex molecules. Fig. 3.7 summarizes in a kind of flow diagram the chemical evolution in space with its two important branches: formation of biomolecule precursors and that of refractory organic compounds.

All radiation induced interactions build up and destroy complexity at the same time. Figures 3.8 and 3.9 show schematically the hot atom chemistry pathways in the "hydrocarbon world" which is ordered according to hydrogen content and the number of C atoms per molecule. Starting from simple molecules such as CH_4, all kinds of alkanes, alkanes, aromatic, and polycyclic aromatic hydrocarbons are formed and finally, even amorphous carbon.

On the other hand, starting with complex molecules such as PAHs gives rise to formation of CH_4, alkanes, etc., and likewise carbonaceous matter. Thus, there is no one-way street to complexation, but rather a circle process such as formulated

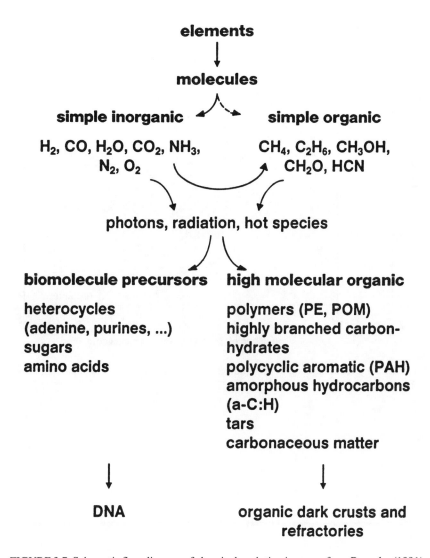

FIGURE 3.7. Schematic flow diagram of chemical evolution in space, from Roessler (1991).

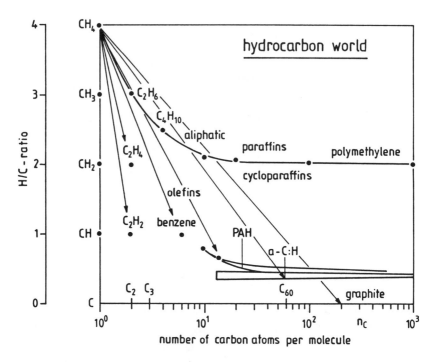

FIGURE 3.8. Product formation by suprathermal reactions in solid CH_4 in the framework of the "hydrocarbon world," from Roessler (1992a).

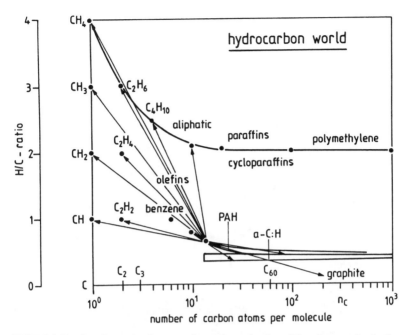

FIGURE 3.9. Product formation by suprathermal reaction in solid anthracene in the framework of the "hydrocarbon world," from Roessler (1992a).

in Fig. 3.10 what is called by the authors a "swing." It seems even somewhat questionable whether it is still allowed to call CH_4 a primitive and amorphous carbon a complex body. In this framework it seems that suprathermal nonequilibrium reactions favor somewhat more the build-up branch and that radiolysis and photolysis act more in the direction of destruction. This is suggested from a comparison of effects of VUV photolysis and 17 MeV ion irradiation of solid organic compounds such as polymers, paraffins, PAHs, and kerogen (Mahfouz et al., 1992).

But even in VUV photolysis circle reactions were observed (e.g., for polyoxymethylene POM in Fig. 3.11). What is discussed here for hydrocarbons will hold equally for organic compounds including nitrogen, oxygen, and sulphur which may lead to precursors of biomolecules. It also will be valid for a domain which has not so much been studied hitherto, i.e., the interaction of CHON atoms with the mineral part of extraterrestrial bodies such as silicates, oxides, carbides, etc., and furthermore for the reactions of hot Si and P atoms. The build-up of alkanes and alkenes by interaction of hydrogen ions and atoms with graphite as well as the oxidation by oxygen ions has been studied in the keV range.

The circle or swing reactions are not only typical for radiolytic interactions. Table 3.11 plots the H/C, O/C, and N/C ratios of some typical molecules involved, the so-called carbocentric element ratio indices (CERI) (Roessler, 1993). When plotting them in a H/C, O/C, N/C diagram shown in Fig. 3.12 the swing between syntheses and metabolism becomes obvious. Both abiotic (in particular, radiolytic)

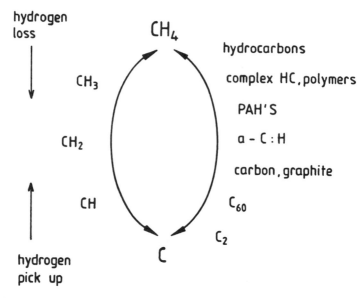

FIGURE 3.10. Circle reactions ("swing") in the hydrocarbon system, from Roessler (1992a).

and biological synthesis and destruction or metabolism processes work at the same time to a medium state of complexation. It has to be mentioned that biological synthesis and destruction are mainly due to electronic processes, whereas most of the radiation-induced processes are ionic (e.g., protic) or atomic ones (radical, hot atoms).

3.3   In Situ Measurements

Having reviewed now the various processes controlling the formation of organic material in space, we now look into the results of the first—and so far only— missions, to comet P/Halley which carried instruments to perform mass spectrometry of the comet dust and gas.

3.3.1   Overview

In March 1986 a fleet of spacecraft approached comet P/Halley. Three of them carried a dedicated payload and achieved an approach close enough to return detailed data on cometary phenomena: Vega 1 and 2, and Giotto.

Among the big surprises were the low albedo of the nucleus surface, the relative high abundance of CO gas, and the organic component of the dust, espe-

TABLE 3.11. Carbocentric Element Ratio Indices (CERI), from Roessler (1993).

Name	Formula	H/C	O/C	N/C
Carbon dioxide	CO_2	0	2	0
Cyan	$(CN)_x$	0	0	1
Formaldehyde, POM, sugars	$(H_2CO)_x$	2	1	0
HCN and its polymers	$(HCN)_x$	1	0	1
Urea	NH_2CONH_2	4	1	1
Pyrimidine	$C_4H_4N_2$	1	0	0.5
Uracil	$C_4H_4O_2N_2$	1	0.5	0.5
DNA	$C_{39}H_{46}O_{23}N_{15}P_4$	1.18	0.59	0.38
RNA	$C_{39}H_{46}O_{27}N_{15}P_4$	1.18	0.69	0.38

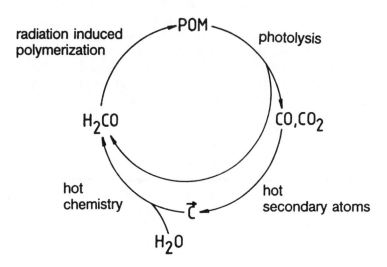

FIGURE 3.11. Circle reactions ("swing") in CH_2O systems, from Roessler (1992a).

cially its intimate intermixture with the mineral component down into the smallest submicron-sized grains.

The instruments returned a great wealth of data, and many of these direct observations still await our understanding. After all, it is still unresolved whether the dark cometary surface is made of dust and boulders not having escaped the nucleus' gravity or rather a layer of ices modified by cosmic radiation into refractory organic material.

As to the chemical composition, the infrared spectrometers IKS on Vega (Moroz et al., 1987), the neutral and ion mass spectrometers NMS and PICCA on Giotto, and the dust mass spectrometers PUMA and PIA on Vega and Giotto, respectively, provided the data. Earth bound observations showed the presence of many organic fragments as neutrals or ions, e.g. McFadden et al. (1987), as well as the existence of discrete CN jets, e.g., A'Hearn and Festou (1990).

Cometary dust plays a key role for the explanation of this phenomenon. It is the solid component of the comet that can carry volatile material away from the nucleus. Consequently it is considered the location of the extended source needed to understand the CO profile measured in the gas phase (Eberhardt et al., 1987). Since only Giotto approached the nucleus closely enough, Clark et al. (1987) in their first analysis of the PIA data, found that the occurrence of mass spectra dominated by C and O was indeed limited to the same area.

The PUMA and PIA mass spectra have allowed determination of the elemental composition of cometary dust (Kissel et al., 1986 a, b; Jessberger et al., 1988). The measurements suggest that cometary matter is of solar system composition, with an enrichment of the light elements compared to CI-Chondrites which were, prior to Halley's dust, the most primitive material available.

Schulze and Kissel (1992) have tried to interpret the anorganic part of the spectra and found the particles to be mostly monomineralic. The most frequent minerals are Fe poor Mg-silicates and Ni enriched Fe-sulfides.

A few mass spectra revealed an unexpected feature of the carbon isotopy, a component dominated by ^{12}C (see Solc et al., 1987; Jessberger and Kissel, 1991).

These were the type of results easily expected from the mass spectrometers, since the spacecraft passed the nucleus of comet P/Halley at high relative speeds of $69 - 78$ km s^{-1}. Laboratory tests at such speeds are limited to femtogram particles. Impact ionization models predict atomic ions only with a substantial fraction carrying multiple charges. Molecular ions were not expected even though Mayer et al. (1986) and Kissel and Krueger (1987a) have shown at lower impact speeds, that ion desorption is one of the important ion formation mechanisms to be considered.

The flight data did set an upper limit to the occurrence of multiply charged ions of even Ca^{2+} to less than one part in a thousand. This indicates that the temperatures at which ionization would reach equilibrium are considerably lower than anticipated. This fact in turn left a chance to find traces of any molecular constituents of the particle. Kissel and Krueger (1987b) have searched mass spectra, selected for sufficient apparent mass resolution, for molecular ions. We present their work in somewhat more detail, as it is still the only direct mass spectrometric evidence for

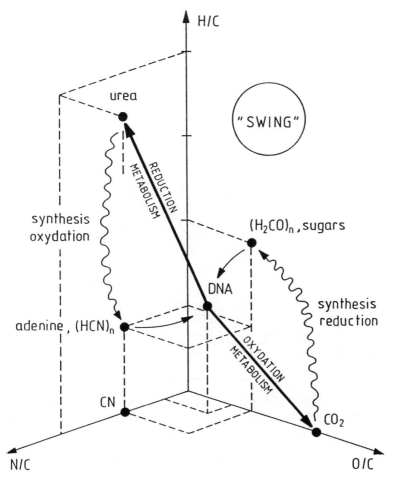

FIGURE 3.12. CERI diagram (seen from −1, −1, −1 direction) of DNA and some of its metabolic products and synthesis components, from Roessler (1993).

organic matter in comets.

The results concerning the physical properties of the dust particles may be described as follows. The mass spectra and energy distributions are compatible with a typical dust particle as shown in Fig. 3.13. A somewhat fluffy ($\rho_p \approx 1$-3 g cm^{-3}) mineral core, more or less chondritic, is embedded in a mantle of organic refractory material, which is even more fluffy ($\rho_p \approx 0.3$-1 g cm^{-3}).

When a dust particle impinges on the target a shock wave travels through the particle. This shock wave causes an emission of atomic and molecular species from its surface, some of which will be ionized up to energy levels corresponding to the shock velocity. Further, the directed momentum flow of the shock wave is dissipated in the entire residual volume of the particle and the nearby target region, producing a plasma which recombines almost totally during expansion, releasing

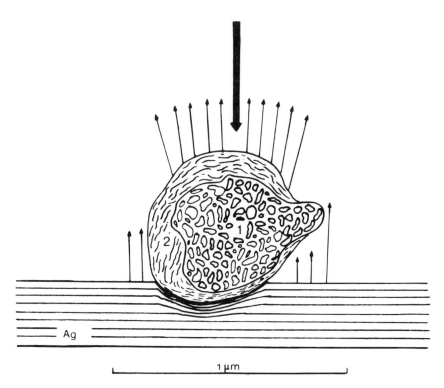

FIGURE 3.13. Dust particle incident on Ag target. Ion type and energy distribution show that most dust particles consist of a fluffy silicate core (1), often covered by also relatively fluffy refractory (and icy) organic material (2). When interacting with high velocity, target and particle materials are compressed and shock waves travel through target and particle causing the emission of atomic and molecular ions from their surfaces due to collision cascades. Later, when the material is compressed into a plasma state, it expands again into the surrounding vacuum, recombination leaves only a small fraction of singly charged ions from the particle's interior.

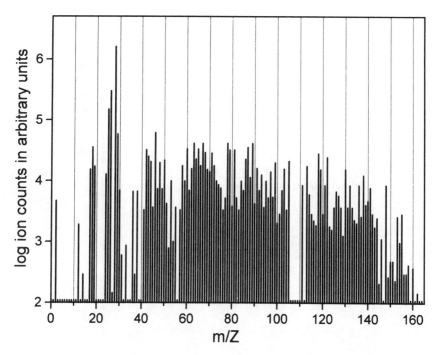

FIGURE 3.14. Cumulative mass spectrum of the molecular ions from 43 evaluated mass spectra.

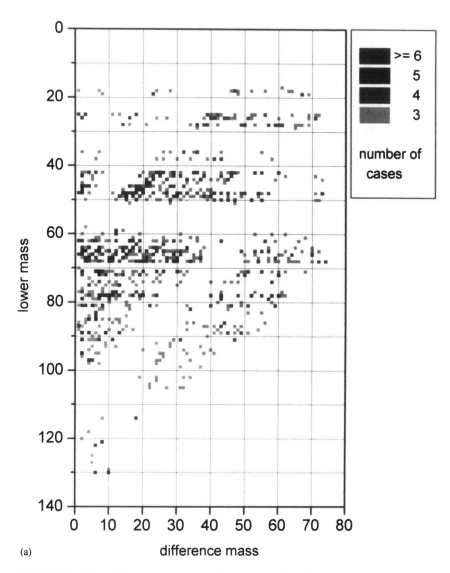

(a)

FIGURE 3.15. Coincidence diagrams of molecular mass lines. The contributions of atomic ions to the mass spectra have been subtracted. Subsequently all pairs of mass lines occurring mass peaks are represented by dots. The dots are darker the more frequent the coincidences show up. Thus the darker dots mark chemically significant mass pairs, pointing to reaction- and/or decomposition partners (as Fig. 3.16) in the dust itself (or during ion formation). (a) Shows the data as difference masses to be added to the lower mass peak. The upper mass lines show up on diagonal lines.

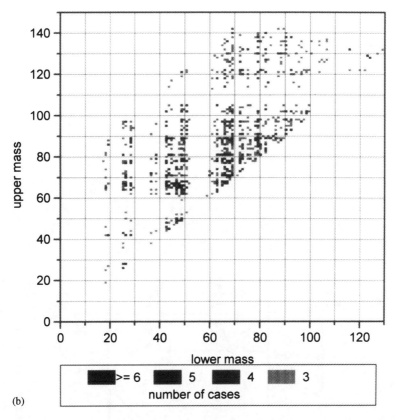

(b)

FIGURE 3.15. Coincidence diagrams of molecular mass lines. The contributions of atomic ions to the mass spectra have been subtracted. Subsequently all pairs of mass lines occurring mass peaks are represented by dots. The dots are darker the more frequent the coincidences show up. Thus the darker dots mark chemically significant mass pairs, pointing to reaction- and/or decomposition partners (as Fig. 3.16) in the dust itself (or during ion formation). (b) Shows the same data as (a). Here the occurrence of the upper mass partner is plotted against the lower mass number. Constant differences lie on diagonal lines. In both plots the empty areas correspond to the range not evaluated because of possible interference of major atomic mass lines.

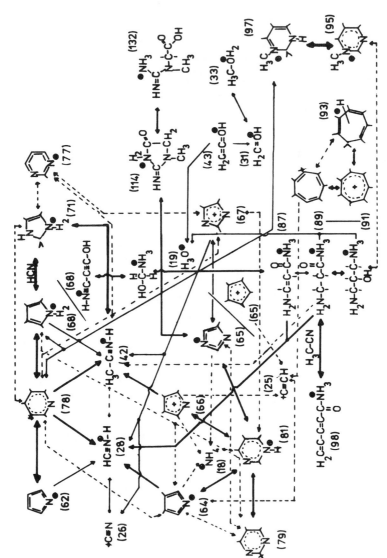

FIGURE 3.16. Reaction and decomposition scheme between various species. The ions formed in this scheme are consistent with the frequently occurring mass line coincidences in Fig. 3.15. It is also consistent with organic molecular ion formation as it occurs with other techniques of fast dissipation of energy near solid surfaces.

TABLE 3.12. Mean Abundances of the Main Elements (relative to Si = 100) in the Core, Mantle, and Icy Phase. Errors are within a Factor of Two except for H for Which the Measurement Accuracy is Uncertain.

Element	Abundance	Element	Abundance
Organic mantles		Chondritic cores	
H	400	C	100
C	500	O	300
N	20	Na	2
O	100	Mg	70
S	10	Al	5
		Si	100
Water ice		S	40
H	300	Ca	4
O	150	Fe	70

only a certain fraction of singly charged atomic ions. In this model one can deduce not only the density but also the mass of the incident particle, if the Y_p and Y_t numbers are known. Masses between 2×10^{-15} g and 10^{-11} g have been measured so far, most of them in the region between 10^{-12} g and 10^{-13} g. The absolute values of the model have been fitted to laboratory experimental points. From this procedure one can estimate the systematic error of the density determination to 50%, whereas that of the particle mass is within an order of magnitude.

3.3.2 Impact Formation of Molecular Organic Ions

We want to concentrate on the composition of the organic refractory part of the dust particles. The elemental composition of the core and mantle of the dust particles as measured with PUMA-1 and evaluated with our sputtering-type model is given in Table 3.12. The mineral core has been analyzed and found to be chondritic by Jessberger et al. (1986) within a factor of 2, on average (which was also the estimate of the systematic error of this method) and the composition of the particles as a whole was found to be practically solar (except hydrogen). However, the abundance variations of individual grains are large. As compared to SIMS from organically contaminated surfaces, as measured by Benninghoven and Sichtermann (1978), the ratio of the number of molecular ions to that of atomic ions in the dust particle impact is reduced by at least a factor of 30. Consequently, the molecular mass spectral lines play only a minor role in the dust mass spectra and have to be discriminated against several types of background lines, thus requiring a refined statistical analysis.

The smaller relative intensity of molecular ions compared to SIMS may be due to the much higher mechanical momentum flow in the impact process. Nevertheless, the fact that only singly charged ions occur leads us to the conclusion that all molecular ions emitted are also electronically cold. Their formation processes are thus governed by protic reactions and a lot of skeletal rearrangements due to their high vibrational heat as produced by the shock wave. So one may expect organic molecular ions from the outermost layers of the dust particles to have considerable intensity only if they possess the following properties:

1. even-electron (nonradical) species (due to the nonexcited electronic system);

2. stable for at least 0.1 ms (MS flight time);

3. considerable Bronsted basicity of that related species which is expected to form a positive ion by proton addition; and

4. considerable polarity (charge centers) to maintain a sufficiently low enthalpy of formation of such an ion.

These properties are known to be relevant for all such related processes of molecular ion formation (Krueger, 1983). Such types are known to be already present in the matrix and so this process is called "desorption." Bearing in mind, however, that skeletal rearrangements and fragmentation occur during desorption, it cannot be determined whether a certain ion species is a degradation product of a molecule present in the dust, or whether it is a molecular ion of a precursor molecule formed by reactions taking place in the solid phase of the comet nucleus. Nevertheless, the general class of molecules forming the organic part of the dust could be determined.

For this purpose a total of 43 mode-0 spectra of PUMA-1 have been selected for their suitability. The known atomic lines and the obvious background have been subtracted. Line patterns not representing a normal isotope distribution of atomic ions have been interpreted as mixtures of atomic ions of elements with normal isotope distributions and molecular ions. Mass ranges which are dominated by major elemental lines have been excluded from processing. At a first glance these single molecular ion mass spectra appear confusing, as one does not know the extent to which stochastic background lines contribute. The only interesting phenomenon seen in these single spectra is the "cationization" (Schueler et al., 1981), well known with all surface ionization phenomena: silver ions from the target react with polar molecules from the dust. Target contamination is excluded, because the neutral gas stream during the encounter sputter-cleans the target surface (Krueger, 1984a). The characteristic Ag doublet is found near (Ag+17), or (Ag+18), near (Ag+27) and (Ag+30), and thus the molecules possibly involved in this process are interpreted as being ammonia, water, hydrocyanic, and formic acid, respectively. To learn more about the organic component, further analysis is needed.

A cumulative spectrum has been produced by adding up the linear counts of each m/Z over all individual spectra and plotting the logarithms of these sums. The result is shown in Fig. 3.14. The widths of the K, Ca, Fe, and Ag lines make these regions unidentifiable leading to the related "holes" in the cumulative spectrum. Another method of producing a cumulative spectrum is to sum the square roots of the count rates, thus considering each line with its statistical weight. Moreover, the frequency distribution of the occurrence of each line regardless of its intensity has also been calculated. The results are not too different, except for the ranking of the intensities. The major m/Z-lines that are attributed to molecular ions were: 18, 25, 26, 28, 29, 42, 43, 44, 60, 63, 65, 67, 68, 71, 78, 79, 81, 87, 89, 118, 122, 130, 138, and 148. (The mass determination above the Ag peak may be unsure by 1 AMU.)

These cumulative spectra have been compared with SIMS spectra taken from targets in order to recognize possible contributions due to any remaining contamination. The lines m/Z 13(CH^+), 25(C_2H^+), 29(C_2H^{3+}), 53(C_4H^{5+}), 65(C_5H^{5+}), 67(C_5H^{7+}), 77(C_6H^{5+}), 89(C_7H^{5+}), and 91(C_7H^{7+}) are typical contamination products, which thus cannot be attributed to dust species without further examination. (The frequent contamination ions C_2H^{3+} (27) and C_2H^{7+} (31) make the quantification of Al and P unsure; their values are thus to be taken as upper value limits only.)

A next step of analysis was done for the presence of lines from related ion species. The probability that two lines are present in the same mass spectrum has an almost Poisson distribution if the lines themselves are randomly distributed over the mass spectra independently from each other. On the other hand, if decomposition processes of molecules occur during ion formation, mother–daughter relations are represented as line coincidences of the related ion species, and their mass differences give the mass of the neutral elimination products. Also, the union of small molecules and the breakdown of larger molecules within the dust particle before impact cause related species to form ions simultaneously and to show up in the same mass-spectrum. Furthermore, reaction equilibria may be applicable between certain molecular species in the dust, the related ions of which are therefore expected to occur simultaneously in the mass spectrum. Thus, a frequency distribution of the line coincidences will exhibit such mechanisms if one considers only those coincidences which occur much more frequently than expected from simple Poisson distribution.

In Fig. 15(a) and (b) this frequency distribution is plotted. Each crosspoint of ordinate (lower mass number m_1) and abscissa (mass difference number Δm; upper mass number $m_2 = m_1 + \Delta m$, diagonal) marks a possible coincidence of m_1 and m_2. If, in a statistically large number of spectra, such a coincidence occurs, this crosspoint is marked by a dot. Its size gives the frequency of the coincidences and thus its statistical significance. The smallest dot size means $n = 3$ spectra contain this coincidence (significance $s = 50\%$); the next $n = 4$, $s = 80\%$; the next $n = 5$, $s = 92\%$; and the largest size means $n \geq 6$, $s \geq 96\%$. In the next section we report a consistent set of a few molecular species only, which can explain nearly all coincidences with $n \geq 4$.

Furthermore, we may sum all coincidences of fixed m_1 (ordinate) or m_2 (diagonal!), thus resulting in the mass spectra of the ions involved in the coincidences. Contrary to the still confusing cumulative mass spectrum of Fig. 3.14, only a few mass lines seem to be of real chemical importance, in particular, m/Z = 18, 26, 28, 42, 46, 48, 64, 65, 66, 68, 71, 78, 81, 89 AMU. On the other hand, we may sum all coincidences of fixed m (abscissa) giving a frequency distribution of neutral eliminate products. For example, m = 2, 4, 10, 14, 17, 29, 41, 43 AMU are clearly important, whereas m = 18 and 28 AMU seldom occur. Ion pairs sometimes arising at m/Z = 17–19, 35–37, 53–55, 71–73, 89–91 are in part attributed to cluster ions OH.$(H_2O)_n^+$ and $H_3O.(H_2O)_n^+$ (n = 0, 1, 2, ...), respectively, due to water ice residues.

3.3.3   Results

The ion chemistry of desorption is distinct from that of electron impact ionization in the gas phase. Whereas electron impact produces radical ion species which may decompose into radical or nonradical ions by single molecule decomposition, desorption ions are mostly nonradical ones which are preformed in the condensed (in this case, dust) phase through multi-molecular reaction processes. A further single molecule decomposition in the gas phase is possible; the related mechanisms obey the rules of gas-phase nonradical ions and decompose predominantly into more stable nonradical ions again. However, in TOFMS with a reflector the daughter species are observed only with lifetimes of the mother $<10^{-8}$ s, and the mother species with $>10^{-5}$ s, contrary to quadrupole- or magnet-MS. As a result, most of the desorption ions observed with TOFMS are electronically and, to a lesser extent, also vibrationally cold species. Consequently, m/Z = 18 is by no means H_2O^+ and m/Z = 28 is neither CO^+, nor N_2^+, as would be the case with gas-phase electron impact. With desorption m/Z = 18 is known as the ammonia ion NH_4^+; and m/Z = 28 is most likely HC=NH$^+$. Such ions are a frequently occurring species in the condensed phase as well as in desorption, being described as the proton-transfer result of a neutral species with Bronsted-basic properties in a polar environment. For instance, the hydrocyanic acid HCN is a weak base, exhibiting an amphoteric character and thus forming H_2CN^+; however, it is a stronger acid, forming most likely the carbanion CN$^-$ which we do not observe because it has a negative charge. It is less likely, due to its electronically unsaturated property, to form the carbenion N=C$^+$, but this was found to a lesser extent. The presence of the ions related to ammonia and hydrocyanic acid does not necessarily mean that these substances are present in the dust. However, it does point to the presence of nitrogen-containing hydrocarbons. The first one especially indicates the presence of amines, the second one nitriles, imines, and heterocyclic aromates. Water or hydrates would form H_3O^+ (m/Z = 19), and carbohydrates COH^+ (m/Z = 29) or COH_3^+ (m/Z = 31), which are found to a much lower extent. Carbohydrides form preferably C_2H^+ (m/Z = 25), especially if unsaturated, otherwise they form $C_2H_3^+$ (m/Z = 27, indistinguishable from Al), but only with low probability due to its low polarity.

A very important mass number in the spectra, m/Z = 42, is attributed to the acetonitryl ion $H_3C — C=NH^+$. Like the hydrocyanic cation m/Z = 28, this ion does not only indicate the presence of acetonitrile but also the same class of substances mentioned above.

With m/Z = 44 the imino ion $H_3C — CH=NH_2^+$ as well as the H_2CN^+ (due to the low basicity of the (iso-)cyanic acid H_2CNO^+ or HOCN, respectively) may both contribute, and with m/Z = 46, 48 further reduction products of both ions may be present. (However, coincidences between m/Z = 46, 48, 50 indicate the presence of titanium atomic ions which were not considered before analysis.) The organic molecular part of m/Z = 48 can only be explained as the amino methanol ion $HO - CH_2 - NH_3^+$ by coincidence analysis, as it tends to disintegrate under water elimination to the imino methane ion $H_2C = NH^+$ (m/Z = 30). In general, their lower intensities indicate that they are only a small proportion of the oxygen-containing ion types.

The same is true when looking at the neutral eliminated products. An elimination of water from a molecular ion m_2 leading to m_1, with $m_2 - m_1 = \Delta m = 18$ is not found to contribute significantly. Also the elimination of CO ($\Delta m = 28$), well known for ketones, esters, furanes, or chinones does not contribute. On the contrary, the elimination of NH_3 ($\Delta m = 17$, ammonia), $H_2C = NH$ ($\Delta m = 29$, methylimine), that of $\Delta m = 41$ (acetonitrile $H_2C - C \equiv N$), and $\Delta m = 43$ (either $H_3C - CH = NH$ (ethanimine) or tautomerically $H_2C = CH - NH_2$ (ethenamine) are found to be frequent reactions. This all points to the presence of unsaturated N-containing carbohydrides.

The different oxidation-reduction states of these unsaturated compounds correspond to the common values of $\Delta m = 2$ and 4, interpreted as being elimination (oxidation) processes of one or two H_2 molecules, forming -enes, dienes, or -ines. So we are able to present a consistent interpretation of the main low mass lines, coincidences and mass differences. The presence of unsaturated nitrogen-containing species like nitriles, aldimines, enamines and possibly their polymerization products is assured. The latter ones are the main types of species which are able to explain all the major coincidences as shown in Fig. 3.16: pyrrole, pyrazole/imidazole, pyridine, pyrimidine, and its derivatives are very likely to be present in the dust.

Due to the low fraction of nitrogen relative to carbon found in the elemental composition, it may be surprising at first sight not to find the unsaturated carbohydrides to a larger extent. However, their probability of forming ions in desorption is a factor of ≈ 100 lower than the related nitrogen-containing ones, because they lack a charge center and thus have only a low basicity. Furthermore, they are consequently more volatile, so they may be evaporated out of the dust before impact. Nevertheless, we conclude that a considerable fraction of alkenes, alkynes, or cyclic, especially aromatic, molecules (like benzene, toluene) may well be present, as the coincidences of m/Z = 89, 91, 93 indicate, which may be correlated with the most stable tropyl-ion (91) and its oxidation product dehydrotropyl as well as the reduction product the cycloheptatriene (93) ion.

With larger m/Z the interpretation becomes ambiguous. The presence of oxygen atoms in the molecules may well occur more frequently with larger molecules, but this has not been proved. With a lack of oxygen there are even some indications from multiple coincidences in a few spectra that a molecule as large as adenine may be present. Its quasi-molecular ion (m/Z = 136) is completely oxygen-free, but as an aminopurine it belongs to the class of heterocyclic aromates discussed above.

It is therefore consistent that no indications for alcohol or sugars as N-free carbohydrates have been found, nor α-amino acids ($R - CH - NH_2 - COOH$), although the latter ones are known to produce large yields of ions, easily detectable due to their common elimination of formic acid (m = 46). So the biologically important alpha-amino acids are, if they are formed at all, at least a factor of 30 less abundant than the biologically important pyrimidines and purines.

It is puzzling that the presence of oxygen in the organic molecules was not found to a larger extent, although there is much oxygen within the dust, as seen from the elemental composition. Possibly, most of the oxygen is bound in the silicates, and also present as frozen H_2O and CO. However, we found indications that (Ag-cationized) formic acid was present, and possibly also some other N-free carboxylic acids as acetic or oxalic acid may well be formed. Also methanal and/or ethanal may be present. Their efficiencies of producing positive ions are fairly low. With a strong acidic behavior they are expected to show up in the negative ion spectrum, which we did not measure. The same is true for S-containing species, which tend to form more negative ions from thiols, thials, and so on. The lack of coincidences for m = 34 yielding a possible H_2S elimination analogous to that of water do not point to the presence of such species. The substance classes that are probably present in comet Halley's dust are summarized in Table 3.13.

3.3.4   Possible Implications

One major aspect of the interest in cometary organic material is its relevance for the formation of life on the early Earth. Clark (1988) has used the PUMA results and described that a comet impacting on Earth would leave a pond in which the organic material would lead to biogenic molecules and ultimately to the evolution of first self-reduplicating entities: a prerequisite for the formation of life on the Early earth.

The hypothesis that the organic part of the dust of a comet consists of spermiae by F. Hoyle and N.C. Wickramasinghe (1986) has been disproved at least for that of comet Halley as analyzed by Vega 1. First of all, dried cells contain at least some fraction of Na and K: we found only less than 1 per 10^6 in the organic dust particles. No indications have been found for a significant presence of oxygen-rich compounds such as sugars and peptides, although the latter molecules should show up in the mass spectra if present. The value m/Z = 31 is found to be of minor importance. It is due to phosphorus in the mineral core only. However, nucleic acids should yield phosphorus, but we do not see it in the organic fraction sufficiently enough; also sugars are known to produce m/Z = 31 as a CH_2OH^+ ion.

Nevertheless, we do think that the question of the origin of life in the context of primordial matter as found in comets has become even more exciting. We stress that the substance classes, which we believe are present in the cometary dust are highly reactive especially in warm water. Imagine that some mechanism brings such substances into contact with liquid water. The unsaturated carbohydrides may add water molecules thus reacting to give carbohydrates such as sugars. The nitrogen-containing species may react to give bases of nucleosides, if not already present. The mineral core may serve as the phase boundary necessary to give a local concentration gradient, and also as a source for dissolved phosphoric acid. With

TABLE 3.13. Types of Organic Molecules as Inferred from the Mass Spectra.*

C—H— Compounds

(Only high-molecular probable due to volatility; hints only to unsaturated)

$HC{\equiv}C(CH_2)_2CH_3$	Pentyne
$HC{\equiv}C\ (CH_2)_3CH_3$	Hexyne
$H_2C{=}CH{-}CH{=}CH_2$	Butadiene
$H_2C{=}CH{-}CH_2{-}CH{=}CH_2$	Pentadiene

Cyclopentene,
  cyclopentadiene

Cyclohexene,
  cyclohexadiene

Benzene, toluene

C—N—H— Compounds

(Mostly of high extensity; also higher homologues possible)

$H{-}C{\equiv}N$	Hydrocyanic acid
$H_3C{-}C{\equiv}N$	Ethanenitrile (acetonitrile)
$H_3C{-}CH_2{\equiv}N$	Propanenitrile
$H_2C{=}N{-}H$	Iminomethane
$H_3C{-}CH{=}NH$	Iminoethane
$H_2C{=}CH{-}NH_2$	Aminoethene
$H_2C{=}CH{-}CH{=}NH$	Iminopropene

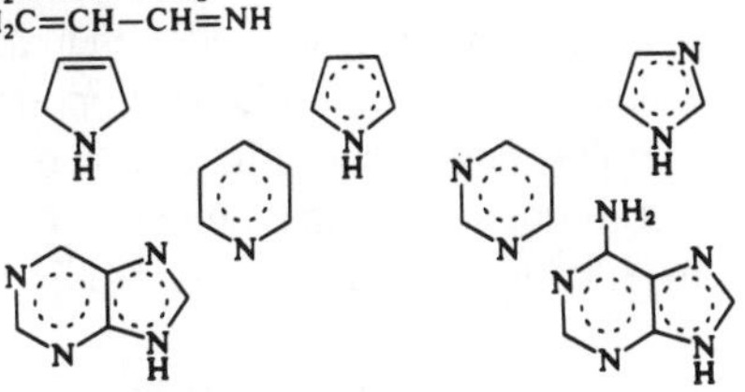

Iminoethane / Aminoethene (tautomeric)

Pyrroline, pyrrole, imidazole

Pyridine, pyrimidine

(and derivatives)

Purine, adenine

C—O—H— Compounds

(Only very few hints to existence)

$HC{=}OH$	Methanal (formaldehyde)
$H_3C{-}C{=}OH$	Ethanal (acetaldehyde)
$HCOOH\ H_3C{-}COOH$	Methanoic (formic) and
	ethanoic (acetic) acid (?)

C—N—O—H— Compounds

(Amino-, Imino-, Nitrile of -ole, -ale, -keto- only probable with
  higher C-numbers of -anes, -enes, and -ines or cyclic aromates)

$N{\equiv}C{-}OH\ \ O{=}C{=}NH$	(Iso-) cyanic acid (?)
$N{\equiv}C{-}CH_2{-}OH$	Methanolnitrile
$HN{=}CH{-}CH{=}O$	Methanalimine

Methanolnitrile / Methanalimine (tautomeric)

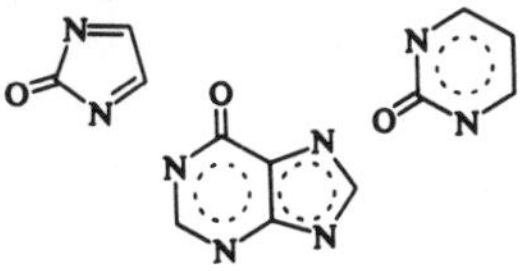

Oxyimidazole, oxypyrimidine

Xanthine

* Additionally, structure isomers are possible. Several types may form tautomers, mesomers, and conformational isomers. Thus the molecules given here serve only as examples of the class of substances possibly present in the organic component of the dust. We are not really sure if oxygen containing species are present (see text).

such a model, the chemical prerequisites and also the requirements of nonlinear nonequilibrium thermodynamics are fulfilled to trigger, for example, hypercyclic behavior (Eigen and Schuster, 1977). High affinity, high concentration gradient, and localization of the reactions have been considered (Nicolis and Prigogine, 1977; Krueger, 1984) to trigger the self-organization of nucleic acids. This is known to be possible without amino acids at this early stage (Eigen and Schuster, 1977) because the replication of RNA is not necessarily associated with transcription (Lohrmann et al., 1980), when Zn^{2+}-catalysts are present, as is the case with the dust.

Krueger and Kissel (1989) consequently favor another scenario in which the intimate mixture of organic material and dust present the key point. The organic material as described above has large amounts of latent free energy. The dust exposes a very large specific surface area with catalytic properties. As the comet nucleus after entry through the atmosphere dissolves in the seas, water enters the voids of the dust particles and favors reactions specific for the biogenesis pond. The dust also provides a certain compartilization, which keeps the reaction products close together long enough to achieve the high complexity of biogenic molecules. As all this happens in the ocean, any entities emerging from this environment could feed on organic material already formed on Earth by various processes and therefore present in the early seas and waste products can be dispersed in that environment.

The data available up to now have been extensively discussed and used to derive properties of cometary material. In most cases statements could be supported as to what type of material is not present in comets rather than solid proofs of what material is definitely there. It has, however, become clear that cometary organic material is of a complex nature indeed. This situation would naturally call for more and dedicated data.

NASA had set up a rendezvous mission to a comet, the Comet Rendezvous and Asteroid Flyby Mission (CRAF) designed to spend several years in the neighborhood of a short-period comet. In order to provide more and better data a secondary ion time-of-flight mass spectrometer has been developed, featuring a mass range of up to 3000 Da and a mass resolution of about $m/dm = 10,000$. This instrument, the Cometary Matter Analyzer, would work with both positive and negative secondary ions and provide elemental, isotopic, and molecular composition of collected cometary dust. It is described in greater detail by Zscheeg et al. (1992) and has been designed to fit on the CRAF spacecraft and would well complement the data provided by other on-board instruments such as a scanning electron microscope (SEMPA), a gas chromatograph (CIDEX), a neutral gas mass spectrometer (NGIMS), which were very well suited to provide new high quality data on cometary dust, and to help resolve the open issues. Unfortunately, the CRAF mission was canceled in January 1992. On a brighter note, the European Space Agency and NASA are collaborating in the Rosetta mission, planned for launch in 2003 (McKay, this volume).

3.4 References

A'Hearn, M.F. and Festou, M.C. (1990), Physics and Chemistry of Comets. In W.F. Huebner (Springer Verlag, Heidelberg), pp. 69–112.

Benninghoven, A. and Sichtermann, W.K. (1978), Detection, identification and structural investigation of biologically important compounds by secondary ion mass spectrometry, *Analyt. Chem.*, **50**, 1180–1184.

Clark, B.C. (1988), Primeval procreative comet pond. *Origins Life*, **18**, 209–238.

Clark, B.C., Mason, L.W., and Kissel, J., 1987, Systematics of the CHON and other light-element particle populations in Comet P/Halley. *Astron. Astrophys.*, **187**, 779–784.

Draganic, I.G., Draganic, Z.D., and Vujosevic, S. (1984), Some radiation-chemical aspects of chemistry in cometary nuclei. *Icarus*, **60**, 464–475.

Duley, W.W. and Williams, D.A. (1984), *Interstellar Chemistry* (Academic Press, London).

Eigen, M. and Schuster, P. (1977), *The Hypercycle* (Springer-Verlag, Berlin).

McFadden, L.A., A'Hearn, M.F., Feldman, P.D., Roettger, E.E., Edsall, D.M., and Butterworth, P.S. (1987), Activity of comet P/Halley 23-25 March, 1986: IUE Observations *Astron. Astrophys*, **187**, 333–338.

Goldanski, V.I. (1984).

Greenberg, J.M. (1978). In J.A.M. McDonnell (ed.), *Cosmic Dust* (Wiley, New York), pp. 187–294.

Greenberg, J.M. (1984), The structure and evolution of interstellar grains. *Sci. Am.*, **250**, 124–135.

Heyl, M. and Roessler, K. (1992), In J.P. Adloff et al. (eds.), *Handbook of Hot Atom Chemistry* (Kodansha, Tokyo and Verlag Chemie, Weinheim), pp. 602–624.

Hoyle, F. and Wickramasinghe, N.C., pre-print no. 122, University College, Cardiff.

Huebner, W.F. (1987), First polymer in space identified in Comet Halley. *Science*, **237**, 628–630.

Huebner, W.F., Boice, D.C., and Sharp, C.M. (1987), Polyoxymethylene in Comet Halley. *The Astrophysical Journal*, **320**, 149–152.

Huebner, W.F.(1990), *Physics and Chemistry of Comets*, (Springer-Verlag, Berlin).

Jessberger, E.K., Kissel, H., Fechtig, H., and Krueger, F.R. (1986), *Eur. Space Agency Spec. Publ.*, **249**, 27.

Jessberger, E.K., Christoforidis, A. and Kissel, J. (1988), Aspects of the major element composition of Halley's dust. *Nature*, **332**, 691–695.

Jessberger, E.K. and Kissel, J. (1991). In R. Newburn, M. Neugebauer (eds.), *Comets in the Post-Halley Era (II)* (Kluwer Academic Publishers, Dordrecht), pp. 1075–1092.

Kaiser, R.I. (1991), *Chemische Prozesse durch Zyklotronionen in festem Methan* (Report JL-2492, Research Centre KFA Jülich, Germany).

Kaiser, R.I., Lauterwein, J., Maller, P. and Roessler, K. (1992), *Nucl. Instr. Meth.*, **B 65**, 463–467.

Kaiser, R. I. and Roessler, K. (1992), Cosmic ray modification of organic cometary matter simulated by cyclotron irradiation. *Ann. Geophysicae*, **10**, 222–225.

Kaiser, R.I. (1993), *MeV-Ionen induzierte chemische Reaktionen in festem Methan, Ethen and Ethin* (Report Jl- 2856, Research Centre KFA Jülich, Germany).

Kissel, J., Sagdeev, R.Z., Bertaux, J.L., Angarov, V.N., Audouze, J., Blamont, J.E., Bachler, K.V., Hoerner, H., Inogamov, N.A., Khomorov, V.N., Knabe, W., Krueger, F.R., Langevin, Y., Levasseur- Regourd, A.C., Managadze, G.G., Podkolzin, S.N., Sharipo, V.D., Tabaldyev S.R., and Zubkov, B.V. (1986a), Encounters with Comet Halley - The First Results). *Nature*, **321**, 280–282.

Kissel, J., Brownlee, D.E., Bachler, K., Clark, B.C., Fechtig, H., Gran, E., Hornung, K., Igenbergs, E.B., Jessberger, E.K., Krueger, F.R., Kuczera, H., McDonnell, J.A.M., Morfill, G.E., Rahe, J., Schem, G.H., Sekanina, Z., Utterback, N.G., Vilk, H.J., and Zook, H. (1986b), Encounters with Comet Halley - The First Results. *Nature*, **321**, 336–337.

Kissel, J. and Krueger, F.R. (1987a), Ion formation by impact of fast dust particles and comparison with related techniques. *Appl. Phys.*, **A 42**, 69–85.

Kissel, J. and Krueger, F.R. (1987b), The organic component in dust from comet Halley as measured by the PUMA mass spectrometer on board Vega 1 *Nature*, **326**, 755–760.

Korth, A., Krueger, F.R., Mendis, D.S., and Mitchell D.L. (1990), In C.I. Lagerkvist et al. (eds.), *Asteroids, Comets, Meteors III*, (Uppsala), pp. 373.

Korth, A., Marconi, M.L., Mendis, D.A., Krueger, F.R., Richter, K.A., Lin, R.P., Mitchell, O.L., Andersen, K.A., Carlson, C.W., Réme, H., Savaud, J.A. and d'Uston, C. (1989), Probable detection of organic-dust-borne aromatic $C_3H_3^+$ ions in the coma of comet Halley. *Nature*, **337**, 53–55.

Krueger, F.R. (1983), Thermodynamics of ion formation by fast dissipation of energy at solid surfaces. *Naturforschung*, **38a**, 385–394.

Krueger, F.R. (1984), *Physik und Evolution* (Parey, Berlin).

Krueger, F.R. (1984a), *Eur. Space Agency Spec. Publ.*, **224**, 49.

Krueger, F.R. and J. Kissel (1989), Biogenesis by cometary grains – thermodyamic apsects of self-organization. *Origins Life Evol. Biosphere*, **19**, 87–93.

Krueger, F.R., Korth, A., and Kissel, J. (1991), The organic matter of comet Halley as inferred by joint gas phase and solid phase analyses. *Space Science Reviews*, **56**, 167–175.

Lecluse, C. (1993), *Fractionnement isotopique des éléments légers au cours de la formation du systéme solaire* (Ph.D. thesis, Université Paris VII).

Lohrmann, R., Bridsen, P.K., and Orgel, L.E. (1980), Efficient metal-ion catalyzed template-directed oligonucleotide synthesis. *Science*, **208**, 1464.

Mahfouz, R.M., Sauer, M., Atwa, S.T., Kaiser, R.I., and Roessler, K. (1992). *Nucl. Instr. Meth.*, **B 65**, 447–451.

Mayer, F.J., Krueger, F.R., and Kissel, J. (1986). In A. Benninghoven (ed.), *Proceedings in Physics*, **9** (Springer-Verlag, Berlin), pp. 169.

Moroz,V.I., Combes, M., Bibring, J.P., Coron, N., Crovisier, J., Encrenaz, T., Crifo, J.F., Sanko, N., Grigoryev, A.V., Bockelée-Morvan, D., Gispert, R., Nikolsky, Y.V., Emerich, C., Lamarre,J.M., Rocard, F., Krasnoplosky, V.A., and Owen, T. (1987), Detection of parent molecules in comet P/Halley from the IKS-Vega experiment. *Astron. Astrophys.*, **187**, 513–518.

Nicolis, G. and Prigogine, I. (1977), *Self-Organization in Non-equilibrium Systems* (Wiley, New York).

Patnaik, A., Roessler, K., and Zádor, E. (1989), Modification of simple organic solids in space -Energetic carbon interactions with solid methane. *Adv. Space Res*, **9**(6), 49–52.

Patnaik, A., Roessler, K., and Zádor, E. (1990), *Radiochimica Acta*, **50**, 75–85.

Rettig, T.W., Tegler, S.C., Pasto, D.J., and Mumma, M.J. (1992), Comet outbursts and polymers of HCN. *Astrophys. J.*, **398**, 293–298.

Roessler, K. (1987), Polycyclic Aromatic Hydrocarbons and Astrophysics In A. Léger, L. d'Hendecourt, and N. Boccara (eds.), *Polycyclic Aromatic Hydrocarbons and Astrophysics* (Reidel, Dordrecht), 173–176.

Roessler, K. and Eich, G. (1987), In *Amorphous Hydrogenated Films, E-MRS*, **XVII** (Les Editions de Physique, Paris), pp. 167–175.

Roessler, K., Eich, G., Patnaik, A., and Zador, E. (1990), Polycyclic aromatic hydrocarbons via multicenter reactions induced by solar radiation, *Lunar and Planetary Sci.*, **XXI**, 1035–1036.

Roessler, K. (1991), In E. Bussoletti and G. Strazzulla (eds.), *Solid State Astrophysics* (North Holland, Amsterdam), pp. 197–266.

Roessler, K. (1992a), *Nucl. Instr. Meth.*, **B 65**, 55–66.

Roessler, K. (1992b), In J.P. Adloff (eds.), *Handbook of Hot Atom Chemistry* (Kodansha, Tokyo and Verlag Chemie, Weinheim), pp. 602–624.

Roessler, K. (1993), *Aristoteles und die Weltraumsimulation* (Script, Inauguration Lecture, Mathematisch-Naturwissenschaftliche Fakultät, Universität Münster, 8. Dec. 1993).

Schulze, H. and Kissel, J. (1992), Chemical heterogeneity and mineralogy of Halley's dust. *Meteoritics*, **27**, 286–287.

Schutte, W., Allamandola, L.J. and Sandford, S.A. (1993a), Formaldehyde and organic molecule production in astrophysical ices at cryogenic temperatures. *Science*, **259**, 1143–1145.

Schueler, B., Feigl, P., Krueger, F.R., and Hillenkamp, F. (1981), Cationization of organic molecules under pulsed laser induced ion generation. *Org. Mass Spectrom*, **16**, 502–506.

Solc, M., Vanysek, V., and Kissel, J. (1987), Carbon-isotope ratio in PUMA 1 spectra of P/Halley dust. *Astron. Astrophys.*, **187**, 385–387.

Stöcklin, G. (1969), *Chemie heißer Atome* (Verlag Chemie, Weinheim) (Chimie des atomes chauds, Masson et Cie, Paris, 1972).

Strazzulla, G. and Johnson, R.E. (1991), Irradiation effects on comets and cometary debris. In R.L. Newburn, M. Neugebauer and J. Rahe (eds.), *Comets in the Post- Halley Era Vols. I-II* (Dordrecht, Boston), pp. 243–275.

Whipple, F.L. (1950), A comet model. I. The acceleration of Comet Encke. *Astrophys. J.*, **111**, 375–394.

Winnewisser, G. and Armstrong, J.T. (eds.) (1989), *The Physics and Chemistry of Interstellar Molecular Clouds* (Springer, Berlin).

Zscheeg, H., Kissel, J., Natour, Gh., and Vollmer, E. (1992), COMA - An additional space experiment for in situ analysis of cometary matter. *Astrophysics and Space Science*, **195**, 447–461.

4

Polymers and Other Macromolecules in Comets

W.F. Huebner and D.C. Boice

ABSTRACT Prebiotic molecules derive from abiotic organic molecules, radicals, and ions. Organic molecules pervade the universe from the low temperatures in interstellar clouds to temperatures as high as a few 1000 K in circumstellar envelopes. Here we review the role of organic molecules that condensed at low temperatures before or during comet formation in the early history of the solar system. New laboratory data and spacecraft encounters and ground-based observations of carbon-rich volatile and dust components of comet comae provide a broad database for the investigation of these organic molecules. Probable icy organic constituents of the nucleus and of complex organic particles, which are the most likely candidates for the distributed sources of gas-phase organic species in the coma, will be discussed. There is broad agreement that many organic molecules observed in the coma originate from the dust that must have existed in the solar nebula at the time and place of comet formation. We show that complex organic molecules found in comets may be a source of the prebiotic molecules that led to the origins of life and may include some biologically important compounds that did not form by abiotic synthesis on Earth.

4.1   Introduction

Modern cells are based on organic chemistry and evolution builds on a plethora of pre-existing organic molecules. It is therefore reasonable to assume that life is a consequence of organic chemistry. However, the surface of the primitive Earth lacked the necessary abundance of organics. This then raises the question of how life began on Earth.

Molecules, in particular a large variety of carbon-containing molecules, exist in the gas phase at high temperatures in the atmospheres of cool stars and at low temperatures in circumstellar envelopes, interstellar clouds, atmospheres of planets and some of their satellites, and the comae of comets. They must have also existed in the outer regions of the solar nebula where interstellar silicate grains that had formed in the circumstellar envelopes of stars survived. In some cold regions of the nebula, where densities were sufficiently high, gases condensed on interstellar grains to form ices and organic mantles. This was the case before and during comet formation. Comets, meteoroids, asteroids, and planets also contain inorganic matter, such as silicates. Here we will restrict ourselves to the discussion of the organic molecules. In the inner solar system the temperatures were too high for volatiles, including water and organic molecules, to condense during planet formation. Only the rock-forming constituents condensed at these high temperatures. Why then are the organic molecules of *comets* of interest to the origins of life?

Prebiotic molecules derive from abiotic organic molecules, radicals and ions. The Earth was formed 4.6 Gyr ago, but the time for the origin of life on Earth remains uncertain. A physical record of astronomical and geological data provides information on the character of the Earth's surface up to the epoch during which life began. The primitive Earth provided a suitable environment for the *evolution* of life, but lacked water and organic materials as well as processes to initiate molecular chirality and to amplify it to chiral homogeneity that is necessary for self-replications and the *origins* of life. Table 4.1 compares the relevant relative abundances of bioelements (H, C, N, O, P, and S) to the rock-forming elements in the Earth's lithosphere and in comets. Although oxygen appears to be very abundant in the lithosphere, about 90% of it is chemically bound to the rock-forming elements to form oxides, mostly SiO_2 and MgO. These oxides are organized into more complex molecules or minerals such as enstatite ($MgSiO_3$, the magnesium-rich component of orthopyroxene, $(Mg, Fe)SiO_3$) and fosterite (Mg_2SiO_4, the magnesium-rich component component of the mineral olivine, $(Mg, Fe)_2SiO_4$), which predominate in the lithosphere. This leaves little oxygen for organic molecules. The lack of elements to form organic molecules is even more severe if one considers the ration of C to the sum of rock-forming elements Mg, Al, Si, Ca, and Fe. For the lithosphere this ratio is 10^{-3} while for comets it is about three. Only outer solar system bodies (including comets) are particularly rich in water and organic molecules. There exists ample evidence of meteoroid, asteroid, and comet bombardment of planets and moons during the early history of the solar system but, of that group, only comets come from the outer solar system. This led Oró (1961) to suggest that, in very early times, comets may have collided with the Earth to deliver the organic material and water necessary for the origins of life. The possibility of cometary transport has been investigated recently by Chyba et al. (1990). Chyba (1990a, b) maintains that planets could have acquired their oceans from late-accreting impacts of comets. If the conjecture by Bonner (1991, 1992) turns out to be correct, that terrestrial chirality of biological molecules must have had an extraterrestrial origin, then comets will provide the only efficient means to transport these molecules to Earth. Studies by Engel et al. (1990) showed that amino acids isolated from the Murchison meteorite, contained about an 18% excess of L-alanine over D-alanine in which the ^{13}C content of each enantiomer indicated an extraterrestrial origin. Comets may have an even higher enantiomeric excess since they have been much less exposed to heating which accelerates racemization. Comets also carry much more water than asteroids and the abundance of organic material in comet P/Halley was about twice as high as in the Orgueil meteorite which in turn was richer in organics than the Murchison meteorite. Thus, comets may have not only extinguished life during the Cretaceous–Tertiary boundary period of geological time by causing the extinction of dinosaurs (among many other species) but they may also be responsible for making life on Earth possible.

We distinguish between prebiotic and organic molecules in the sense that all molecules containing carbon (except some simple ones such as CO, CO_2, carbonates, cyanides, and cyanates) are called organic. If organic molecules also contain hydrogen and possibly oxygen, nitrogen, sulfur, or phosphorus, then they have the necessary elements to be prebiotic molecules, but this is not a sufficient condition for them to be of biological nature. Many organic molecules produced in the laboratory or identified in meteorites are racemic mixtures of isomers of biological molecules. They lack biological activity, which is related to self-reproduction. Racemic mixtures are optically inactive, but separable into lavorotary and dextrorotary isomeric forms, a property associated with all amino acids, except the simplest one, glycine. Presently, there are no observational data available about the chiral structure of interstellar or cometary molecules.

4.2   The Mass Ratio of Dust-to-Gas Emission in Comets

The detailed composition of the nucleus of a comet is not known. The nucleus is only a few kilometers in sise and has never been probed directly (see, e.g., Keller, 1990). When a comet is close to the Sun, the ices of the nucleus sublimate (evaporate) under the influence of solar radiation. As they escape from the negligible gravitational attraction of the nucleus they entrain dust, thereby forming the very large cometary atmosphere, or coma. Because of the constantly expanding and escaping gas and dust, the visible coma of an active comet, before it becomes too faint to be seen, is of the order of 10^5 km in size, very large compared to the nucleus and about thirty times larger than the Earth. Solar fluoresence of several minor gas species, such as CN, C_2, C_3, NH, NH_2, etc., and sunlight reflected by the coma dust obscure the nucleus. The nucleus may be visible in reflected sunlight only when the coma is weak, i.e., when the comet is far from the Sun, but then its surface composition cannot be determined because the body is far too small and too far from the Earth to be resolved. Thus, until we have an extended mission to investigate a comet nucleus in situ, all compositional information about the nucleus must be deduced from the composition of the coma. As the coma freely expands, it undergoes chemical changes initiated by dissociations and ionizations from solar UV radiation. We can infer the original molecular species in the nucleus by working back from the coma species identified through remote sensing or in situ spacecraft measurements. However, this is imprecise; the law of increasing entropy prevents us from a time-reversal of detailed chemical kinetics calculations. Thus, we must guess the composition of the gas as it sublimates from the nucleus and calculate its chemical evolution until we find a composition that matches the observations of the coma (see, e.g., Schmidt et al., 1988). Such a solution is, however, not unique. In addition, coma species appear to have two origins: one component is known to come from the sublimation of frozen gases in the surface layers of the nucleus, while another, smaller component derives from the disintegration and sublimation of the coma dust (Boice et al., 1990; Eberhardt et al., 1987; Meier et al., 1993). The coma dust is known to have an organic component and may also have volatile

TABLE 4.1. Relative Element Abundances of the Earth's Lithosphere and Comets.

Element	Lithosphere	Comets, $\chi = 1$
H	0.029	0.5464
He	—	—
C	0.00037	0.1137
N	0.00006	0.0132
O	0.602	0.2834
F	0.00051	—
Ne	—	—
Na	0.025	0.0010
Mg	0.025	0.0099
Al	0.062	0.0007
Si	0.205	0.0183
P	0.00079	—
S	0.00023	0.0071
Cl	0.00011	—
Ar	—	—
K	0.0089	—
Ca	0.019	0.0006
Ti	0.0037	—
Cr	0.000039	0.0001
Mn	0.00035	—
Fe	0.018	0.0052
Co	—	—
Ni	0.000025	0.0004
Total	1.0000	1.0000

inclusions in a porous matrix. As observed in comet P/Halley, the dust is spread throughout the coma, acting as a distributed source for some gaseous species. A further complication is that even if we could determine the relative abundance of molecules in the coma just above the nucleus, it would only approximate the relative abundances of the frozen constituents in the nucleus. The reason is that solar heat has depleted the most volatile constituents on the surface. Heat penetrates into the nucleus where it evaporates the more volatile constituents which then diffuse through the nucleus layers, in part outward to become detectable in the coma, in part inward to recondense in cooler places. This has been experimentally verified by Spohn and Benkhoff (1990) in comet simulation experiments. Thus the abundance of volatiles in the coma is different than their abundance in the nucleus.

Even though the *molecular* composition of the ice and dust in the nucleus is difficult to ascertain, the *elemental* abundances can be determined with some accuracy. Thus we will examine these abundances as additional clues. Because the dust contains semivolatile components (and perhaps even volatile inclusions such as water ice) that contribute as a distributed source to the coma gas, we must know the mass ratio of the dust-to-gas release of the comet and the fraction of the dust that is organic to perform a meaningful analysis.

The dust consists of silicates and organic material, known as CHON (so labeled because it is composed of carbon, hydrogen, oxygen, and nitrogen). The mass ratio of dust-to-gas release, χ, in comets is difficult to determine, but typically ranges between $0 < \chi < 2$. For recent analyses of the dust release rates see Singh et al. (1992) and others cited in this chapter. The uncertainty with χ lies with the particle size distribution. There are few particles of large size, making their abundance difficult to measure. Nevertheless, these large particles carry most of the mass of the dust. In comet P/Halley the ratio was found to lie between $1 < \chi < 2$ (see, e.g., McDonnell et al., 1991). For comets P/Kopff and P/Tempel 2, which are potential targets of comet missions, $\chi < 0.1$. The in situ measurements of comet P/Halley have provided us with the most complete data sets and have shed new light on ground-based observations of succeeding comets. Therefore, in the discussion that follows, we will use primarily results from P/Halley and some of the most recent comets.

4.3   The Molecular Abundances

In Table 4.2, we compile the molecules that have been detected in comets. They have been identified by their spectra in the UV, visible, IR, and radio range of the spectrum and in a few cases by in situ mass spectrometry during spacecraft flybys of comet P/Halley. We will say more about their relative abundances and the associated uncertainties later in this section. First, it is interesting to compare them with identified gas-phase molecules detected in the interstellar medium as presented in Table 4.3. This table is updated from the work of Irvine et al. (1991) who also present another, smaller table of interstellar species identified in dark clouds.

It will be noted that all of the neutral cometary molecules (except CO_2, S_2, and the neutral cometary radicals NH_2 and NH) are also identified interstellar molecules. CO_2 and S_2 are symmetric species that have no dipole moment and, therefore, no pure rotational spectra by which most interstellar molecules are identified. On the other hand, S_2 does have a fine-structure transition, but a search for it in the interstellar medium turned out to be negative. From the presence of CO_2H^+ (protonated CO_2) in the interstellar medium, one can conclude that CO_2 is present. Also solid interstellar CO_2 has been identified by d'Hendecourt and Jourdain de Muizon (1989). Interstellar NH has been identified in the UV in diffuse clouds with background stars (Meyer and Roth, 1991).

Because NH and NH_2 are light molecules, their rotational transitions in the submillimeter region of the spectrum for which observational capabilities are being

TABLE 4.2. Identified Cometary Molecules.

Simple hydrides, oxides, sulfides and related molecules

H_2O	CO	CS	C_2
	CO_2	S_2	
		H_2S	

Nitriles, acetylene derivatives and related molecules

C_3 HCN CH_3CN?

Aldehydes, alcohols and related molecules

H_2CO CH_3OH

Cyclic molecules

Ions

CH^+	CO^+	N_2^+	H_2O^+
C_3^{+*}	CO_2^+	NH_4^{+*}	H_3O^{+*}
C_3H^{+*}			H_3S^{+*}
$C_3H_3^{+*}$			$CH_3OH_2^{+*}$

Radicals

CH	CN	NH_2
NH		
OH		

Notes. Gas phase only.
      * Indicates indirect detection by ion mass spectrometers.
      ? Claimed but not yet confirmed.

TABLE 4.3. Identified Interstellar Molecules.

Simple hydrides, oxides, sulfides, and related molecules

H_2	CO	NH_3	CS	$NaCl^*$
HCl	SiO	SiH_4^*	SiS	$AlCl^*$
H_2O	SiO_2	C_2	H_2S	KCl^*
	OCS	CH_4^*	PN	AlF^*
	HNO			

Nitriles, acetylene derivatives, and related molecules

C_3^*	HCN	H_3CNC	CH_3CH_2CN	$H_2C_2H_2^*$
C_5^*	HC_3N	CH_3CN	CH_2CHCN	HC_2H
C_2O?	HC_5N	CH_3C_3N	HNC	
C_3O	HC_7N	CH_3C_2H	HNCO	
C_5O	HC_9N	CH_3C_4H	HNCS	
C_3S	$HC_{11}N$	CH_3C_5N?		
C_4Si^*	CHC_2HO			

Aldehydes, alcohols, and related molecules

H_2CO	CH_3OH	OHCHO	H_2CNH	CH_2C_2
H_2CS	CH_3CH_2OH	CH_3OCHO	H_3CNH_2	CH_2C_3
CH_3CHO	CH_3SH	CH_3OCH_3	H_2NCN	
NH_2CHO	$(CH_3)_2CO$?	CH_2CO		

Cyclic molecules

C_3H_2	SiC_2^*	$c\text{-}C_3H$

Ions

CH^+	HCO^+	$HCNH^+$	H_3O^+?
HN_2^+	COH_2^+	SO^+?	HOC^+?
	HCS^+		H_2D^+?

Radicals

OH	C_2H	CH	HCO	C_2S
CH	C_3H	C_3N	NO	NS
NH	C_4H	CH_2CN	SO	SiC^+
CH_2?	C_5H	CP^*		
NH_2	C_6H			

Notes. Gas phase only.

    * Indicates detection only in the envelopes around evolved stars.

    ? Claimed but not yet confirmed.

TABLE 4.4. Abundance Estimates for Ices in Comet Nuclei.

H_2O	0.85
CO	0.04
CO_2	0.03
H_2CO	0.02
CH_3OH	0.02
N_2	0.01
H_2S, HCN, NH_3, CH_4, CS_2, H_2CO_2, etc.	0.03

developed only now. As a very recent consequence of this development, the first identification of NH_2 in the interstellar medium has just been reported by van Dishoeck et al. (1993). Therefore, there is no reason to expect that NH is not present in the interstellar medium. Thus all molecules in comets, with the exception of S_2 that may be related to comet dust, appear to overlap with interstellar molecules.

In contrast to the neutral molecular species, there are many cometary radicals and ions that have not been identified in the interstellar medium. This is not surprising. Photo processes are the primary mechanisms for dissociation and ionization and the solar radiation field is very different from the interstellar radiation field.

The molecular similarities seem to indicate a close affinity between comets and interstellar clouds, but we must be cautious and not misinterpret such a conclusion since the relative abundances of species must be considered also. If is difficult to understand how comets might have formed at the very low densities in the interstellar medium. On the other hand, interstellar molecules may have survived or reformed in the outer parts of the solar nebula where densities may have been high enough to form comets. Nevertheless, the list of interstellar molecules may be taken as a guide to identify new species from observed, but still unassigned, cometary spectral lines. Assimilating the available comet observations, model results, and the arguments presented above, Table 4.4 lists our best estimate for the average abundance of the icy composition of comet nuclei based directly on coma observations.

However, the actual abundances of the nucleus could be different from that of the coma. The data come from a variety of sources and many, allegedly "normal" comets, although the in situ P/Halley measurements strongly influenced the relative abundances. As was suspected for many years, it is now clear that water is the dominant ice.

It must also be kept in mind that comet nuclei are inhomogeneous and that there may be compositional differences between comets. To be specific, all variations are associated with detections in the coma. Since the coma is the result of gas emission from an active area, which is only a small part of the surface of a comet nucleus, we must be very careful in the interpretation of these variations. Some authors (see, e.g., Cochran, 1987; Fink, 1991; Budzien and Feldman, 1992) claim that the variations are due to compositional differences of the nuclei, possibly related to the place and time of comet formation. On the other hand, the differences may be only local, due to the inhomogeneous nature of a comet nucleus (Mumma and Reuter, 1989; Sanzovo et al., 1993). It is known that comets that enter the solar system for the first time can be bright at large heliocentric distances, but their brightness increases do not live up to expectations as they approach the Sun. This relative dimming has been ascribed to the sublimation of a volatile frosting on their surface, but may also be caused by the sublimation of volatiles in the layers just below the surface. Part of these volatiles escapes to make the comet appear bright, while another part diffuses inward to recondense at deeper layers in the nucleus, thereby creating a volatile-depleted zone near the surface. Other than this, no clear compositional difference has yet been found between short- and long-period comets.

The best known variation in comets is the mass ratio of dust-to-gas release rates. Some comets appear to be very dusty ($\chi > 0.5$), while others appear to be almost dust free ($\chi < 0.1$). Again, we must view such a simple interpretation with caution. A "dust-free" comet may actually contain large particles that carry most of the mass but are difficult to detect because of their small number, or a very large number of submicrometer-sized particles that are difficult to detect because they are inefficient light scatterers. However, such an argument only shifts the emphasis from dusty versus dust-free comets to comets with different dust particle size distributions.

We know almost nothing about the ratio of organic (CHON) to silicate dust components in comets other than P/Halley. The ratio of CO to H_2O is also variable, based on UV observations of CO and observations of CO^+, the main plasma tail ion. Some comets have a very weak plasma tail or none at all, while others have a very dominant plasma tail. Comets Morehouse (1908 III) and Humason (1962 VIII) are examples of comets that had unusually strong and active CO^+ tails.

Finally, there is the variation of relative abundances of some minor species like CN and C_2. In most comets, the ratio of production rates for C_2/CN varies from 1.2 to 1.5, but in some comets, like P/Giacobini–Zinner, this ratio is only about 0.3. This could be linked to the dust-to-gas mass ratio, which was also low for P/Giacobini–Zinner. Fink (1991) reported the unusual observations of comet Yanaka (1988r) at 0.91 AU heliocentric distance that showed strong emissions of oxygen and NH_2, but no trace of either C_2 or CN, the two species that are among the strongest in most comets. However, it should not necessarily be interpreted that comet Yanaka (1988r) was carbon poor. The carbon may have been present in the organic component of the dust particles (Greenberg et al., 1993).

We should also note the coexistence of oxidized and reduced species in comets. Interstellar clouds are rich in CO and N_2. It is only natural to assume that the solar nebula was also rich in these species. On the other hand, the regions of the solar nebula where the giant planets formed were probably rich in CH_4 and NH_3, because the temperatures and densities were high enough to convert CO to CH_4 and N_2 to NH_3. It is now believed that the abundances of CH_4 (Boice et al., 1990) and NH_3 (Wyckoff, 1990) in comets may be very low. This is consistent with expectations since it would be difficult for comets to escape from the giant planet subnebulae, but the conclusion is based primarily on data from only one comet, namely, comet P/Halley. If the low CH_4 and NH_3 abundances can be confirmed in other comets, then comets must have formed outside of the subnebulae of the giant planets, possibly in the trans-neptunian region of the solar nebula where they are guaranteed a longer lifetime against ejection from the solar system and can replenish the Oort cloud (see, e.g., Rickman and Huebner, 1990).

It is quite possible that the ratios of CH_4/CO and NH_3/N_2 vary from comet to comet. The ratio of reduced-to-oxidized species may be a measure of meta-morphosis toward chemical equilibrium, related to the place of origin of comets. Oort-cloud comets may be more similar to the interstellar composition (i.e., richer in oxides) while short-period comets may be more similar to other solar system bodies (i.e., containing more reduced species). This would leave us with an apparent inconsistency, because P/Halley is a short-period comet while being rich in oxides. However, P/Halley has another characteristic which is unusual for short-period comets; its orbit is retrograde and highly inclined with the ecliptic. One may therefore speculate that it is a captured Oort cloud comet (Weissman and Campins, 1993).

4.4   The Atomic Abundances

Various authors have analyzed the elemental abundances of the gas-phase species and the organic particles of comet P/Halley (e.g., Geiss, 1988; Delsemme, 1991). Their conclusions are that the relative abundances of most elements in that comet are solar with two notable exceptions: hydrogen and nitrogen are depleted. Table 4.5 presents the relative elemental abundance in comets (primarily based on P/Halley) and compares it to that of the Sun. The important quantities to compare are the ratios of C:N:O:S, which we discuss below.

The reason for the nitrogen discrepency is similar to that for hydrogen. Of all molecules, two of the most stable are CO and N_2. They are thought to be abundant in interstellar clouds (Prasad and Huntress, 1980). The conversion in the solar nebula of CO to CH_4, which is also very volatile, and N_2 to NH_3 was very limited (except in the regions of the giant planet subnebulae). Nitrogen is less reactive than carbon or oxygen and tends to form N_2. Like H_2, N_2 and CO are too volatile at low temperatures to condense if comets were formed at temperatures between 25 K and 50 K. Only small amounts of CO and N_2 (and probably an even smaller fraction of CH_4) were trapped in water ice during its condensation. The solar ratio of N:C is approximately 1:3, thus if the probability of trapping N_2 is about the same as for CO, the ratio of N_2:CO in comets should be about 1:6. Considering that carbon is very reactive, while nitrogen is not, some hydrocarbons formed in circumstellar envelopes of stars where they condensed on grains and eventually found their way, via the solar nebula, into comets. Thus the carbon content in comets relative to N may be enhanced making N_2:CO in comets closer to 1:5. N_2 is difficult to detect by ground- or space-based spectroscopy of comets (however, N_2^+ has been observed in comet plasma tails). In addition, N_2 shares the same mass

TABLE 4.5. Relative Element (Number) Abundances of Comets and the Sun.

Element	Ice	Dust	Ice and dust, $\chi = 1$	Ice and dust, $\chi = 2$	Sun
H	0.5929	0.4810	0.5464	0.5273	0.92048
He	—	—	—	—	0.07835
C	0.0570	0.1934	0.1137	0.1370	0.00030
N	0.0154	0.0100	0.0132	0.0122	0.00008
O	0.3347	0.2114	0.2834	0.2621	0.00061
Ne	—	—	—	—	0.00008
Na	—	0.0024	0.0010	0.0014	—
Mg	—	0.0238	0.0099	0.0140	0.00002
Al	—	0.0016	0.0007	0.0009	—
Si	—	0.0439	0.0183	0.0258	0.00003
S	—	0.0171	0.0071	0.0101	0.00001
Ca	—	0.0015	0.0006	0.0009	—
Ti	—	0.0001	—	0.0001	—
Cr	—	0.0002	0.0001	0.0001	—
Mn	—	0.0001	—	0.0001	—
Fe	—	0.0124	0.0052	0.0073	0.00004
Co	—	0.0001	—	0.0001	—
Ni	—	0.0010	0.0004	0.0006	—
Total	1.0000	1.0000	1.0000	1.0000	1.0000

channel as CO, making in situ mass spectroscopic identification very difficult. Thus, no exact value is available for N_2:CO in the ice of a comet nucleus, but the above ratio appears to be reasonable. Thus N_2:CO is small in comets and nitrogen is underabundant relative to its ratio in the Sun.

In the gas phase, carbon is also underabundant in the coma. However, when the carbon contained in the organic particles is added, assuming $\chi \approx 1$ or 2, then the relative carbon abundance in comet P/Halley is restored to the solar value (Grün and Jessberger, 1990). Oxygen is sufficiently abundant in the water ice so that the organic contribution to the oxygen reservoir is negligibly small. Nitrogen is also underabundant in the organic particles so that its relative abundance in comets remains below the solar value. It should be mentioned that sulfur in comets is most likely bound to iron, but some sulfur is also contained in the organic particles and its relative abundance in comets is very close to the solar value. Phosphorus, on the other hand has not been detected in comets, but its relative solar abundance is so low that it may have eluded detection in comets.

4.5   The Organic Component of Dust Particles

An important discovery made by the spacecraft encounters with comet P/Halley was that of organic materials consisting of carbon, hydrogen, oxygen, and nitrogen ("CHON") in dust particles. In this section two aspects of the organic component of the dust particles are discussed: the solar phase and the heavy, gas-phase species that have been detected in the coma of comet P/Halley, presumably related to the solid phase. If prebiotic molecules exist in comets, they will most likely be in the polycondensates of the organic dust component. Thus, we will concentrate on the solid phase first.

An early analysis of P/Halley dust particles by Kissel et al. (1986a, b) and Clark et al. (1987) indicated three major groups of particles: pure silicates, pure organics, and mixed. However, when Lawler and Brownlee (1992) reanalyzed the particle mass spectrometer data of Kissel et al. (1986a, b), selecting only those sets with high dynamic range and a minimum of defects, they found that in all particles the organic and silicate components are interdispersed at submicrometer scales. The peak in the ratio of distributions of organic material, represented by the amount of carbon, to rock-forming elements Mg, Si, Ca, and Fe, lies between 0.4 and 4. From the more symmetric distribution of carbon to rock-forming elements in large particles and the very skew distribution with a steep slope for low ratios of C: (Mg + Si + Ca + Fe), they concluded that volatile organic material sublimates; sublimation being more effective in small particles than in large ones. Clearly, particles for which the ratio of CHON to rock-forming elements is 4 or larger, approach the pure organic component discussed earlier by Kissel et al. (1986a, b) and Clark et al. (1987). For these particles the discussion by Huebner and Boice (1992), about the implications of the six compositional subgroups containing almost exclusively: (1) H and C; (2) H, C, and N; (3) H, C, and O; (4) C and O; (5) H and O; and (6) H, C, N, and O; to the composition of CHON particles, remains valid.

The detailed molecular composition of the organic component of the dust particles is still mostly unknown. However, many species have been suggested for the solid organic phase. Kissel and Krueger (1987) have suggested *unsaturated linear molecules*, including pentyne, hexyne, butadiene, pentadiene, hydrocyanic acid, ethanenitrile (acetonitrile), propanenitrile, iminomethane, iminoethane, aminoethane, iminopropene, methanal (formaldehyde), ethanal (acetaldehyde), methanoic (formic) acid, ethanoic (acetic) acid, isocyanic acid, methanolnitrile, and methanalimine, and *ring molecules*, including cyclopentene, cyclopentadiene, cyclohexene, cyclohexadiene, benzene, toluene, pyrroline, pyrrole, imidazole, pyridine, pyrimidine, purine, adenine, oxyimidazole, oxypyrimidine, and xanthine. Several of these species are potential prebiotic molecules.

Several solid phase constituents of organic particles have been suggested. Polycyclic Aromatic Hydrocarbons (PAH) have spectral properties that are consistent with the 3.3 μm feature detected in many comets (Allamandola et al., 1987). Hydrogenated Amorphous Carbon (HAC) also has spectral properties consistent with that feature (Colangeli et al., 1990). Tholins, a class of complex organic heteropolymers thought to be widely distributed throughout the solar system, have optical properties (wavelength-dependent complex index of refraction) that are consistent with P/Halley dust scattering (Khare et al., 1984). Polyoxymethylene (POM, or oligomerized formaldehyde) dissociation and ionization products are consistent with the mass spectra obtained with the heavy ion analyzer RPA-2 (Rème Plasma Analyzer) of the Positive Ion Cluster Composition Analyzer (PICCA) (Huebner, 1987), the Plasmag-1 instrument on the Vega-2 spacecraft, and some distributed sources of H_2CO and CO in P/Halley (Boice et al., 1990). HCN oligomer (a short chain of a polymer) has interesting properties that make its dissociation products consistent with distributed sources of NH_2, CN, C_2, and C_3 in the coma. Carbon suboxide C_3O_2 has been suggested also as a component of the organic dust (Huntress et al., 1991). Further analysis of the PICCA data has led to the identification of C_3H^+ (Marconi et al., 1989), $C_3H_3^+$ (cyclopropenyl, Korth et al., 1989), and H_3S^+ (Marconi et al., 1990). These authors argue that the dominant sources of these ions are the organic particles, although the specific parents of these ions remain open to conjecture. The last three detailed studies confirm our model predictions for these and other molecular ions that remain likely candidates for unexplained peaks in the mass spectra (Wegmann et al., 1987).

In the gas phase, we must consider: (1) neutral species from distributed sources as evidenced by the neutral mass spectrometer (NMS) on the Giotto spacecraft; and (2) heavy ions as detected by the PICCA instrument on the Giotto spacecraft and the electrostatic analyzer CRA of the Plasmag-1 instrument on the Vega-2 spacecraft. Jet-like features have been observed in the emission bands of CN, C_2, C_3, and possibly also in NH and NH_2. Although no parents have been identified for these radicals, polyaminocyanomethylene, an HCN chain oligomer (see, e.g., Matthews and Ludicky, 1986), is an excellent candidate for all of these species. Distributed sources have also been identified by the NMS on the Giotto spacecraft for CO (Eberhardt et al., 1987) and in the ion mode of the NMS for H_2CO (Meier et al., 1993). Meier et al. confirm the earlier suggestions of Boice et al. (1990) that the distributed source of CO is derived from a distributed source of polymerized H_2CO, although at a lower abundance. POM has been identified in laboratory ice experiments (Schutte et al., 1992). It is worth mentioning at this point that Oró et al. (1959) used POM as the first molecule to investigate the origin of life. Korth et al. (1990) argue for alkyls, aliphatic alcohols, ethers, aldehydes, ketones, acids, esters, alcohols, polyols, thiols, sulfides, dienes, alkenes, and aromatics to explain the mass spectrum of the PICCA instrument.

4.6   Laboratory Experiments

The organic components of comet dust could not be analyzed with the instruments on the spacecraft to Halley's comet. Laboratory experiments are thus very important to provide clues and give guidance for future comet missions. Such experiments with frozen gases relevant to comet formation scenarios yield complex organic molecules and oligomers. Three groups of experiments should be mentioned specifically: energetic particle and photon irradiation of ice mixtures, gradual warming of ice mixtures produced at very low temperatures, and investigations of specific materials.

To the first group belong the experiments of Goldanskii et al. (1973), Brown et al. (1978, 1980, 1982), Pirronello et al. (1982), Johnson et al. (1982, 1983a, b, 1987), Lanzerotti et al. (1982, 1985, 1986), Starzulla et al. (1984), Pirronello (1985), Khare et al. (1989), and Sandford and Allamandola (1990). Mansueto et al. (1989) used HCl to initiate polymerization of formaldehyde at temperatures between 10 K and 77 K. They showed that the typical chain length of the POM oligomer that formed on warming is 6.8 ± 1.3 in samples rich in formaldehyde.

To the second group belong the experiments of Greenberg (1982), Allamandola et al. (1988) and Schutte et al. (1992, 1993a, b). Considering the limited knowledge about comets and their origin, both groups appear to be relevant for the formation of complex organic polycondensates that may have formed on grains before and during comet formation, and later on the surfaces of comet nuclei, accumulating and building up an insulating crust.

The third group of investigations deals with specific materials. Of all the complex molecules mentioned above, the strongest case that can be made for a chain oligomer existing in comets is polymerized formaldehyde (Huebner and Boice, 1992). Although other chain oligomers such as polyaminocyanomethylene (see, e.g., Cruikshank et al., 1991), polyetheylene and polyethylene oxide have been mentioned, the evidence is much weaker. Chain molecules can more easily account for decay products with increasing cometocentric distance, because they are easier to destroy than ring molecules.

4.7   Conclusions

Polymers and other complex organic molecules, possibly including the first chemical building blocks of life, exist in comets and may have been carried by them, like the wind carrying pollen, to fecundate the Earth. While it is true that polymers are destroyed by UV radiation, they can also form on the surface of comets under the influence of UV and cosmic radiation. Since it is too unlikely that very complex molecules formed during the 76 year period of Halley's comet to replenish those lost during the previous apparition through sublimation of the surface, there is general agreement that oligomers and other complex organic molecules must have existed in solar nebula grains before comets formed and were incorporated into their nuclei during comet formation (Greenberg, 1989; Huebner, 1987; Korth et al., 1990). Inside the nucleus, they are well protected from cosmic and UV radiation. The source of complex prebiotic molecules is more likely to exist in the organic component of the cometary dust particles than in the icy constituents of the nucleus. One element of the six, crucial for the origins of life (H, C, N, O, S, and P), phosphorus (or a phosphor-containing species) has not yet been identified in comets, but assuming that it has the same low relative abundance in comets as in the Sun, it may have escaped detection. It is well established that comet dust contains many complex organic molecules that are potential precursors to the origins of life, but their identity is not clearly established. Presently, there are no observational data about the chiral structure of interstellar or cometary molecules. Basic physico-chemical data, including dissociation and ionization rate coefficients, are needed for complex organic molecules in solid and gas phases. Missions to comets planned for the next decade will be able to answer several of the questions raised here.

Acknowledgments: We are grateful for support from the NASA Planetary Atmospheres Program grants Nos. NAGW-2205 and NAGW-2370.

4.8   References

Allamandola, L.J., Sandford, S.A., and Wopenka, B., 1987, Interstellar polycyclic aromatic hydorcarbons and carbon in interplanetary dust particles and meteorites. *Science*, **237**, 56–59.

Allamandola, L.J., Sandford, S.A., and Valero, G.J., 1988, Photochemical and thermal evolution of interstellar/precometary ice analogs. *Icarus* **76**, 225–252.

Alvarez, L.W., Alvarez, W., Asaro, F., and Michel, H.V., 1980, Extraterrestrial cause for the Cretaceous-Tertiary extinction. *Science*, **208**, 1095–1108.

Boice, D.C., Huebner, W.F., Sablik, M.J., and Konno, I., 1990, Distributed coma sources and the CH_4/CO ratio in Comet Halley. *Geophys. Res. Lett.*, **17**, 1813–1816.

Bonner, W.A., 1991, The origin and amplification of biomolecular chirality. *Origins Life Evol. Biosphere*, **21**, 59–111.

Bonner, W.A., 1992, Terrestrial and extraterrestrial sources of molecular homochirality. *Origins Life Evol. Biosphere*, **21**, 407–420.

Brown, W.L., Lanzerotti, L.J., Poate, J.M., and Augustyniak, W.M., 1978, "Sputtering" of ice by MeV light ions. *Phys. Rev. Lett.* **40**, 1027–1030.

Brown, W.L., Augustyniak, W.M., Brody, E., Cooper, B., Lanzerotti, L.J., Ramirez, A., Evatt, R.E., and Johnson, R.E., 1980, Energy dependence of the erosion of H_2O ice films by H and He ions. *Nucl. Instr. Methods*, **170**, 321–325.

Brown, W.L., Augustyniak, W.M., Simmons, E., Marcantonio, K.J., Lanzerotti, L.J., Johnson, R.E., Boring, J.W., Reimann, C.T., Foti, G., and Pirronello, V., 1982, Erosion and molecular formation in condensed gas films by electron loss of fast ions. *Nucl. Instr. Methods*, **198**, 1–8.

Budzien, S.A., and Feldman, P.D., 1992, Upper limits to the S_2 abundance in several comets observed with the International Ultraviolet Explorer. *Icarus*, **99**, 143–152.

Chyba, C.F., 1990a, Impact delivery and erosion of planetary oceans in the early inner Solar System. *Nature*, **343**, 129–133.

Chyba, C.F, 1990b, Extraterrestrial amino acids and terrestrial life. *Nature*, **348**, 113–114.

Chyba, C.F., Thomas, P.J., Brookshaw, L., and Sagan, C., 1990, Cometary delivery of organic molecules to the early Earth. *Science*, **249**, 366–373.

Clark, B.C., Mason, L.W., and Kissel, J., 1987, Systematics of the CHON and other light-element particle populations in Comet P/Halley. *Astron. Astrophys.*, **187**, 779–784.

Cochran, A.L., 1987, Another look at abundance correlations among comets. *Astron. J.*, **93**, 231–238.

Colangeli, L., Schwehm, G., Bussoletti, E., Fonti, S., Blanco, A., and Orofino, V., 1990, Hydrogenated amorphous carbon grains in Comet Halley? *Astrophys. J.*, **348**, 718–724

Cruikshank, D.P., Allamandola, L.J., Hartmann, W.K., Tholen, D.J., Brown, R.H., Matthews, C.N., and Bell, J.F., 1991, Solid C≡N bearing material on outer Solar System bodies. *Icarus* **94**, 345–353.

Delsemme, A.H., 1991, Nature and history of the organic compounds in comets: An astrophysical view. In *Comets in the Post-Halley Era*. R.L. Newburn, Jr., M. Neugebauer and J. Rahe, eds., Vol. 1, pp. 377–428.

d'Hendecourt, L.B., and Jourdain de Muizon, M., 1989, The discovery of interstellar carbon dioxide. *Astron. Astrophys.*, **223**, L5–L8.

Eberhardt, P., Krankowsky, D., Schulte, W., Dolder, U., Lämmerzahl, P., Berthelier, J.J., Woweries, J., Stubbeman, U., Hodges, R.R., Hoffman, J.H., and Illiano, J.M., 1987, The CO and N_2 abundance in Comet P/Halley. *Astron. Astrophys.* **187**, 481–484.

Engel, M.H., Macko, S.A., and Silfer, J.A., 1990, Carbon isotope composition of individual amino acids in the Murchison meteorite. *Nature*, **348**, 47–49.

Fink, U., 1991, Comet Yanaka (1988r): A new class of carbon poor comet. *Science*, **257**, 1926.

Geiss, J., 1988, Composition in Halley's Comet: Clues to origin and history of cometary matter. *Rev. Mod. Astron.*, **1**, 1–27.

Goldanskii, V.I., Frank-Kamenetskii, M.D., and Barklov, I.M., 1973, Quantum low-temperature limit of a chemical reaction rate. *Science* **182**, 1344–1345.

Greenberg, J.M., 1982, What are Comets made of? A Model Based on Interstellar Dust. In *Comets*. L.L. Wilkening, ed., University of Arizona Press, pp. 131–163.

Greenberg, J.M., 1989, Comet nuclei as aggregated interstellar dust. Comet Halley results. In *Evolution of Interstellar Dust and Related Topics*. A. Bonetti, J.M. Greenberg, S. Aiello, eds., North-Holland, Amsterdam, pp. 383–395.

Greenberg, J.M., Singh, P.D., de Almeida, A.A., 1993, What is new about the new Comet Yanaka (1998r)? *Astrophys. J. Lett.* **414**, L45–L48.

Grün, E., and Jessberger, E., 1990, Dust. In *Physics and Chemistry of Comets*. W.F. Huebner, ed. Springer-Verlag, New York, pp. 113–176.

Huebner, W.F., 1987, First polymer in space identified in Comet Halley. *Science*, **237**, 628–630.

Huebner, W.F., and Boice, D.C., 1992, Comets as a source of prebiotic molecules. *Origins Life Evol. Biosphere*, **21**, 299–315.

Huntress Jr., W.T., Allen, M., and Delitzky, M.L., 1991, Carbon suboxide in Comet Halley? *Nature*, **352**, 316–318.

Irvine, W.M., Ohishi, M., and Kaifu, N., 1991, Chemical abundances in cold, dark interstellar clouds. *Icarus*, **91**, 2–6.

Johnson, R.E., Lanzerotti, L.J., and Brown, W.L., 1982, Planetary applications of ion induced erosion of condensed gas frosts. *Nucl. Instr. Methods*, **198**, 147–158.

Johnson, R.E., Brown, W.L., and Lanzerotti, L.J., 1983a, Energetic charged particle erosion of ices in the Solar System. *J. Phys. Chem.*, **87**, 4218–4220.

Johnson, R.E., Boring, J.W., Reimann, C.T., Barton, L.A., Sieveka, E.M., Garrett, J.W., Farmer, K.R., Brown, W.L., and Lanzerotti, L.J., 1983b, Plasma ion-induced molecular ejection on the Galilean satellites: Energies of ejected molecules. *Geophys. Res. Lett.*, **10**, 892–895.

Johnson, R.E., Cooper, J.F., and Lanzerotti, L.J., 1987, Radiation Formation of a Non-Volatile Crust. In *20th ESLAB Symposium on the Exploration of Halley's Comet*, B. Battrick, E.J. Rolfe, and R. Reinhard, eds., ESA Publication SP-250, Vol. II, pp. 269–272.

Keller, H.U., 1990, The Nucleus. In *Physics and Chemistry of Comets*. W.F. Huebner, ed., Springer-Verlag, New York, pp. 13–68.

Khare, B.N., Sagan, C., Arakawa, E.T., Suits, F., Callcot, T.A., and Williams, M.W., 1984, Optical constants of organic tholins produced in a simulated Titanian atmosphere: From soft x-ray to microwave frequencies. *Icarus*, **60**, 127–137.

Khare, B.N., Thompson, W.R., Murray, B.G.J.P.T., Chyba, C.F., Sagan, C., and Arakawa, E.T., 1989, Solid organic residues produced by irradiation of hydrocarbon-containing H_2O and H_2O/NH_3 ices: Infrared spectroscopy and astronomical implications. *Icarus*, **79**, 350–361.

Kissel, J., Sagdeev, R.Z., Bertraux, J.L., Angarov, V.N., Audouze, J., Blamont, J.E., Büchler, K., Evlanov, E.N., Fechtig, H., Fomenkova, M.N., von Horner, H., Inogamov, N.A., Khromov, V.N., Knabe, W., Krueger, F.R., Langevin, Y., Leonas, V.B., Levasseur-Regourd, A.C., Managadze, G.G., Podkolzin, S.N., Shapiro, V.D., Tabaldyev, S.R., and Zubkov, B.V., 1986a, Composition of Comet Halley dust particles from Vega observations. *Nature*, **321**, 280–282.

Kissel, J., Brownlee, D.E., Büchler, K., Clark, B.C., Fechtig, H., Grün, E., Hornung, K., Igenbergs, E.B., Jessberger, E.K., Krueger, F.R., Kuczera, H., McDonnell, J.A.M., Morfill, G.M., Rahe, J., Schwehm, G.H., Sekanina, Z., Utterback, N.G., Völk, H.J., and Zook, H.A., 1986b, Composition of Comet Halley dust particles from Giotto observations. *Nature*, **321**, 336–337.

Kissel, J., Krueger, F.R., 1987, The organic component in dust from Comet Halley as measured by the PUMA mass spectrometer on board Vega 1. *Nature*, **326**, 755–760.

Korth, A., Marconi, M.L., Mendis, D.A., Krueger, F.R., Richter, A.K., Lin, R.P., Mitchell, D.L., Anderson, K.A., Carlson, C.W., Rème, H., Sauvard, J.A., d'Uston, C., 1989, Probable detection of organic-dust-borne aromatic $C_3H_3^+$ ions in the coma of Comet Halley. *Nature*, **337**, 53–55.

Korth, A., Krueger, F.R., Mendis, D.A., and Mitchell, D.L., 1990, Organic ions in the coma of Comet Halley. In *Asteroids, Comets, Meteors III*. C.-I. Lagerkvist, H. Rickman, B.A. Lindblad and M. Lindgren, eds., pp. 373–377.

Lanzerotti, L.J., Brown, W.L., Augustyniak, W.M., Johnson, R.E., and Armstrong, T.P., 1982, Laboratory studies of charged particle erosion of SO_2 ice and applications to the frosts of Io. *Astrophys. J.*, **259**, 920–929.

Lanzerotti, L.J., Brown, W.L. and Johnson, R.E., 1985, Laboratory studies of ion irradiations of water, sulfur dioxide and methane ices. In *Ices in the Solar System*. J. Klinger, D. Benest, A. Dollfus and R. Smoluchowski, eds., Reidel, Dordrecht, pp. 317–335.

Lanzerotti, L.J., Brown, W.L. and Johnson, R.E., 1986, Astrophysical implications of ice sputtering. *Nucl. Instr. Methods Phys. Res.*, **B14**, 373–377.

Lawler, M.E. and Brownlee, D.E., 1992, CHON as a component of dust from Comet Halley. *Nature*, **359**, 810–812.

Mansueto, E.S., Ju, C.Y., and Wight, C.A., 1989, Laser-initiated polymerization of solid formalehyde. *J. Phys. Chem.*, **93**, 2143–2147.

Marconi, M.L., Korth, A., Mendis, D.A., Lin, R.P., Mitchell, D.L, Rème, H., and d'Uston, C., 1989, On the possible detection of dust-borne C_3H^+ ions in the coma of Comet Halley. *Astrophys. J. Lett.*, **343**, L77–L79.

Marconi, M.L., Mendis, D.A., Korth, A., Lin, R.P., Mitchell, D.L., and Rème, H., 1990,

The identification of H_2S^+ with the ion of mass per charge (m/q) 35 observed in the coma of Comet Halley. *Astrophys. J. Lett.*, **352**, L17–L20.

Matthews, C.N. and Ludicky, R., 1986, The dark nucleus of Comet Halley: Hydrogen cyanide polymers. In *20th ESLAB Symposium of Halley's Comet*. B. Battrick, E.J. Rolfe, and R. Reinhard, eds., ESA report SP-250, Vol. 2, pp. 273–277.

McDonnell, J.A.M., Lamy, P.L., and Pankiewicz, G.S., 1991, Physical properties of cometary dust. In *Comets in the Post-Halley Era*. R.L. Newburn, Jr., M. Neugebauer, and J. Rahe, eds., Kluwer Academic, Dordrecht, pp. 1043–1073.

Meier, R. Eberhardt, P., Krankowsky, D., and Hodges, R.R., 1993, The extended formaldehyde source in Comet P/Halley. *Astron. Astrophys.* **277**, 677–690.

Meyer, D.M. and Roth, K.C., 1991, Discovery of interstellar NH. *Astrophys. J.*, **376**, L49–L52.

Mumma, M.J. and Reuter, D.C., 1989, On the identification of formaldehyde in Halley's comet. *Astrophys. J.*, **344**, 940–948.

Oró, J., 1961, Comets and the formation of biochemical compounds on the primitive Earth. *Nature*, **190**, 389–390.

Oró, J., Kimball, A., Fritz, R., and Master, F., 1959, Amino acid synthesis from formaldehyde and hydroxylamine. *Arch. Biochem. Biophys.*, **85**, 115–130.

Pirronello, V., 1985, Molecule formation in cometary environments. In *Ices in the Solar System*. J. Klinger, D. Benest, A. Dollfus, and R. Smoluchowski, eds. Reidel, Dordrecht, pp. 261–272.

Pirronello, V., Brown, W.L, Lanzerotti, L.J., Marcantonio, K.J., and Simmons, E.H., 1982, Formaldehyde formation in a H_2O/CO_2 ice mixture under irradiation by fast ions. *Astrophys. J.*, **262**, 636–640.

Prasad, S.S. and Huntress, Jr., W.T., 1980, A model for gas phase chemistry in interstellar clouds: I. The basic model, library of chemical reactions, and chemistry among C, N, and O compounds. *Astrophys. J. Suppl. Ser.*, **43**, 1–35.

Rickman, H. and Huebner, W.F., 1990, Comet Formation and Evolution. In *Physics and Chemistry of Comets*. W.F. Huebner, ed., Springer-Verlag, Berlin, pp. 245–303.

Sandford, S.A. and Allamandola, L.J., 1990, The volume- and surface-binding energies of ice systems containing CO, CO_2, and H_2O. *Icarus*, **87**, 188–192.

Sanzovo, G.C., Singh, P.D., and Huebner, W.F., 1993. Dust colors, dust release rates and dust-to-gas ratios in Comet Bowell (1992 I), Bradfield (1979 X), Brorsen-Metcalf (1989 X), Giacobini-Zinner (1985 XIII), Levy (1990 XX) and Stephan-Oterma (1980 X), *Astron. J.* Submitted.

Schmidt, H.U., Wegmann, R., Huebner, W.F., and Boice, D.C., 1988. Cometary gas and plasma flow with detailed chemistry. *Comp. Phys. Comm.*, **49**, 17–59.

Schutte, W., Allamandola, L.J., and Sandford, S.A., 1992. Formation of organic molecules by formaldehyde reactions in astrophysical ices at very low temperatures. In *IAU Symposium No. 150, Astrochemistry of Cosmic Phenomena*. P.D. Singh, ed., Kluwer Academic, Dordrecht, pp. 29–30.

Schutte, W., Allamandola, L.J., and Sandford, S.A., 1993a. Formaldehyde and organic molecule production in astrophysical ices at cryogenic temperatures. *Science*, **259**, 1143–1145.

Schutte, W., Allamandola, L.J., and Sandford, S.A., 1993b. An experimental study of the organic molecules produced in cometary and interstellar ice analogs by thermal formaldehyde reactions. *Icarus*, **104**, 118–137.

Singh, P.D., de Almeida, A.A., and Huebner, W.F., 1992. Dust release rates and dust-to-gas mass ratios of eight comets. *Astron. J.* **104**, 848–858.

Spohn, T. and Benkhoff, J., 1990. Thermal history models for KOSI sublimation experiments. *Icarus*, **87**, 358–371.

Strazulla, G., Calcagno, G., and Foti, G., 1984. Build-up of carbonaceous materials by fast protons on Pluto and Triton. *Astron. Astrophys.*, **148**, 441–444.

Van Dishoeck, E.F., Jansen, D.J., Schilke, P., and Phillips, T.G., 1993. Detection of the interstellar NH_2 radical. *Astrophys. J. Lett.*, **416**, L83–L86.

Wegmann, R., Schmidt, H.U., Huebner, W.F., and Boice, D.C., 1987. Cometary MHD and Chemistry. *Astron. Astrophys.*, **187**, 339–350.

Weissman, P.R., and Campins, H., 1993. Short-period comets. In *Resources of Near Earth Space*. J. Lewis, M.S. Matthews, and M. Guerrieri, eds. University of Arizona Press, Tucson, AZ, pp. 569–617.

Wyckoff, S., 1990. Ammonia and nitrogen abundances in comets. In *Workshop on Observations of Recent Comets* (1990). W.F. Huebner, P.A. Wehinger, J. Rahe, and I. Konno, eds. pp. 28–33.

5
Numerical Models of Comet and Asteroid Impacts

P.J. Thomas and L. Brookshaw

ABSTRACT Numerical simulation techniques can be applied to the collision of large organic-rich objects (comets and carbonaceous chondrite asteroids) with the early Earth. Results from these simulations imply that it is possible for a fraction of the extraterrestrial organic material to survive the high temperatures occurring during the impact (and thus contribute prebiotic material to the early Earth). Recent models for atmospheric passage, however, predict that the fate of such candidate impactors is an airburst capable of pyrolyzing the entire organic inventory of the comet or asteroid.

5.1 Introduction

There is now strong evidence (Delsemme, this volume; Chyba and Sagan, this volume) that a signficant fraction of the Earth's inventory of water and carbon was delivered to the Earth by accretion of cometary and asteroidal material during the period of Heavy Bombardment period ending 3.8 Gyr ago.

In this chapter we discuss the role that large (> 1 m) impactors play as candidates for delivery of prebiotic organic molecules to the early Earth. It is clear (Kissel et al., this volume; Huebner and Boice, this volume) that organic material is common in comets and C-type asteroids. Laboratory analyses of CI and CII carbonaceous chondrite meteorites reveals them to be 3–5% by mass organic heteropolymer (Wilkening, 1978). Mass spectrometry performed during the flyby of comet Halley by the Giotto and Vega spacecraft indicates that Halley dust is $\sim$33% organic by mass, while the gas fraction of the comet is $\sim$14% organic by mass. Assuming a standard gas/dust ratio of 1:1, we obtain an overall cometary organic fraction of $\sim$25% organic by mass.

Detection of organic molecules in carbonacous chondrite meteorites led Chamberlin and Chamberlin (1908) to propose in a seminal paper that prebiotic organic molecules might have been delivered to the Earth by impacts of "planetesimals." Later spectroscopic detection of C- and N- radicals in cometary comae led Oró (1961) to suggest comets as possible organic impact deliverers.

It is clear that the mechanism of delivery of organic molecules to the Earth via large impacts faces multiple difficulties. Impactors must traverse an atmosphere, plausibly an order of magnitude denser than today's (Walker, 1986) at hypersonic speeds. Much of the object may be ablated as a result. Aerodynamic forces may produce sufficient deformation for catastropic fragmentation ("airburst") to occur, in a similar manner to the Tunguska explosion of 1908 (Chyba et al., 1993). If an impactor survives to the surface, a highly energetic explosion will result. The extremely high temperatures resulting from the impact would seem to rule out the possibility of any extraterrestrial organic molecules surviving impact. It was for these reasons that Clark (1988) invoked "an improbable, fortuitous soft-landing of a cometary nucleus" as a delivery event.

We may estimate the energetics of a large cometary impact as follows (this discussion is modified from Chyba et al. (1990)). Consider a comet colliding with the Earth at a velocity of 18 km s^{-1}, in the lowest quartile of short-period (SP) comet impact velocities (Chyba, 1990). The kinetic energy per unit mass of this comet is $\sim 1.6 \times 10^5$ kJ kg^{-1} which, assuming a latent heat appropriate to ice (~ 4 kJ kg^{-1} K^{-1}), corresponds to a temperature of $\sim 40,000$ K. While it is true that this temperature is far too high for organics to survive (by a factor of 20 to 40), this calculation ignores several factors that significantly lower the effective impact temperatures. Aerobraking (slowing of the impactor by atmospheric drag) reduces impact velocity, while detailed impact hydrocode calculations predict that the largest fraction of the energy of a impactor with a density less than its target will be partitioned into heating the target and the kinetic energy of ejecta (O'Keefe and Ahrens, 1982). Finally, it is possible that some organic material would be contained in ejected material that would not be exposed to the highest temperatures of impact.

It is clear that a detailed numerical model is required to provide a reasonably physical model of the complicated dynamics of large-scale collisions. Such a model must contain data obtained from impact experimentation, especially concerning the determination of a realistic equation of state (EOS).

We will discuss the extent to which previous organic survivabilty estimates from numerical models (Chyba et al., 1990) are altered by considering asteroid impacts and results of fully three-dimensional impact models, as well as the sensitivity of these conclusions to model assumptions.

5.2   Atmospheric Passage of Impactors: Ablation

Not all comets and asteroids incident on the upper atmosphere will reach the surface intact. During the traversal of the atmosphere a bow shock forms in front of the bolide. The bow shock consists of compressed and heated atmospheric gases. The heated gases radiate their energy to the surface of the bolide which melts and vaporizes. The melted and vaporized material is entrained in the surface airflow and is quickly carried away.

A conservative assumption is that the organic molecules in the ablated material is destroyed. Ablated material would be exposed directly to the hot, shocked atmosphere surrounding the bolide: typical temperatures in this region are $\sim 25,000-30,000$ K (Biberman et al. 1980).

Atmospheric passage of the impactor can be modeled by a finite difference numerical code that represents atmospheric drag and ablation (Chyba et al., 1990).

The equations governing aerodynamic entry of the impactor are as follows:

$$\frac{dh}{dt} = -V\cos\theta, \tag{5-1}$$

$$\frac{dV}{dt} = -\frac{1}{2}C_D\rho V^2\frac{A}{m} + g\cos\theta, \tag{5-2}$$

$$\frac{d\cos\theta}{dt} = \frac{g}{V}(1 - \cos^2\theta), \tag{5-3}$$

$$\frac{dm}{dt} = -\frac{1}{2}C_H\rho V^3\frac{A}{\zeta}, \tag{5-4}$$

where h is altitude, V is velocity, θ is the angle of the trajectory relative to vertical, ρ is the atmospheric density, A is the frontal area of the impactor, g is the gravitational acceleration, and m is the impactor mass. The extent of ablation of the impactor is determined specifically by the drag coefficient C_D, the heat of ablation ζ, and the heat transfer coefficient C_H. The numerical values of these parameters are functions of impactor shape, composition, mass and velocity. For spherical impactors in the terrestrial atmosphere we select $C_H = 0.01$ and $C_D = 0.92$ (O'Keefe and Ahrens, 1982). Values of ζ depend on the material of the impactor composition and the specific mechanism of ablation. Chyba et al. (1990) followed the suggestion of Passey and Melosh (1980) that ζ can be obtained for iron meteorites by averaging the heats of fusion and vaporization. This approach yields $\zeta = 5$ MJ kg^{-1}, the same value as for ordinary chondrites (Baldwin and Shaeffer, 1971). This chondritic value for ζ, combined with observationally derived values for $\sigma = C_H/C_D\zeta$ (Ceplecha, 1977) can then be used to determine ζ values of 3.2 MJ kg^{-1} and 1.6 MJ kg^{-1} for carbonaceous chondrite and cometary meteors, respectively.

Statistics for contemporary SP comet orbits, plausibly similar to primordial cometary impactors, show that 25% of collisions with Earth occur at or below 18 km s^{-1}, while another 50% occur with velocities between 18 and 23 km s^{-1} (Chyba et al., 1990). By comparison, the median impact velocity for Earth-crossing asteroids is 15 km s^{-1}.

Application of this atmospheric entry model for an atmosphere similar in density to that of present-day Earth (surface pressure of 1 bars) predicts that cometary impactors with initial velocities of 23 km s^{-1} and radii$\approx$100 m have their speeds reduced by approximately 10%. Asteroidal impactors have their speeds reduced by a smaller factor, because of their greater density. Large impactors (radii > 1 km) have their initial speeds decreased by only a negligible amount before impact with the surface.

However, the early Earth's atmosphere was plausibly much denser than the contemporary atmosphere (Walker, 1986; Kasting, 1990), significantly aiding aerobraking. If we consider a primordial CO_2 atmosphere with a surface pressure of 10 bars, with appropriate greenhouse-enhanced surface temperature and scale height we find that 100 m comets entering the atmosphere at an incidence angle of 45° (statistically the most common value) with initial velocities of 23 km s^{-1} are aerobraked to final impact speeds of 5.9 km s^{-1}. Asteroidal impactors are not slowed to impact speeds less than ~ 10 km s^{-1} (Chyba et al., 1990).

5.3   Hydrodynamic Simulation of Impact

The dynamics of impact with the terrestrial surface are modeled using Smoothed Particle Hydrodyamics (SPH) (Monaghan, 1985). This fully Lagrangian approach divides the material in the problem into a set of mass particles which are used to calculate density, pressure, internal energy, and their derivatives when needed. This approach is a powerful one for turbulent flow, avoiding problems associated with tangled grids or limiting boundary conditions. In addition, it can straighfor-wardly be extended from one and two dimensions to a fully three-dimensional representation of the collision.

SPH was developed to simulate nonaxisymmetric phenomena in astrophysics, and most examples of its use can be found in the astrophysics literature. SPH has been successfully applied to the study of gas dynamics, the stabilitry of binary stars, stellar collisions, fragmentation of interstellar clouds, collisions of inter-stellar clouds, disks and rings around stars, radio jets, motion near black holes, supernovae, and impact calculations. References to these applications and more can be found in Monaghan (1992).

5.3.1   Selection of an Appropriate Equation of State

Numerical hydrocodes normally calculate the density ρ and the specific thermal energy E (or temperature T) in one cycle. The pressure P and the pressure gradient are then calculated for the next cycle, or time step. Typically, a material-specific Equation of State (EOS) is required to relate these three thermodynamic quantities (see Eliezer et al. (1986) for a comprehensive introduction to EOS theory). The EOS is material specific.

Until recently the most common form of the EOS used in impact hydrocodes was the Tillotson EOS (TEOS) (Tillotson, 1962; Melosh, 1989). This equation of state consists of parametric equations which have the same form for all materials. The Tillotson equation governing compression has the form:

$$P = \left(a + \frac{b}{(E/E_0\eta^2) + 1} \right) E\rho + A\mu + B\mu^2, \tag{5-5}$$

where $\eta = \rho/\rho_0$ is the normalized density and $\mu = \eta - 1$. The constants a, b, and A are derived a priori to fit experimental data for the given material, and then E_0 and B are then adjusted to provide the best ρ, E, P surface.

In practice, the TEOS has a considerable advantage over other EOS forms be-cause of its simplicity. A major drawback of the TEOS, however, is that temperature is not explicitly determined and can only be calculated using knowledge of the spe-cific heat (which for many materials is a strong function of temperature). This is an important deficit, because a determination of organic survivability requires the solution of temperature-dependent rate equations for pyrolysis.

A second computational technique is to obtain EOS data in tabular form. Tables of values, with density and specific thermal energy as independent variables and pressure as the dependent variable, are used for each material. The tables are constructed using analytic and semi-analytic codes that would be impossible to incorporate directly into a hydrocode. The Los Alamos SESAME database is an example of tabular equations of state for about 100 materials (SESAME, 1983).

The third form of EOS determination is via a semianalytic code such as ANEOS (Thompson and Lauson, 1972). Unlike simple parametric equations, ANEOS uses valid physical approximations in different regimes. To generate a thermodynamically complete and consistent EOS (which both SESAME and Tillotson fail to do) ANEOS formulates all thermodynamic variables in terms of the Helmholtz free energy.

In practice, the TEOS has a considerable advantage over the other EOS forms because of its simplicity. A major drawback of the TEOS, however, is that temperature is not explicitly determined and can only be calculated using knowledge of the specific heat (which for many materials is a strong function of temperature). This is an important deficit, because a determination of organic survivability requires the solution of temperature- dependent rate equtions for pyrolysis. Accordingly, both the SESAME and ANEOS equations of state are preferred for determinations of organic survivability.

Constructing an EOS for comets and carbonaceous chondrites is a difficult task. The composition of comet nuclei is unknown in detail and while the composition of carbonaceous chondrites is known from studies of meteorites, they are highly heterogeneous and the response of meteoritic material to shock loading is currently unknown. Comets are most easily modeled by assuming that they are composed of solid water ice. This is no doubt an over-simplifying assumption, but an unknown composition of dust, water ice, clathrates and voids cannot be modeled. The SESAME database supplies a number of water ice tables covering different density and temperature ranges.

To construct an EOS of a material requires knowledge of the material's response to shock loading. Though examples of carbonaceous chondrites exist, the material properties needed to construct an EOS are unknown. To model carbonaceous chondrites we resort to using a terrestial rock with well-known properties. The rock of choice is serpentinite (a mixture of the three forms of serpentine: chrysotile, antigorite and lizardite, and impurities). The formula for pure serpentine is $Mg_3Si_2O_5(OH)_4$. Table 5.1 compares the composition of naturally occurring serpentine with carbonaceous chondrite. Using the known properties of serpentinite under shock loading (Tyburczy et al., 1991) the parameters for the ANEOS EOS can be constructed.

5.3.2   Results of Impact Simulations

We have analyzed a wide range of collisions of comets and carbonaceous chondrite asteroids on basalt and ocean surfaces. The step-by-step evolution of a typical

TABLE 5.1. A Comparison (by Weight) of Serpentinite and Carbonaceous Chondrite Composition.

Component	Serpentine	Carbonaceous chondrite
FeS	—	6.5—18.4
SiO_2	37.5—43.4	22.5—33.4
TiO_2	—	0.07—0.12
Al_2O_3	0.38—4.68	1.65—2.93
MnO	0.00—0.27	0.19—0.23
FeO	0.00—9.24	9.45—25.43
MgO	39.3—40.0	15.8—24.0
CaO	0.12—0.32	1.22—2.64
Na_2O	0.09—0.28	0.22—0.75
K_2O	0.00—0.28	0.04—0.14
P_2O_5	—	0.23—0.41
H_2O+	11.11—15.77	1.40—19.89
H_2O-	0.10—15.77	0.18—19.89
Cr_2O_3	0.00—0.42	0.33—0.52
NiO	0.00—0.30	0.00—1.71
C	—	0.46—4.83

model, for an ocean impact, is illustrated in Fig. 5.1. The Earth is represented by a semicircular segment of particles, limited in size by computer storage requirements.

Our criterion for survivability of organic compounds is that only those collisions resulting in some fraction of the impactor remaining heated to $T < 1600$ K for <1 s contribute to the terrestrial organic inventory. In all of our simulations where the organic fraction is predicted to survive, peak temperatures have declined in a timescale much smaller than 1 s. We therefore assume that peak temperatures for any individual particle also decline on the same timescale.

The results of our SPH simulations predict that no significant organic fraction survives impact for plausible speeds for carbonaceous chondrite impactors or for comets hitting basalt surfaces. For the case of comets hitting an ocean surface at speeds less that 10 km s^{-1}, a fraction of the impactor is exposed to temperatures below 1800 K. The results for these simulations are shown in Fig. 5.2. The reason for this survivability is that the greater specific heat and lower density of a water target acts to partition a larger fraction of the impactor kinetic energy into thermal energy of the target (i.e. the ocean) than for a basalt surface. These results are independent of the depth of the ocean or the size of the impactor, provided that the ocean is at least an impactor diameter deep. The results are only mildly affected by impact angle (in the direction of slightly lower temperatures experienced), as shown in Figure 5.2b.

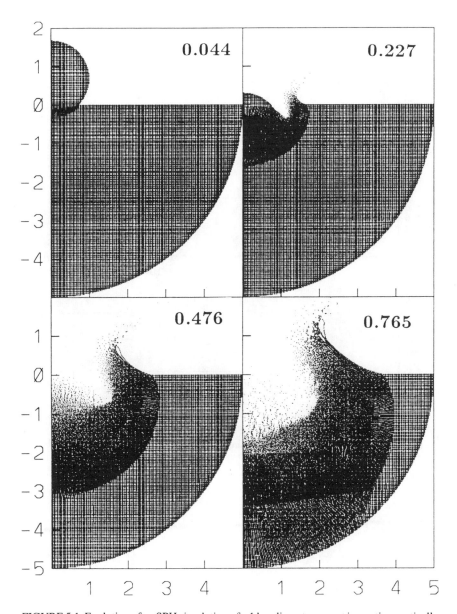

FIGURE 5.1. Evolution of an SPH simulation of a 1 km diameter comet impacting vertically on a 3 km deep ocean overlying a basalt seabed at 7.5 km s^{-1}. Times from first contact are shown in seconds, and axis scales are in kilometers. In the last frame the rebound of the shock from the seabed is visible.

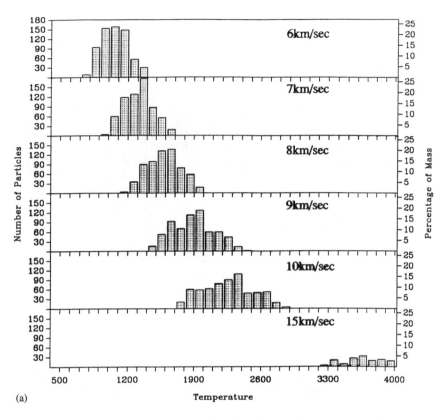

FIGURE 5.2. (a) Histograms representing fractions of comet impactors exposed to various peak temperatures (in kelvins) for impacts into an ocean at speed varying from $6 - 10 \, \text{km s}^{-1}$.

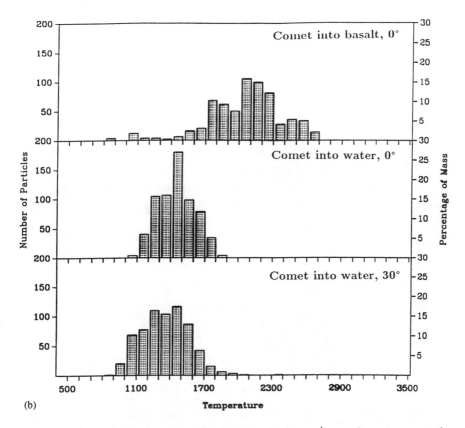

(b)

FIGURE 5.2. (b) Histograms representing fractions of 7.5 km s^{-1} comet impactors exposed to various peak temperatures for various impact scenarios.

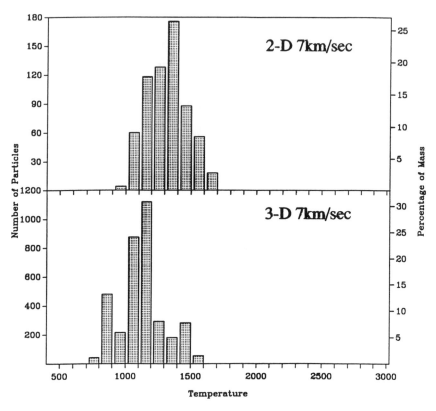

FIGURE 5.3. The effect on peak temperatures of three-dimensional geometry on an SPH simulation. Both histograms are for a comet hitting an ocean at 0° incidence angle.

Survival temperatures of interest for key organics in these simulations (given the timescale of 1 s) are 1800 K for HCN and 1200 K for kerogen pyrolysates and H_2CO. Decarboxylated amino acids will survive temperatures less than 1100 K. The SPH simulations show that, for impact speeds on an ocean less than 7 km s^{-1} (the aerobraked impact speed of a $\sim$200 m comet in a 10 bar atmosphere), the entire comet is exposed to temperatures lower than 1800 K and $\approx$30% of the comet is exposed to temperatures lower than 1200 K.

With some reduction in resolution, we have run models using full three-dimensional symmetry. This has the effect of reducing peak temperatures by as much as an order of magnitude, and exposing a much smaller fraction of the impactor to the highest temperatures during impact, as shown in Fig. 5.3. For a three-dimensional simulation of a comet–ocean impact at 7 km s^{-1}, the entire comet is exposed to temperatures lower than 1600 K, and $\approx$75% of the comet is exposed to temperatures lower than 1200 K, siginficantly enhancing our estimate of organic survivability.

5.4   Airbursts and Organic Survivability

Atmospheric explosions ("airbursts") of hypersonic impactors occurs because the impactor's material strength is exceeded by the differential internal pressure between the stagnation pressure in the bow shock at the leading face ($\sim \rho V^2$) and the vacuum in the wake at the trailing face. Subsequent deformation of the impactor increases atmospheric drag forces, leading to a "run-away" of deformation and subsequent disintegration. However, an impactor will only fragment during atmospheric passage if sufficient time is available for this pressure difference to be established. This time is given by the diameter of the body divided by the speed of sound of the appropriate material. Typical atmospheric passage times for impactors with $V_0 \sim 18$ km s^{-1} are ~ 4 s. Therefore, given sound speeds ~ 500 m s^{-1}, we find that only objects with diameters < 2 km will fragment. If we assume $V_0 \sim 25$ km s^{-1}, fragmentation will be limited to objects smaller than 1.4 km.

If the differential pressure across the impactor exceeds the material strength the bolide will begin to deform substantially and will eventually fragment. For an object traveling at 20 km s^{-1} the stagnation pressure at 45 km is approximately 1 MPa. Plausible strengths for comet and carbonaecous chondrite impactors are 0.1 MPa and 1 MPa, respectively (Chyba et al., 1993).

A model for the dynamics of an airburst was applied to the 1908 explosion over Tunguska, Siberia, by Chyba et al. (1993). We follow their analysis here. This model assumes the impactor to be a "cubical cylinder," with a diameter equal to its height. The cylindrical shape makes calculation of internal hydrostatic stresses a simple process. The impactor velocity vector lines along its axis of symmetry, and once the differential pressure across its leading and trailing edges exceeds its strength, deforms in the manner of an inviscid fluid. The rate of deformation is determined by a force balance equation with the pressure differential. The effect of the deformation is to increase the cross-sectional area (A in equation (2)), increasing aerobraking. As the impactor deforms to become a squatter version of itself, the deformating pressure increases, producing a catastrophic fragmentation, in which most of the original impactor kinetic energy is deposited at a small range of altitudes.

We have modified the airburst code of Chyba et al. (1993) to incorporate denser primordial CO_2 atmospheres (with surface pressures of 10–20 bars, as described above) to examine whether airbursts are an important factor in the survivability of organic molecules from comet impactors in the ~ 100 m size range. Calculated heights and energies of airbursts for these objects are illustrated in Figs. 5.4 and 5.5.

Our calculations indicate that airbursts occur at heights in excess of 40 km for ~ 100 m comet impactors. Energies associated with the airbursts are in the range of 50–100 megatons (MT) of TNT (1 MT = 4.18×10^{15} J). Airburst energies are somewhat lower in a 20 bar atmosphere because pre-airburst aerobraking occurs to a greater extent than for a 10 bar atmosphere. Carbonaceous chondrite asteroids airburst at lower altitude, because of their greater density and strength: airburst altitudes for ~ 100 m carbonaceous chondrite impactors are in the range of 20–

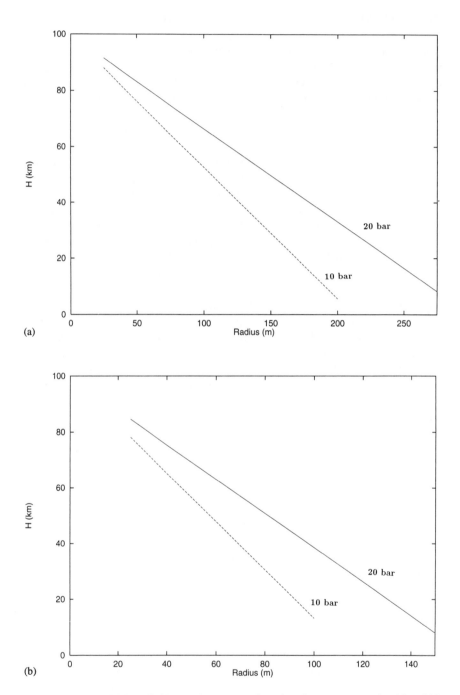

FIGURE 5.4. (a) Heights of airbursts for comets of varying diameters traversing 10 and 20 bar CO_2 atmospheres. Initial velocity = 25 km s^{-1}, angle of incidence = 45°. (b) Heights of airbursts for carbonaceous chondrite asteroids. Initial velocity = 15 km s^{-1}, angle of incidence = 45°.

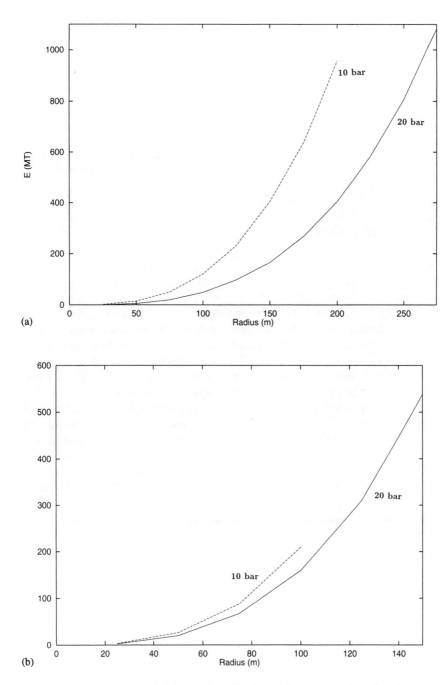

FIGURE 5.5. (a) Energies of airbursts (in MT of TNT) for comets of varying diameters traversing 10 and 20 bar CO_2 atmospheres. Initial velocity = 25 km s^{-1}, angle of incidence = 45°. (b) Energies of airbursts for carbonaceous chondrite asteroids. Initial velocity = 15 km s^{-1}, angle of incidence = 45°. As can be seen from Fig. 5.4, asteroids with radius > 100 m reach the surface without airbursting.

40 km.

A well-recorded instance of organic material surviving an airburst is the fragmentation of a CI carbonaceous chondrite at an altitude of 30 km above Revelstoke, Canada, in 1965 (Levin and Bronshten, 1986). Photomicrographs of the resulting fragments, up to 0.5 cm in size, showed no heat damage within an exterior layer <0.7 mm thick, suggesting that such airbursts could allow organic inclusions to survive. The energy released in the Revelstoke airburst was of the order of $\sim 10^{-2}$ MT (Krinov, 1966; Folinsbee et al., 1967). Since this energy is three orders of magnitude smaller than the airburst energies calculated for ~ 100 m comet and asteroid impactors, it seems unlikely that a large fraction of the organic molecules present in an impactor of this size would survive pyrolysis from energy released by the airburst.

5.5   Conclusions

While it seems possible to make a case that organic molecules were successfully delivered to the prebiotic Earth via impact, the efficiency of such delivery is low: only impactors of a limited size range ~ 100 m impacting oceans would have experienced sufficiently low-impact temperatures to avoid pyrolyzing their organic fraction. Although the energy released in the Revelstoke airburst is three orders of magnitude smaller than the calculated energy release from that calculated above for ~ 100 m comet and asteroid impactors, it demonstrated that organic inclusions within rocky fragments can survive pyrolysis in low-energy airbursts: this result may not be irrelevant even for the extreme energies considered here, although a conservative approach would of course be to assume that effectively all extraterrestrial organic molecules contained *within* comets and asteroids are pyrolyzed, either by ablation, airburst, or impact. The detailed partitioning of post-airburst energy, and its likely effect on organic molecules in comets and asteroids remains an interesting problem to be analyzed, perhaps using insights gained from the modeling of the collision of the fragments of Comet Shoemaker–Levy 9 with Jupiter in July 1994 (see, e.g., Zahnle and Mac Low, 1994).

The fate of organic-rich extraterrestrial dust is distinctly separate from the scenarios discussed here, and is discussed by Chyba and Sagan in the next chapter.

Acknowledgments: We wish to thank Chris Chyba and Chris McKay for useful discussions.

5.6   References

Baldwin, B. and Y. Shaeffer (1971), Ablation and breakup of large meteoroids during atmospheric entry, *J. Geophys. Res.*, **76**, 4653.

Biberman, L.M., S. Ya. Bronin, and M.V. Byrkin (1980), *Acta Astronaut.*, **7**, 53–65.

Ceplecha, Z. (1977), Meteoroid populations and orbits. In *Comets, Asteroids and Meteorites* (A.H. Delsemme, ed.). University of Toledo, Toledo, OH.

Chamerberlin, T.C. and R.T. Chamberlin (1908), Early terrestrial conditions that may have favored organic synthesis, *Science*, **28**, 897–911.

Chyba, C.F., P.J. Thomas, L. Brookshaw, and C. Sagan (1990), Cometary delivery of organic molecules to the early Earth, *Science*, **249**, 366–373.

Chyba, C.F., P.J. Thomas, and K.J. Zahnle (1993), The 1980 Tunguska explosion: Atmospheric disruption of a stony asteroid, *Nature*, **361**, 40–44.

Clark, B.C. (1988), Primeval procreative comet pond, *Origins Life*, **18**, 209.

Eliezer, S., A. Ghatak, and H. Hora (1986), *An Introduction to Equations of State: Theory and Applications*. Cambridge University Press, Cambridge.

Folinsbee, R.E., J.A.V. Douglas, and J.A. Maxwell (1967), Revelstoke, a new Type I carbonaceous chondrite, *Geochim. Cosmochim. Acta*, **31**, 1625–1635.

Kasting, J.F. (1990), Bolide impacts and the oxidation state of carbon in the Earth's early atmosphere, *Origins Life*, **20**, 199–231.

Krinov, E.L. (1966), *Giant Meteorites*. Pergamon, Oxford.

Levin, B.Y. and V.A. Bronshten (1986), The Tunguska event and the meteors with terminal flares, *Meteoritics*, **21**, 199–215.

Melosh, H.J. (1989), *Impact Cratering: A Geologic Process*. Oxford University Press, New York.

Monaghan, J.J. (1985), Particle methods for hydrodynamics, *Comput. Phys. Rep.*, **3**, 71–124.

Monaghan, J.J. (1992), *Ann. Rev. Astrophys.*, **30**, 543–574.

O'Keefe, J.D. and T.J. Ahrens (1982), Cometary and meteorite swarm impact on planetary surfaces, *J. Geophys. Res.*, **87**, 6668–6680.

Oró, J. (1961), Comets and the formation of biochemical compounds on the primitive Earth, *Nature*, **190**, 389–390.

Passey, Q.R. and H.J. Melosh (1980), Effects of atmospheric breakup on crater field formation, *Icarus*, **42**, 211–233.

SESAME '83 (1983), Report on the Los Alamos Equation-of-State Library, LALP-83-4. Los Alamos National Laboratory, Los Alamos, NM.

Thompson, S.L., and H.S. Lauson (1972), Improvements in the Chart-D radiation-hydrodynamic code III: Revised analytic equations of state. Sandia National Laboratory Report RR-71 0714.

Tillotson, J.H. (1962), Metallic equation of state for hypervelocity impact. General Atomic Report GA-3216.

Tyburczy, J.A., T.S. Duffy, T.J. Ahrens, and M.A. Lange (1991), Shock wave equation of state of serpentine to 150 GPA: Implications for the occurrence of water in the earth's lower mantle, *J. Geophys. Res.*, **96**, 18011.

Walker, J.C.G. (1986), Carbon dioxide on the early Earth, *Origins Life*, **16**, 117–127.

Wilkening, L.L. (1978), Carbonaceous chondritic material in the Solar System, *D. Naturwiss.*, **65**, 73.

Zahnle, K. and M.-M. Mac Low (1994), The collision of Jupiter and comet Shoemaker–Levy 9, *Icarus*, **108**, 1–17.

6

Comets as a Source of Prebiotic Organic Molecules for the Early Earth

C.F. Chyba and C. Sagan

ABSTRACT Life on Earth originated during the final throes of the heavy bombardment, in which the Earth–Moon system, as well as the rest of the inner solar system,
was subjected to an intense bombardment of comets and asteroids. This bombardment may have rendered the Earth's surface inhospitable for life for hundreds of
millions of years subsequent to terrestrial formation. It may also have delivered to
the Earth's surface the bulk of the current terrestrial volatile inventory, in the form
of a late-accreting impact veneer. Delivering intact prebiotic organic molecules of
interest for the origins of life is much more difficult. However, several mechanisms
seem likely to have been delivering exogenous organics to the surface of the Earth, or
shock-synthesizing them in impacts. In an early carbon dioxide-rich terrestrial atmosphere, these mechanisms would have quantitatively rivaled or exceeded terrestrial
organic synthesis in situ. In an early reducing (methane-rich) atmosphere, the exogenous sources would have been quantitatively unimportant compared to atmospheric
production.

6.1   The Lost Record of the Origin of Life

Isotopic dating of the terrestrial atmosphere and mantle suggests that the Earth
was formed about 75–100 Myr subsequent to the formation of primitive meteorites
(Swindle et al., 1986). The oldest high-precision meteorite formation date is from
the Allende CV chondrite, which yields a $^{207}Pb/^{206}Pb$ model age of 4.559 ± 0.004
Gyr (Tilton, 1988). So it seems that the Earth was nearly complete by about 4.5
Gyr ago.

Almost the entire surface of the Earth is geologically much younger than this.
The ocean floors are typically less than 100 Myr old, and most continental crust
post-dates the Archean period, which ended 2.5 Gyr ago (Veizer, 1983). Of the
extant Archean crust, rocks as old as 3.5 Gyr are found only in Western Australia,
South Africa, and Greenland.

The Pilbara Block in the Warrawoona Volcanics of Western Australia are sedimentary rocks nearly 3.5 Gyr old that contain both microscopic fossils (Schopf
and Walter, 1983; Schopf, 1993) and fossil stromatolites—macroscopic structures
formed by sediment-trapping microbes (Walter, 1983). The ~3.5 Gyr old On-
verwacht Group of South Africa also appears to contain microfossils, although
this identification is less certain (Schopf and Walter, 1983). The oldest terrain of

sedimentary origin on Earth, the ~3.8 Gyr old Isua metasediments in Southwestern Greenland, have been metamorphosed past the point where organic-walled fossils could be expected to remain (Robbins and Iberall, 1991). However, even here there is controversial evidence for biologically mediated carbon isotope fractionation (Schidlowski, 1988), suggesting photosynthesizing life may have existed by this time.

We therefore find ourselves in the frustrating position that the oldest terrestrial rocks which could contain evidence for early life appear to contain it. Therefore our knowledge of the environment of early Earth during the period of the origin of life is severely curtailed: there is virtually no remaining geological record of terrestrial conditions prior to the existence of a biosphere.

This is why comparative planetology is so important for understanding early Earth and the origin of life. We know from radioactive dating of lunar samples that much of the surface of the Moon is older than the oldest terrestrial sedimentary rocks. Crater counts suggest that much of the southern hemisphere of Mars, as well as the surface of Mercury, is similarly ancient. Each of these surfaces shows evidence of having withstood a massive bombardment early in Solar System history (Strom, 1987). The Earth, being an extremely active planet geologically, no longer retains its record of these events. As a result, studying the cratering records of other nearby worlds, but especially that of the Moon, provides the most direct route to understanding terrestrial conditions at the time of the origins of life.

6.2   The Uninhabitable Habitable Zone

Life on Earth depends critically on a handful of "biogenic" elements, such as carbon, nitrogen, and hydrogen. An examination of the distribution of these elements in our Solar System reveals them to be abundant in the outer Solar System, but comparatively rare in the region of the terrestrial planets. McKay (1991) has illustrated this graphically for the case of the element carbon. Fig. 6.1 shows the ratio of carbon atoms to heavy atoms (all atoms heavier than He) throughout the solar system. The inner solar system is extremely depleted in carbon, compared to solar abundances. Moving outward from the Sun, it is not until we reach the C-type asteroids in the main asteroid belt that the carbon abundance begins to approach that of the Sun.

Yet it is only at heliocentric distances less than the distance to the asteroid belt that planetary surface temperatures are high enough for the existence of liquid water (barring tidal heating mechanisms). Liquid water is the sine qua non of life as we know it (Horowitz, 1986), and those terrestrial environments in which liquid water is most rare are those closest to being sterile—indeed, the harshest deserts on Earth may actually *be* sterile outside of specific, protected environments (McKay, 1986, 1991).

We may define a star's habitable zone as that region where liquid water can exist for extended periods of time (Kasting et al., 1993). The zone's inner boundary is that distance at which a runaway greenhouse effect occurs (as in the case of Venus

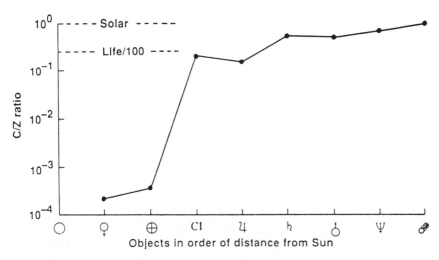

FIGURE 6.1. Ratio of carbon (C) atoms to heavy (Z) atoms (atoms heavier than helium) throughout the Solar System (from McKay, 1991).

in our Solar System). The outer boundary is that distance at which carbon dioxide condenses out of an atmosphere and is no longer available as a greenhouse gas (in our Solar System, around the orbit of Mars). In the context of the solar habitable zone, the distribution of biogenic elements presents an apparent dilemma: the elements essential to life are comparatively rare exactly where the temperatures make liquid water possible.

This dilemma only worsens when Solar System formation models are taken into account. The problem is that the biogenic elements are among the most volatile elements. Delsemme (1992, and this volume) has reviewed progress in thermochemical equilibrium and chemical kinetics models of the accretion disk out of which our planetary system formed. Such models have consistently found temperatures in the neighborhood of 1000 K in the region of terrestrial accretion, making it difficult to provide Earth with its observed inventories of water, nitrogen, and carbon (Lewis, 1974; Lewis et al., 1979; Cameron, 1983; Prinn and Fegley, 1989). Delsemme finds similar temperatures by using the transition between volatile-poor S-type asteroids and comparatively volatile-rich C-type asteroids to peg the accretion disk's temperature to 450 ± 50 K at 2.6 AU, then extrapolating inward to the region of the terrestrial planets. At temperatures near 1000 K, water, carbon, nitrogen, and other volatiles remain almost entirely in the gas phase over the range of pressures likely to have existed in the accretion disk. Volatiles are therefore almost entirely absent from the dust grains out of which the planet-forming planetesimals aggregate. From the point of view of the elements needed for the origin of life, the habitable zone is, initially at least, nearly uninhabitable. In particular, the Earth may have formed almost entirely devoid of biogenic elements.

However, Wetherill (1990) has shown that the latter stages of planetary formation in the inner Solar System were marked by collisions of large planetary embryos scattered across considerable heliocentric distances. In this way, Earth may have acquired an initial complement of volatiles through early radial mixing. Moreover, the latter stages of outer planet formation should have led to the scattering of extremely volatile-rich planetesimals (comets; see the uppermost point in Fig. 6.1) throughout the Solar System (Fernández and Ip, 1983; Shoemaker and Wolfe, 1984; Delsemme, 1992). Hartmann (1987, 1990) has argued that spectral observations of Solar System satellites, as well as the preponderance of CM clasts among the foreign fragments in polymict meteoritic breccias, provide evidence for an intense scattering of C-type asteroids during the first $\sim 10^8$ yr of Solar System history.

But it is unclear whether volatiles accreted by Earth prior to 4.4 Gyr ago would have been retained. The hypothesized Moon-forming impact may have stripped the Earth of whatever terrestrial atmosphere existed prior to that event (Cameron, 1986). In any case, terrestrial water present prior to core formation at ~ 4.4 Gyr ago ($\sim 10^8$ yr after the Earth's formation (Stevenson, 1983, 1990; Swindle et al., 1986; Tilton, 1988)) should have been efficiently destroyed by reacting with metallic iron according to $Fe + H_2O \rightarrow FeO + H_2$; large quantities of hydrogen produced in this way may have removed other degassed volatiles by hydrodynamic escape (Dreibus and Wänke, 1987, 1989). It is possible that only those volatiles accreted by Earth subsequent to ~ 4.4 Gyr ago would have been able to contribute to the planet's extant volatile inventory. Icy planetesimals scattered from the accretion regions of Uranus and Neptune, with scattering timescales of hundreds of Myr, may in this case have been more likely contributors to the present terrestrial volatile inventory than icy planetesimals scattered from the jovian and saturnian regions, whose scattering timescales were perhaps as short as tens of Myr (Ip, 1977; Fernández, 1985).

6.3   A Procrustean Model for Lunar Cratering

Because of the virtual absence of a terrestrial geological record of the period prior to 3.5 Gyr ago, we look to the other worlds of the inner Solar System for insights. The Moon is by far the most useful for this purpose, for it alone has been sufficiently sampled to allow a tentative history of its cratering record to be developed, based on absolute radiometric dating of returned lunar samples.

Our goal in this discussion is to make use of the observed lunar cratering record to estimate the amount of mass accreted by the Moon as a function of time prior to around 3.5 Gyr ago. This accretion may then be extrapolated to the Earth, using the ratio of the two worlds' gravitational cross sections to determine how much more mass Earth would have accreted than the Moon. Ultimately, we will be interested in the fraction of this terrestrially accreted mass that consisted of intact organic molecules. These would have represented an exogenous contribution to the terrestrial prebiotic organic inventory.

We begin by recalling the essentials of impact cratering. (The best recent review of this topic is that of Melosh (1989).) Consider a crater excavated by an impacting body of mass m colliding at velocity v. The initially excavated, or transient, crater diameter D_i is related to the mass m according to Schmidt and Housen (1987):

$$m = \gamma v^{-1.67} D_i^{3.80}, \tag{6-1}$$

where v is impactor velocity, and γ is a constant that depends on surface gravity, impactor and target densities, and impactor incidence angle θ (taken to be the most probable angle of collision, 45°):

$$\gamma = 0.31 g^{0.84} \rho^{-0.26} \rho_t^{1.26} (\sin 45°/\sin \theta)^{1.67}. \tag{6-2}$$

MKS units are assumed throughout this discussion. In (6-2), $\rho_t = 3000$ kg m^{-3} is the lunar crustal density (Ryder and Wood, 1977), $\rho = 2200$ kg m^{-3} is taken as the density of a typical impacting asteroid, and $g = 1.67$ m s^{-2} is the gravitational acceleration at the lunar surface (Melosh, 1989). These values give

$$\gamma = 1.6 \times 10^3 \text{ kg s}^{-1.67} \text{ m}^{-2.13}. \tag{6-3}$$

The final crater diameter D_f is enlarged over the initial transient crater diameter D_i, due to subsequent crater collapse. McKinnon et al. (1990) argue that simple craters on the Moon near the simple–complex transition (which they define, on the basis of depth–diameter data, to be $D_c \approx 11$ km) are ~15–20% wider than the original transient craters. McKinnon et al. (1990) write

$$D_f = k D_i^{1.13}, \tag{6-4}$$

where

$$k = \kappa D_c^{-0.13}. \tag{6-5}$$

The requirement that D_f be 17.5% (the mean of 15% and 20%) larger than D_i when $D_f = D_c$, combined with (6-4) and (6-5), gives $\kappa = (1.175)^{1.13} = 1.2$, whence

$$D_f = 1.2 D_c^{-0.13} D_i^{1.13}, \tag{6-6}$$

where $D_c = 11$ km.

Fig. 6.2 shows the dependence of cumulative lunar crater density on surface age, using data from the Basaltic Volcanism Study Project (BVSP, 1981). Also shown is an analytical fit to these data, which is discussed below. Elsewhere one of us has compared (Chyba, 1990a) other authors' fits to this, as well as an alternate lunar data set from Wilhelms (1984). The data, for craters bigger than a diameter D, are well modeled by the equation

$$N(t, D) = \alpha[t + \beta(e^{t/\tau} - 1)](D/4000 \text{ m})^{-1.8} \text{ km}^{-2}, \tag{6-7}$$

where t is in billions of years (Gyr), and α and β are determined by two-dimensional χ^2 minimization.

Equation (6-7) is the mathematical statement that cratering has been roughly constant for about the past 3.5 Gyr, while increasing exponentially into the past prior to that time. Since Fig. 6.2 is a cumulative crater plot, its data are fit by the

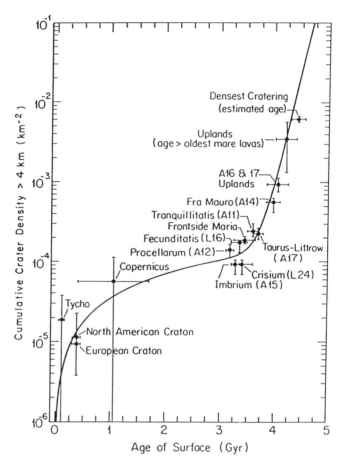

FIGURE 6.2. Analytical fit (6-7) to cumulative lunar crater density as a function of surface age, with a decay time constant $\tau = 144$ Myr (100 Myr half-life). From Chyba (1991).

integral (over time t) of this constant plus exponential, giving a sum of a linear and an exponential term. The logarithmic ordinate in Fig. 6.2 results in the term linear in t extrapolating to $-\infty$ as t goes to 0.

Choices of the decay constant τ in the literature have ranged from 70 Myr to 220 Myr. However, fitting any single decay "constant" to the cratering flux is a procrustean approximation, as the impactor flux cannot actually have decayed at a constant rate, but must instead have made a transition from rapidly swept-up objects in Earth-like orbits, to objects from comparatively long-lived, slowly decaying orbits (Wetherill, 1977; Hartmann, 1980; Grinspoon, 1988). The fit used here should therefore represent the impactor flux subsequent to the time after which the flux of those impactors with very short ($\tau \sim 20$ Myr) sweep-up timescales has decayed to a level unimportant compared to that of more slowly ($\tau \sim 100$ Myr) decaying populations (Grinspoon, 1988).

We have suggested (Chyba, 1990a; Chyba and Sagan, 1992) choosing a heavy bombardment decay half–life of 100 Myr, or $\tau = 144$ Myr (which gives $\alpha = 3.5 \times 10^{-5}$ and $\beta = 2.3 \times 10^{-11}$ in (6-7)). A ~ 100 Myr half-life has been found for the decrease of the primordial comet flux through the inner Solar System by numerical simulations of the formation of Uranus and Neptune (Fernández and Ip, 1983; Shoemaker and Wolfe, 1984). This choice also has the advantage of lying in the middle range of the values for τ with which the lunar cratering record is reasonably consistent. An independent estimate of τ by Oberbeck and Fogleman (1989), that correlates absolute age estimates for lunar impact basins from a crustal viscosity model (Baldwin, 1987a) with crater counts for these basins (Baldwin, 1987b), yields $\tau = 150$ Myr. Finally, a choice of τ consistent with a 100 Myr decay half-life is in excellent agreement with lunar and terrestrial geochemical constraints, as discussed in the next section.

6.4   Lunar Cratering and Geochemical Evidence

Equations (6-1), (6-6), and (6-7) may be used to obtain an estimate of the total mass delivered to the Earth or Moon subsequent to some time t. These estimates may then be compared with lunar and terrestrial geochemical data, which provide an independent check on the cratering model developed here.

Median asteroid collision velocities for the Moon and Earth are 12 km s^{-1} and 15 km s^{-1}, corresponding to a velocity at infinity of about 12 km s^{-1} (Chyba et al., 1995). Equations (6-1) and (6-6) yield

$$m = 0.54 \gamma v^{-1.67} D_c^{0.44} D_f^{3.36} = 8.0 \times 10^{-3} D_f^{3.36}, \qquad (6\text{-}8)$$

an equation relating the final crater diameter observed on the Moon to the mass (and impact velocity) of the projectile. Equations (6-7) and (6-8) may be combined to give the number of objects of mass $> m$ that have impacted the Moon as a function of time t:

$$n(>m, t) = \psi [t + \beta(e^{t/\tau} - 1)] m^{-b} \, \text{kg}^b, \qquad (6\text{-}9)$$

where $b = (1.8/3.36) = 0.54$, and $\psi = 3.4 \times 10^8$.

The total mass $M(t)$ incident on the Moon subsequent to some time t in impactors with masses in the range m_{min} to m_{max} is given by the integral:

$$M(t) = \int_{m_{max}}^{m_{min}} m[\partial n(> m, t)/\partial m]\, dm, \qquad (6\text{-}10)$$

which yields

$$M(t) = 3.9 \times 10^8[t + 2.3 \times 10^{-11}(e^{t/\tau} - 1)](m_{max}^{1-b} - m_{min}^{1-b})\ kg^b. \qquad (6\text{-}11)$$

Scaling net impact flux from the Moon to Earth requires accounting for Earth's larger gravitational cross-section. A planet's gravitational cross-section is $\sigma = \pi R_g^2$, where the gravitational radius R_g for a planet of physical radius R and escape velocity v_{esc} is given by

$$R_g = R[1 + (v_{esc}/v_\infty)^2]^{1/2}. \qquad (6\text{-}12)$$

With $v_\infty = 12\ km\ s^{-1}$, the ratio of the Earth's gravitational cross-section to that of the Moon is then

$$\sigma_\oplus/\sigma_m \equiv \xi = 1.8(R_\oplus/r_m)^2. \qquad (6\text{-}13)$$

The Earth's volume is equivalent to a sphere of radius $R_\oplus = 6371$ km (Stacey, 1977), and the radius of the Moon $r_m = 1738$ km, so that $\xi \approx 24$. To scale to the Earth, $M(t)$ must therefore be multiplied by $\xi \approx 24$.

The correct choice of m_{max} in (6-11) requires some care, and has been discussed at length by Chyba et al. (1995). Here, we merely quote the results of calculating $M(t)$ over a range of uncertainties in m_{max}. One finds $(0.4-1.3) \times 10^{20}$ kg of extralunar material incident upon the Moon subsequent to crustal solidification 4.4 Gyr ago. About half this material should have been retained, the other half being lost to space. This result overlaps the geochemical estimates of Sleep et al. (1989) of the meteoritic component mixed into the lunar crust that yield about $(0.4-1.6) \times 10^{20}$ kg.

Similarly, scaling (6-11) to Earth, *and also* taking account of the statistical probability that the largest impactors incident on Earth were more massive than the largest incident on the Moon, gives an estimate of $\sim 2 \times 10^{22}$ kg of material accumulated by Earth subsequent to 4.4 Gyr ago (Chyba et al., 1995), about the time of terrestrial core formation. (Note that such estimates require a choice of the mass distribution for the largest terrestrial impactors; for reasonable choices the result is dominated by the mass of the single largest impactor.) This result is in good accord with geochemical estimates of post-core formation meteoritic input. These estimates, based on chondritic abundances of highly siderophile elements in the terrestrial mantle (Chou, 1978; BVSP, 1981; Sun, 1984; Dreibus and Wänke, 1989), and depending on assumptions about mixing in the Earth's mantle, lie in the range $(1-4) \times 10^{22}$ kg (Chyba et al., 1995). Any calculations, whether based on the lunar cratering record or on planetary accretion models, that give estimates for total terrestrial mass accretion subsequent to 4.4 Gyr ago in great excess of $\sim 10^{22}$ kg are inconsistent with the geochemical data, and need to be reconsidered.

6.5   Are the Earth's Oceans Extraterrestial?

The mass of the Earth's oceans is 1.4×10^{21} kg. Since comets appear to be more than 40% water ice by mass (Delsemme, 1992), the results of the previous section suggest that the heavy bombardment need have been only ~10% cometary by mass to have delivered the terrestrial oceans as a late-accreting veneer, if all impacting cometary mass were retained. This conclusion is consistent with the observation that the deuterium to hydrogen ratio for comet Halley ($0.6–4.8 \times 10^{-4}$) overlaps that of Earth's oceans (1.6×10^{-4}). Both values are also consistent with meteoritic values, but are an order of magnitude higher than interstellar, jovian, or saturnian values (Eberhardt et al., 1987). Alternatively, if the heavy bombardment were composed entirely of CI carbonaceous chondrites, Earth could have acquired its oceans entirely from an asteroidal source.

This conclusion is complicated by the effects of impact erosion. Sufficiently fast and large impacts produce an expanding vapor plume powerful enough to carry the impactor's material away into space, and erode some of the terrestrial atmosphere in the process (Melosh and Vickery, 1989). About half of cometary impacts with Earth have velocities sufficient to cause impact erosion (so that twice as large a cometary fraction as would otherwise be required is needed to deliver an ocean of water); ~10% of asteroidal collisions do (Chyba et al., 1995). Water condensed into oceans is largely immune to erosion; only volatiles remaining in the vapor phase are significantly at risk to being permanently lost in this way. For the larger worlds of the inner Solar System, accretion of planetary oceans was favored over impact erosion (Chyba, 1990a). For the Moon, the situation was exactly the opposite (Chyba, 1991), whereas Mars seems to have been right on the edge. Melosh and Vickery (1989) have argued that Mars may have lost its early atmosphere through impact erosion.

Cometary delivery of the bulk of the Earth's oceans is quantitatively consistent with the Earth deriving its inventory of carbon from this source as well. Explaining the terrestrial nitrogen abundance in this way is more problematic, but the nitrogen inventory of the Earth's mantle is poorly known, and substantial quantities of N_2 could have been lost due to atmospheric erosion (Chyba et al., 1995). Overall, a consistent picture can be drawn of the Earth's acquisition of its hydrosphere, biogenic volatiles, and highly siderophilic elements as a late-accreting veneer delivered by the heavy bombardment subsequent to 4.4 Gyr ago.

6.6   Mass Flux Through Time

The consistency of lunar cratering extrapolations with lunar and terrestrial geo-chemical constraints suggests that (6-11) may be used as a reasonable model for terrestrial mass accretion subsequent to core formation. Equation (6-11), multiplied by ξ, gives the total mass $M(t)$ incident on Earth in impactors within a certain size range, subsequent to some time t. The quantity of interest in the remainder of this chapter is not $M(t)$, but rather the mass *flux* (kg yr^{-1}) at a particular time in the Earth's past. Define $\dot{m}(t) \equiv \xi[\partial M(t)/\partial t]$ as the terrestrial mass flux from objects within a given mass range (m_{min} to m_{max}) being accreted by Earth at a time t. From Eq.(6-11),

$$\dot{m}(t) = 9.4 f(t)(m_{max}^{1-b} - m_{min}^{1-b}) \text{ kg}^b \text{ yr}^{-1}, \tag{6-14}$$

where

$$f(t) \equiv (1 + 1.6 \times 10^{-10} e^{t/\tau}). \tag{6-15}$$

The number flux of objects with mass greater than m being accreted by Earth as a function of time is $\dot{n}(t) \equiv \xi[\partial n(>m, t)/\partial t]$. From (6-9),

$$\dot{n}(t) = 8.2 f(t)m^{-b} \text{ kg}^b \text{ yr}^{-1}. \tag{6-16}$$

Equation (6-14) may be used to give an expression for the mass flux at time t, $\dot{m}(t)$, in terms of the contemporary mass flux, $\dot{m}(0)$:

$$\dot{m}(t) = \dot{m}(0) f(t). \tag{6-17}$$

It is therefore simple to extrapolate from the contemporary mass flux back to that at time t.

6.7   Cautions and Uncertainties

The lunar cratering model derived here faces numerous uncertainties, which have been reviewed elsewhere (Chyba 1993a). For example, consider the data used in Fig. 6.2. The oldest lunar province for which a radiometric date actually exists (the Apollo 16 and 17 uplands) is only 3.85–4.25 Gyr old; the ages of more heavily cratered provinces can at present only be estimated (BVSP, 1981). Different choices for the decay rates fitted to lunar cratering data (and alternate choices for that data set) can lead to very different conclusions about terrestrial mass influx during the heavy bombardment. However, the more extreme of these choices can be excluded as in contradiction with lunar and terrestrial geochemical data on meteoritic input. The role of impact erosion of atmospheres (Melosh and Vickery, 1989) and surfaces must also be considered (Chyba et al., 1995).

Most fundamentally, however, the entire interpretation of the heavy bombardment as representing exponentially decaying remnants of planetary formation is questioned by those arguing for the occurrence of a lunar cataclysm 3.85 Gyr ago (e.g. Tera et al., 1974; Ryder, 1990; Dalrymple and Ryder, 1993). Hartmann (1995) has reviewed this long-standing controversy, arguing that the appearance of a "cataclysm" is an artifact of an exponentially decreasing early impactor flux, which would have reset radiometric ages of rock fragments near the lunar surface or among meteorites prior to ~4 Gyr ago.

6.8   Endogenous Production of Prebiotic Organics

Before proceeding to a discussion of exogenous sources of prebiotic organic molecules on early Earth, it makes sense first to examine organic production in situ. Only those exogenous sources that deliver organics in quantities comparable to or greater than endogenous production need be considered as potentially important contributors to the terrestrial origins of life (unless a case can be made for an important qualitative difference between prebiotic organics of exogenous and endogenous origins).

Chyba and Sagan (1992) have summarized a number of endogenous sources of organics on early Earth. Critical to these estimates is the nature of the early atmosphere. In a putative early reducing atmosphere (rich in methane, CH_4), organic production from ultraviolet light, lightning, and coronal discharges is high. However, it seems likely—though by no means certain—that the early atmosphere was of an intermediate oxidation state, rich in carbon dioxide, CO_2, rather than CH_4 (Walker, 1986; Kasting and Ackerman, 1986; Kasting, 1993). In such an atmosphere, at least in the likely case that $[H_2]/[CO_2] \leq 0.1$, organic production is far lower than in the reducing case. As long as the exact nature of the early atmosphere is unknown, comparisons are best made to both reducing and intermediate oxidation state atmospheric end-members.

Electrical discharges in a reducing atmosphere would have produced about 3×10^9 kg yr^{-1} of organic molecules, while yielding only 3×10^7 kg yr^{-1} in an intermediate oxidation state atmosphere. Production of organics by UV in the intermediate case would have been about an order of magnitude higher than that for electrical discharge, giving 3×10^8 kg yr^{-1}. The action of long-wavelength UV ($\lambda \leq 2700$ Å) on hydrogen sulfide, H_2S, could, via the generation of superthermal hydrogen atoms, produce organics at a rate 2×10^{11} kg yr^{-1} in a reducing atmosphere, three orders of magnitude more than for an atmosphere of intermediate oxidation state.

An additional important source of organics on early Earth is implied by experimental results for the production of organics by UV light of wavelength $\lambda \leq 1550$ Å acting directly on CH_4/N_2 atmospheres Sagan and Thompson (1984) summarized efficiencies of production in such atmospheres at these wavelengths, and found a production efficiency of 1.2×10^{-8} kg J^{-1}, in good agreement with the early results of Noyes and Leighton (1941). From Zahnle and Walker (1982), the net flux of UV energy below 1500 Å on Earth 4.0 Gyr ago was about 1×10^{20} J yr^{-1}, about three times the current value (Chyba and Sagan, 1992). The production of organics from CH_4 photolysis in the terrestrial atmosphere 4 Gyr ago would therefore have been about 1×10^{12} kg yr^{-1}.

6.9   Impact Delivery of Intact Exogenous Organics

Organic molecules are being delivered to Earth today from a variety of sources. These include those interplanetary dust particles (IDPs) small enough to be gently decelerated in the upper atmosphere, and meteorites small enough to be substantially decelerated during their fall, but large enough to avoid complete ablation (Anders, 1989). Impactors in the 1–100 m range often catastrophically fragment in the atmosphere (Chyba, 1993b), as was the case with the Tunguska asteroid of 1908 (Chyba et al., 1993). More recently, a CI carbonaceous chondrite exploded with an energy of $\sim$20 kilotons over Revelstoke, Canada, in 1965; photomicrographs of recovered millimeter-sized fragments reveal unheated interiors (Folinsbee et al., 1967), within which organics should have survived. Finally, the discovery of apparently extraterrestrial amino acids in Cretaceous/Tertiary (K/T) boundary sediments at Stevns Klint, Denmark (Zhao and Bada, 1989), suggests that large asteroid or comet impacts with Earth somehow result in the deposition of large quantities of organics.

Each of these exogenous sources of organics would have been more abundant on early Earth than is the case today. Here we estimate their quantitative importance, scaling by the lunar impact record according to (6-17). The results may be compared with the estimates of endogenous production just cited.

6.9.1   Interplanetary Dust Particles (IDPs)

Anders (1989) has estimated the flux of intact organic matter reaching Earth's surface in the form of IDPs. Particles in the mass range 10^{-12}–10^{-6} g (or about 0.6 –60 μm in radius for a typical density of 1 g cm^{-3}) are sufficiently gently decelerated in the upper atmosphere to avoid being heated to temperatures where their organics are pyrolyzed. With a terrestrial mass flux of 3.2×10^6 kg yr^{-1} of IDPs in this size range, and given an average IDP organic mass fraction of 10%, Anders found $\dot{m}(0) = 3.2 \times 10^5$ kg yr^{-1} of organics from the contemporary IDP source. Estimates of the IDP mass flux have since been put onto a firmer footing, as a result of a direct count of hypervelocity impact craters on the space-facing end of the Long Duration Exposure Facility (LDEF) satellite (Love and Brownlee, 1993). This work shows the IDP mass flux to be $\sim$3 times higher than the estimate used by Anders, giving an improved value for the contemporary organic carbon flux of $\dot{m}(0) = 1 \times 10^6$ kg yr^{-1}. At 4 Gyr ago, Earth would then have been accreting $\sim$2 $\times 10^8$ kg yr^{-1} of organics from IDPs. This represents a net flux some 5000 times smaller than that due to CH_4 photolysis in an early reducing atmosphere— the IDP source is swamped. However, if the early atmosphere were CO_2-rich, the IDP source is comparable to that produced by UV photolysis at 4 Gyr ago, and would have been the quantitatively greatest source shortly prior to that time.

The contemporary terrestrial mass influx from IDPs appears to have been roughly constant over the past 3.6 Gyr, as shown by the consistency of a variety of data sets sensitive to very different timescales (intercompared in Chyba and Sagan, 1992). How to scale this flux back into the time of the heavy bombardment is unclear, however. Here we have simply scaled according to the lunar cratering record. While this approach has the benefit of simplicity, there are arguments for alternative extrapolations back in time (Chyba and Sagan, 1992).

6.9.2   Interstellar Dust

The Earth accretes dust whenever the Solar System passes through interstellar clouds. Obviously this source of exogenous material is independent of the heavy bombardment. Greenberg (1981) has estimated that during its first 700 Myr, Earth would have passed through four or five such clouds, accreting 10^6-10^7 kg yr^{-1} of organic molecules during each 6×10^5 year-long passage. Even during cloud passage, this source would have been quantitatively unimportant compared with the accretion of Solar System IDPs.

6.9.3   Meteorites

Anders (1989) has estimated that contemporary Earth accretes organic material from meteorites at a rate of ~ 8 kg yr^{-1}, obviously quantitatively negligible compared with the accretion of organics from IDPs.

6.9.4   Catastrophic Airbursts

Comets and asteroids in the 1–100 m diameter range typically explode in the terrestrial atmosphere prior to reaching the surface (Chyba, 1993b). One such object, the several-meter-diameter Revelstoke carbonaceous chondrite, exploded over Canada in 1965 with an energy in the tens of kilotons range (Krinov, 1966; Folinsbee et al., 1967). Photomicrographs of dust recovered from a snow field beneath the explosion reveal unheated interiors (Folinsbee et al., 1967) within which organics should have survived. Therefore it seems possible to deliver intact organics to Earth as a result of catastrophic airbursts in the atmosphere.

Chyba and Sagan (1992) attempted to estimate an upper limit for the net contemporary organic influx from this source, by assuming that all exogenous organics survive the atmospheric explosion. With even this extremely generous assumption, they found a net organic mass influx from this source 1–2 orders of magnitude lower than the flux due to IDPs.

The situation has since become more complicated, with the discovery by Rabinowitz and coauthors (Rabinowitz, 1993; Rabinowitz et al., 1993) that the flux of Earth-colliding asteroids in the 10–50 m range is a factor 10–100 times higher than previously believed. The Earth experiences several Revelstoke-magnitude explosions per year, and a megaton-scale explosion every ~ 20 y. The mass influx from objects in this size range is $\sim 10^7$ kg yr^{-1} (Ceplecha, 1992; Chyba, 1993b), comparable to the net annual mass influx from IDPs. IDPs, however, deliver essentially 100% of their organics intact, so even with this increased flux, it seems likely that organic delivery by airburst will only deliver a fraction of that due to IDPs.

6.9.5   Big Impacts

Clark (1988) has suggested the possibility of a rare, soft landing of a comet on Earth, leading to a rich local mix of exogenous water and organics. However, comets and asteroids larger than several hundred meters in diameter typically pass through the atmosphere without airbursting and impact the surface at hypervelocities (Chyba, 1993b). It is extremely difficult for organics to survive intact the shock heating resulting from these collisions (Chyba et al., 1990, Thomas and Brookshaw, this volume), a conclusion supported by xenon data at the K/T boundary (Anders, 1989), as well as laboratory shock experiments with meteorite samples (Tingle et al., 1992). The apparently extraterrestrial amino acids found at the K/T boundary must therefore be the result of quench synthesis or some other mechanism (Chyba, 1990b). These possibilities will be discussed in Section 6.11 below.

The most robust organics, such as polycyclics, were suspected to be sufficiently stable to survive even big impacts, based on results of thermogravimetric analyses (Chyba et al., 1990). The discovery of extraterrestrial fullerenes in the Sudbury impact structure (Becker et al., 1994) may provide an empirical example of this sort of impact survival. The presence of extraterrestrial helium trapped within these fullerenes has been taken to imply that some portion of the Sudbury impactor must have remained below 1000 °C (Becker et al. 1996). (Note, however, that the timescales for helium release in the relevant experiments appear to be on the order of minutes, much longer than the timescales for peak shock heating in impacts (Chyba et al., 1990; Thomas and Brookshaw, this volume).) Simulations of impacts using full three-dimensional symmetry (Thomas and Brookshaw, this volume) predict peak impact temperatures lower by as much as an order of magnitude than those previously calculated in simulations (Chyba et al., 1990). Yet some recent collision experiments have been taken to suggest that this earlier work may have *underestimated* shock heating resulting from impacts (Fiske et al., 1995). Since the bulk of extraterrestrial organic matter incident on Earth (but not necessarily surviving the ensuing impact) resides in the largest impactors, it remains an important objective to reconcile fully results from numerical simulations, laboratory experiments, and samples from impact sites.

6.10   Atmospheric Shock Synthesis of Organic Molecules

The suggestion that atmospheric shock waves from meteoroids could have synthesized organics on early Earth was first made in the early 1960s by Gilvarry and Hochstim (1963) and Hochstim (1963). The production of amino acids via shock chemistry was experimentally demonstrated by Bar-Nun et al. (1970), and subsequently revised and elaborated (Barak and Bar-Nun, 1975; Bar-Nun and Shaviv, 1975).

Atmospheric organic shock synthesis will be proportional to η, the organic synthesis efficiency (kilograms organic carbon produced per joule of shock energy). η is very strongly dependent on atmospheric composition. As previously discussed in Section 6.8, models for the early terrestrial atmosphere range from strongly reducing to an intermediate oxidation state. We therefore calculate early terrestrial shock synthesis for end-members of the two possibilities.

As summarized by Chyba and Sagan (1992), the net efficiency for organic carbon production in reducing atmospheres (including $CH_4/N_2/H_2O$ and NH_3/CH_4 atmospheres) is $\eta \approx 1 \times 10^{-8}$ kg J^{-1}. Efficiencies for intermediate oxidation state (CO_2-rich) atmospheres are much lower, with $\eta \approx 3 \times 10^{-16}$ kg J^{-1}.

6.10.1   Shocks from Meteors

The small particle accretion rate of the Earth's atmosphere is about 4×10^7 kg yr^{-1} (Love and Brownlee, 1993). These objects deposit 100% of their kinetic energy into the atmosphere. Taking their initial velocity to be 17 km s^{-1} (Love and Brownlee 1993), the annual deposition of energy by meteors to the atmosphere is then $\sim 6 \times 10^{15}$ J yr^{-1}. Following Pollack et al. (1986) and McKay et al. (1988), taking the fraction of this kinetic energy converted into atmospheric shock heating to be $\sim 30\%$, we have a total organic production in the atmosphere due to meteors of $\dot{m}(0) = \eta(2 \times 10^{15}$ J yr$^{-1})$. Assuming meteor flux to scale in time as the lunar cratering record (see the discussion in Chyba and Sagan, 1992), a reducing atmosphere at time t would have experienced an organic production rate by this mechanism of $\dot{m}(t) = (2 \times 10^7$ kg yr$^{-1})f(t)$. A CO_2-rich atmosphere would have experienced a negligible organic production nearly eight orders of magnitude smaller.

6.10.2   Shocks from Airbursts

As previously discussed, the mass influx from small asteroids and comets in the 10–100 m range is $\sim 10^7$ kg yr^{-1}. For a median terrestrial impact velocity of 13 km s^{-1} for these objects (Chyba, 1993b), we have a net deposition of energy into the atmosphere of 8×10^{14} J yr^{-1}, or nearly an order of magnitude below that for meteors. Whether this production rate may be extrapolated back into the time of the heavy bombardment according to (6-17) depends on the origin of the 10–100 m asteroid population. Rabinowitz (1993) notes that the power-law dependence for this population is inconsistent with its evolution from the main-belt population. Perhaps these objects are remnants of short-period comets (Rabinowitz, 1993) or impact ejecta from the Moon. Either of these might be expected to scale in time according to the lunar cratering record. On the other had, if these objects are instead debris from recent collisions of larger asteroids with perihelia near Earth (Rabinowitz, 1993), they represent a transient event and the flux can certainly not reliably be scaled back in time.

6.10.3   Shocks from Giant Impact Plumes

Large impactors shock-process the atmosphere primarily through the rapidly expanding post-impact vapor plumes resulting from their impacts with the terrestrial surface. Shock processing by big impactors during atmospheric passage is comparatively negligible (Chyba and Sagan, 1992). Taking into account the fact that plumes from the largest impacts rise far above the atmosphere, so that the shock processing yield per unit of impact energy decreases as the energy of impact increases (Zahnle, 1990; Kasting, 1990), one can attempt to calculate net organic production from impactors of sizes ranging all the way up to that of the South-Pole Aitken basin; one finds $\dot{m}(t) \approx \eta(10^{16} \text{ J yr}^{-1}) f(t)$ (Chyba and Sagan, 1992), or about a factor of five greater than that due to meteors.

Of course, the effects of the two sources over short timescales are vastly different: whereas shock-processing of the atmosphere due to meteors is constant but at a comparatively low level, shock processing of the atmosphere due to large impacts is dominated by the largest, most infrequent events. Large impacts would have produced extremely large quantities of organics in very short pulses (see, e.g., Oberbeck and Aggarwal, 1992).

6.11   Post-Impact Recombination

A number of authors have suggested that organics may have been synthesized on early Earth by recombination from reducing mixtures of gases resulting from the shock vaporization of asteroids or comets on impact (Oró, 1961; Mukhin et al., 1989; Mckay et al., 1989; Oberbeck et al., 1989; Chyba and Sagan, 1992; Oberbeck and Aggarwal, 1992). The outcome is complicated by possible entrainment of target material and background atmosphere into the vapor plume, effects that are difficult to model and have typically not been taken into account. Kasting (1990) has argued that organic carbon in the hot rock vapor resulting from a large impact would be almost entirely converted to carbon monoxide.

There are some relevant, though limited, experimental data. Barak and Bar-Nun (1975) have demonstrated the shock synthesis of amino acids even when the initial gas mixtures contain large quantities of water and air. Shock vaporization and organic recombination by laser pulse-heating of terrestrial rocks and meteorite samples (Mukhin et al., 1989) yield a wide variety of carbon products of varying oxidation states (see Chyba and Sagan (1992) for a critical discussion of these experiments). The Mukhin et al. (1989) experiments suggest that some 4% of impactor carbon is incorporated into organics in the vapor plume formed subsequent to impact. If these experimental results may in fact be scaled up over the many orders of magnitude required to reach the energies of large impact events, (6-14) could simply be multiplied by 0.04 in order to obtain the organic flux on early Earth due to post-impact recombination.

6.12   Implications of the K/T Organics

The most important datum for an impact source of organics on Earth is the discovery of the nonbiological amino acids, racemic isovaline and α-aminoisobutyric acid (AIB), immediately above and below the clay layer marking the Cretaceous–

Tertiary (K/T) boundary (Zhao and Bada, 1989). These molecules are unlikely to be extraterrestrial organics that survived the impact of the K/T object (Anders, 1989; Chyba, 1990b; Tingle et al., 1992). Possibly they were delivered to the Earth by interplanetary dust evolved off the K/T impactor (Zahnle and Grinspoon, 1990). More probably, they represent material that was synthesized in the post-impact fireball (e.g., Oberbeck and Aggarwal, 1992).

Whatever their origin, their abundance at the boundary may be used as a kind of "ground truth" for organic delivery or synthesis in large impacts. It hardly need be stressed that any attempt to extrapolate to "typical" large impacts from this single event is fraught with peril. Nevertheless, it is interesting to see where such an approach leads.

The AIB/iridium (AIB/Ir) ratio in the K/T boundary layer is ~ 100 (Zhao and Bada, 1989). The mass fraction abundance of Ir in a type I carbonaceous (CI) chondrite is $\sim 5 \times 10^{-7}$ (Alvarez et al., 1980), so that the ratio of the mass of AIB produced in the collision to the mass of the impactor is $\sim 5 \times 10^{-5}$. (This simple model assumes that all the Ir, as well as all the AIB, delivered or produced in the impact was retained, or that the two were lost in equal proportions. This may not have been the case.) Total organic carbon produced is typically ~ 10 times that incorporated into amino acids, so that the ratio of the mass of organics produced in the collision to the mass of the impactor is $\sim 5 \times 10^{-4}$. CI chondrites are typically $\sim 4\%$ (organic) carbon by mass (Wilkening, 1978), suggesting that about 1% of the K/T impactor's carbon was converted into organics by post-impact synthesis. Note this synthesis would have occurred in the presence of a strongly oxidizing background atmosphere. It is striking that this result is so close to that found by Mukhin et al. (1989) in their laser vaporization experiments with meteoritic samples.

In (6-14), the exponent $(1-b) = 0.46$, so that $\dot{m}(t)$ is not strongly dependent on the choice of m_{max}. We take m_{max} to be the mass corresponding to the lunar Imbrium impactor, or, from (6-8), about 2×10^{18} kg for a main rim diameter of 1160 km. Choosing the South Pole–Aitken object (main rim ~ 2200 km) instead of Imbrium in (6-14) yields a result that differs by only a factor of ~ 3.

With $m_{max} = 2 \times 10^{18}$ kg in (6-14), we have $\dot{m}(t) = 2.5 \times 10^9 \phi f(t)$ kg yr^{-1} as the mass of organics synthesized on Earth in post-impact recombination as a function of time, assuming that results from the K/T event may be bravely extrapolated to all other big impacts. Here ϕ is the ratio of the mass of organics produced or delivered in the impact to the mass of the impactor. Note this extrapolation does not require knowledge of the exact mechanism responsible for the K/T organics; it only assumes that the K/T result may be taken as representative of the results of other big collisions as well. Chemistry in an early, less oxidizing atmosphere would likely give higher yields of organics than this calculation suggests. Taking $\phi = 5 \times 10^{-4}$ from the K/T results, $\dot{m}(t) = 1 \times 10^6 f(t)$ kg yr^{-1}, or 2×10^8 kg yr^{-1} 4 Gyr ago.

TABLE 6.1. Major Sources of Prebiotic Organics for Two Candidate Early Earth Atmospheres

Source	Reducing $(kg\ yr^{-1})$	Intermediate $(kg\ yr^{-1})*$
UV photolysis	1×10^{12}	3×10^{8}
Electric discharge	3×10^{9}	3×10^{7}
IDPs	2×10^{8}	2×10^{8}
Shocks from impacts	2×10^{10}	4×10^{2}
Shocks from meteors	4×10^{9}	8×10^{1}
K/T extrapolation	2×10^{8}	2×10^{8}
Totals	1×10^{12}	7×10^{8}

$*[H_2]/[CO_2]=0.1$

6.13   An Inventory of Organic Production on Early Earth

The results of the preceding sections are summarized in Table 6.1, which shows production of organics on Earth 4 Gyr ago (in kg yr^{-1}) from the quantitatively most important sources. Table 6.1 gives these results for both reducing and intermediate oxidation state atmospheres. Rates of production in the intermediate oxidation state case assume $[H_2]/[CO_2] \approx 0.1$, a hydrogen fraction which is probably optimistically high. With lower $[H_2]/[CO_2]$ ratios, endogenous production drops precipitously (Schlesinger and Miller, 1983a, b).

These results assume the exogenous sources scale as the heavy bombardment cratering flux. The exogenous contribution to the terrestrial prebiotic organic inventory may therefore be easily calculated for a different date: since τ from the lunar cratering record corresponds to a half-life of 100 Myr, these fluxes increase, to a good approximation, by a factor of 2 every 100 Myr further into the past.

This discussion has so far ignored the effects of atmospheric erosion. However, only ~10% of asteroid, and ~50% of comet, collisions with the Earth occur at velocities high enough to cause erosion (Chyba, 1991), effects small compared with the other uncertainties in the problem, even if comets made up the bulk of heavy bombardment impactors.

Of course, there may have been other organic production or delivery mechanisms than those considered here. (For example, Zahnle (1986) has proposed a pathway by which an N_2 atmosphere with a CH_4 concentration of $\sim 10^{-4}$ will produce $\sim 10^{10}$ kg yr^{-1} of hydrogen cyanide (HCN) through short-ultraviolet photolysis.) Moreover, the results given here will continue to require revision, just as these results have been revised since the inventory presented by Chyba and Sagan in 1992. Regular revision should protect against a repetition of the subsequent history of the original estimates by Miller and Urey (1959) for organic synthesis by lightning and coronal discharge on early Earth. This landmark paper made use of the best data then available for global electrical discharge rates, those of Schonland (1953). Those data in turn were from experimental work Schonland had performed in South Africa some 25 years earlier (Schonland, 1928), in which he had connected a galvanometer to a potted tree, measured currents passing through it during thunderstorms, then extrapolated. Unsurprisingly, our knowledge of global electrical discharge rates has improved substantially since these original measurements were made. Nevertheless, after their use by Miller and Urey, Schonland's numbers continued to be cited until 1991, when the subject was revisited (Chyba and Sagan, 1991), and it was found that more recent data required the lightning and coronal discharge estimates to be revised downward by factors of 20 and 120, respectively. It should be expected that the results presented here will undergo comparable revision. However uncertain the extrapolations of the exogenous sources may be, the endogenous estimates evidently face comparable dilemmas. (For example, the recent discovery of new, high-altitude lightning phenomena (Fishman et al., 1994; Kerr, 1994; Lyons, 1994) suggests that our knowledge of electrical discharge sources remains far from complete.) Indeed, estimates of electrical discharge energy on early Earth continue to face the currently unquantifiable uncertainty that one can seemingly do no better than assume that discharge rates 4 Gyr ago were equal to those today.

Table 6.1 shows that net organic production and delivery on early Earth would have been more than three orders of magnitude higher in the case of a reducing atmosphere than for an atmosphere of an intermediate oxidation state. For an early reducing atmosphere, exogenous sources would, on average, have been quantitatively swamped by endogenous ones. In the case of a CO_2-rich atmosphere 4 Gyr ago, however, exogenous organics from IDPs, as well as those produced in big impacts (probably through recombination), would have been comparable to those produced by UV photolysis, with the exogenous sources being of greatest relative importance further back in time.

There would have been qualitative differences as well; for example, amino acid production in CO_2-rich atmospheres appears to be almost limited to glycine (Schlesinger and Miller, 1983a), whereas the K/T data imply that post-shock recombination yields a rich array of amino acids as products (in an oxidizing atmosphere). On the other hand, IDPs are radiation-hardened from their stay in interplanetary space, and undergo some heating during atmospheric entry, both of which may decrease their complement of the more fragile organics (Chyba, 1990b).

6.14   Organic Concentrations and Sinks

A common way of assessing the importance of different endogenous sources of organics on early Earth is to calculate concentrations of particular organics in the primitive ocean (e.g., Stribling and Miller, 1987). This is much more difficult to do in the case of exogenous sources, as the nature of organics in IDPs, or even produced by recombination in K/T-like impacts, remains very poorly known. Calculation of concentrations requires knowledge of sink, as well as source, terms. It is obviously difficult to assess sinks when one has little idea which organics one has to consider. Thermal decomposition rates, for example, vary greatly between different types of simple, prebiotically relevant organics (Chyba et al., 1990).

For these reasons, Chyba and Sagan (1992) considered only the simplest sink model, that of the entire ocean cycling through hydrothermal vents on a timescale of $\sim 10^7$ yr, as is the case for the contemporary ocean. If one assumes that all organics thereby cycled are destroyed, one is freed from the necessity of knowing thermal decomposition rates for the different relevant molecules. However, some organics (such as some of those in IDPs) are likely to sink in the oceans and be buried in abyssal clays (e.g., Kyte and Wasson, 1986), while others will float on the ocean surface, forming a kind of primordial "oil slick" (Lasaga et al., 1971), and remaining protected against vent passage.

Nevertheless, if all the organic products in the reducing and intermediate oxidation state atmospheres considered above *were* fully soluble in an early ocean of contemporary mass (1.4×10^{21} kg), and had a mean lifetime of 10 Myr against destruction in mid-ocean vents or subducted plates, the steady-state organic abundance in the ocean 4 Gyr ago would have been $\sim 10^{-2}$ g per g for a reducing atmosphere—an extraordinary concentration—and $\sim 10^{-5}$ g per g for an intermediate oxidation state atmosphere. Although various concentration mechanisms for organics in solution undoubtedly existed, reducing atmospheres seem to be more favorable for the origin of life by some three orders of magnitude.

6.15   Episodic Events

The "catastrophic" exogenous sources would have produced transient, extremely high concentrations of organics in the terrestrial oceans. Consider, for example, an event analogous to the K/T impact. Some 10^4 such events would have occurred during the first 700 Myr of Earth's history (Chyba et al., 1995). Taking the K/T object to have been about 5 km in radius (Alvarez et al., 1980), and assuming it had a CI chondrite density of around 2 g cm^{-3}, its total mass would have been $\sim 10^{15}$ kg. Using the results of Section 6.11 above, this implies that some 5×10^{11} kg of organics would have been produced in that impact, eventually settling out of the atmosphere and into the oceans. This sudden pulse of organic production would, averaged over the year following the impact, be comparable to the net annual organic production in a reducing atmosphere. In an intermediate oxidation state atmosphere, it would utterly swamp "background" organic production by

UV photolysis and IDP delivery by some three orders of magnitude. The most thorough attempt to treat these kinds of transient events for the case of amino acid production, taking into account sinks due to adsorption onto clays and degradation by ultraviolet light, is that of Oberbeck and Aggarwal (1992).

6.16   Prebiotic Organics on the Early Earth

A different way to put the prebiotic organic production and delivery rates calculated here into context is to ask how long would be required to produce a contemporary terrestrial biomass worth of organics. The terrestrial oceanic and total biomasses are $\sim 3 \times 10^{12}$ kg and $\sim 6 \times 10^{14}$ kg, respectively (Hayes et al., 1983). In the case of an early reducing atmosphere, we have the extraordinary result that production by UV photolysis would have provided a terrestrial biomass of organics in only 600 years! For an intermediate oxidation state atmosphere, closer to one million years would have been required to produce a biomass of organics. In either case, early Earth hardly seems to have been at a loss for organics.

Indeed, the organic production rate for an early reducing atmosphere appears almost too high to be plausible. At 4 Gyr ago, the timescale for the complete UV processing of a 1-bar CH_4 atmosphere would have been only $\sim 10^7$ yr. A 1-bar atmosphere on early Earth could therefore only have been sustained if terrestrial volcanism or exogenous sources 4 Gyr ago resupplied CH_4 to the atmosphere at a rate $\sim 10^{12}$ kg yr^{-1}. Terrestrial volcanoes are estimated to release currently as much as $\sim 4 \times 10^{10}$ kg yr^{-1} of carbon into the atmosphere, mostly in the form of CO_2 (Walker, 1977). If carbon were released on a reducing primitive Earth as CH_4, early terrestrial volcanism would need to have been twenty-five times more intense than today to maintain an atmosphere against UV photolysis. If this were not the case, organic production on an early reducing Earth would have been limited not by the availability of UV photons, but rather by the availability of a carbon source. However, estimates of carbon outgassing rates 3 Gyr ago range as high as 6×10^{11} kg yr^{-1} (DesMarais, 1985), with the outgassing rate 4 Gyr ago expected to have been higher still, so carbon resupply may have been able to keep pace with the organic production sink. In this case, organic production integrated over 1 Gyr would have been of the same order as the total estimated terrestrial inventory of carbon in the mantle and crust (Sagan and Chyba, 1996).

How much is enough? Given our current impressive level of ignorance regarding the origin of life on Earth (see, e.g., Chyba and McDonald, 1995), as well as the difficulty in quantifying the likelihood and effectiveness of possible concentration mechanisms, it is difficult to assess what rate of organic production or delivery would have been too little for the terrestrial origin of life. Regardless of the nature of the primitive atmosphere, however, a variety of sources dependent on the heavy bombardment would have maintained a significant production or delivery level of prebiotic organics on the early Earth.

6.17   References

Alvarez, L.W., Alvarez, W.A., Asaro, F., and Michel, H.V. (1980), Extraterrestrial cause for the Cretaceous-Tertiary extinction. *Science* **208**, 1095–1108.

Anders, E. (1989), Pre-biotic organic matter from comets and asteroids. *Nature* **342**, 255–57.

Baldwin, R.B. (1987a), On the relative and absolute ages of seven lunar front face basins. I. From viscosity arguments. *Icarus* **71**, 1–18.

Baldwin, R.B. (1987b), On the relative and absolute ages of seven lunar front face basins. II. From crater counts. *Icarus* **71**, 19–29.

Barak, I. and Bar-Nun, A. (1975), The mechanisms of amino acid synthesis by high temperature shock-waves. *Origins of Life* **6**: 483–506.

Bar-Nun, A. and Shaviv, A. (1975), Dynamics of the chemical evolution of Earth's primitive atmosphere. *Icarus* **24**, 197–210.

Bar-Nun, A., Bar-Nun, N., Bauer, S.H., and Sagan, C. (1970), Shock synthesis of amino acids in simulated primitive environments. *Science* **168**, 470–473.

Basaltic Volcanism Study Project (BVSP) (1981), *Basaltic Volcanism on the Terrestrial Planets*. Pergamon, New York.

Becker, L., Bada, J.L., Winans, R.E., Hunt, J.E., Bunch, T.E., and French, B.M. (1994), Fullerenes in the 1.85-billion-year-old Sudbury impact structure. *Science* **265**, 642–645.

Becker, L., Poreda, R.J., and Bada, J.L. (1996), Extraterrestrial helium trapped in fullerenes in the Sudbury impact structure. *Science* **272**, 249–252.

Cameron, A.G.W. (1983), Origin of the atmospheres of the terrestrial planets. *Icarus* **56**, 195–201.

Cameron, A.G.W. (1986), The impact theory for origin of the Moon. In *Origin of the Moon* (W.K. Hartmann, R.J. Phillips, and G.J. Taylor, eds.), Lunar and Planetary Inst., Houston, pp. 609–616.

Ceplecha, Z. (1992), Earth influx of interplanetary bodies. *Astronomy and Astrophysics* **263**, 361–366.

Chou, C.-L. (1978), Fractionation of siderophile elements in the earth's upper mantle. *Proc. Lunar Planet. Sci. Conf.* **9**, 219–230.

Chyba, C.F. (1990a), Impact delivery and erosion of planetary oceans in the early inner solar system. *Nature* **343**, 129–133.

Chyba, C.F. (1990b), Extraterrestrial amino acids and terrestrial life. *Nature* **348**, 113–114.

Chyba, C.F. (1991), Terrestrial mantle siderophiles and the lunar impact record. *Icarus* **92**, 217–233.

Chyba, C.F. (1993a), The violent environment of the origins of life: Progress and uncertainties. *Geochim. Cosmochim. Acta* **57**, 3351–3358.

Chyba, C.F. (1993b), Explosions of small Spacewatch objects in the Earth's atmosphere. *Nature* **363**, 701–703.

Chyba, C.F. and McDonald, G. (1995). The origin of life in the Solar System: Current issues. *Annu. Rev. Earth Planet Sci.* **24**, 215–249.

Chyba, C. and Sagan, C. (1991), Electrical energy sources for organic synthesis on the early Earth. *Origins Life* **21**, 3–17.

Chyba, C. and Sagan, C. (1992), Endogenous production, exogenous delivery and impact-shock synthesis of organic molecules: An inventory for the origins of life. *Nature* **355**, 125–132.

Chyba, C.F., Thomas, P.J., Brookshaw, L., and Sagan, C. (1990), Cometary delivery of organic molecules to the early Earth. *Science* 249, 366–373.

Chyba, C.F., Thomas, P.J., and Zahnle, K.J. (1993), The 1908 Tunguska explosion: Atmospheric disruption of a stony asteroid, *Nature* **361**, 40–44.

Chyba, C.F., Owen, T.C., and Ip, W.-H. (1995), Impact delivery of volatiles and organic molecules to Earth. In *Hazards Due to Comets and Asteroids* (ed. T. Gehrels) pp. 9–58, University of Arizona Press, Tucson.

Clark, B.C. (1988), Primeval procreative comet pond. *Origins of Life* **18**, 209–238.

Dalrymple, G.B. and Ryder, G. (1993), $^{40}Ar/^{39}Ar$ age spectra of Apollo 15 impact melt rocks by laser step-heating and their bearing on the history of lunar basin formation. *J. Geophys. Res.* **98**, 13085–13095.

Delsemme, A.H. (1992), Cometary origin of carbon, nitrogen and water on the Earth. *Origins of Life* **21**, 279–298.

DesMarais, D.J. (1985), In *The Carbon Cycle and Atmospheric CO_2: Natural Variations Archean to Present.* (eds. E.T. Sundquist and W.S. Broecker), American Geophysical Union, Washington DC, pp. 602–611.

Dreibus, G. and Wänke, H. (1987), Volatiles on Earth and Mars: A comparison. *Icarus* **71**, 225–240.

Dreibus, G. and Wänke, H. (1989), Supply and loss of volatile constituents during the accretion of terrestrial planets. In *Origin and Evolution of Planetary and Satellite Atmospheres* (S.K. Atreya, J.B. Pollack, and M.S. Matthews, eds.), University of Arizona Press, Tucson, pp. 268–288.

Eberhardt, P., Dolder, U., Schulte, W., Krankowsky, d., Lämmerzahl, P., Berthelier, J.J., Woweries, J., Stubbemann, U., Hodges, R.R., Hoffman, J.H., and Illiano, J.M. (1987), The D/H ratio in water from comet P/Halley. *Astron. Astrophys.* **187**, 435–437.

Fernández, J.A. (1985), The formation and dynamical survival of the comet cloud. In *Dynamics of Comets: Their Origin and Evolution* (A. Carusi and G.B. Valsecchi, eds.), Reidel, Dordrecht, pp. 45–70.

Fernández, J.A. and Ip, W.-H. (1983), On the time evolution of the cometary influx in the region of the terrestrial planets. *Icarus* **54**, 377–387.

Fishman, G.J., Bhat, P.N., Mallozzi, R., Horack, J.M., Koshut, T., Kouveliotou, C., Pendleton, G.N., Meegan, C.A., Wilson, R.B., Paciesas, W.S., Goodman, S.J., and Christian, H.J. (1994), Discovery of intense gamma-ray flashes of atmospheric origin. *Science* **264**, 1313–1316.

Fiske, P.E., Nellis, W.J., Lipp, M., Lorenzana, H., Kikuchi, M., and Syono, Y. (1995), Pseudotachylites generated in shock experiments: Implications for impact cratering products and processes. *Science* **270**, 281–283.

Folinsbee, R.E., Douglas, J.A.V. and Maxwell, J.A. (1967), Revelstoke, a new Type I carbonaceous chondrite. *Geochim. Cosmochim. Acta* **31**, 1625–1635.

Gilvarry, J.J. and Hochstim, A.R. (1963), Possible role of meteorites in the origin of life. *Nature* **197**, 624–626.

Greenberg, M.J. (1981), Chemical evolution of interstellar dust–a source of prebiotic material? In *Comets and the Origin of Life* (C. Ponnamperuma, ed.), Reidel, Dordrecht, pp. 111–127.

Grinspoon, D.H. (1988), Large impact events and atmospheric evolution on the terrestrial planets. Ph.D. thesis, University of Arizona.

Hartmann, W.K. (1980), Dropping stones in magma oceans: Effects of early lunar cratering. In *Proceedings of the Conference on the Lunar Highland Crust*, pp. 155–171 (Lunar and Planetary Institute, Houston).

Hartmann, W.K. (1987), A satellite–asteroid mystery and a possible early flux of scattered C-class asteroids. *Icarus* **71**, 57–68.

Hartmann, W.K. (1990), Additional evidence about an early intense flux of C asteroids and the origin of Phobos. *Icarus* **87**, 236–240.

Hartmann, W.K. (1995), Planetary cratering 1. The question of multiple impactor populations: Lunar evidence. *Meteoritics* **30**, 451–467.

Hayes, J.M., I.R. Kaplan, and K.W. Wedeking (1983), Precambrian organic geochemistry, preservation of the record. In *Earth's Earliest Biosphere* (J.W. Schopf, ed.), Princeton University Press, Princeton, NJ, pp. 93–134.

Hochstim, A.R. (1963), Hypersonic chemosynthesis and possible formation of organic compounds from impact of meteorites on water. *Proc. Nat. Acad. Sci.* **50**, 200–208.

Horowitz, N.H. (1986), *To Utopia and Back: The Search for Life in the Solar System.* Freeman, New York.

Ip, W.-H. (1977), On the early scattering processes of the outer planets. In *Comets-Asteroids-Meteorites: Interrelations, Evolution and Origin* (A.H. Delsemme, ed.), University of Toledo Press, Toledo, OH, pp. 485–490.

Kasting, J.F. (1990), Bolide impacts and the oxidation state of carbon in the Earth's early atmosphere. *Origins of Life* **20**, 199–231.

Kasting, J.F. (1993), Earth's early atmosphere. *Science* **259**, 920–925.

Kasting, J.F. and Ackerman, T.P. (1986), Climatic consequences of very high carbon dioxide levels in the Earth's early atmosphere. *Science* **234**, 1383–1385.

Kasting, J.F., Whitmire, D., and Reynolds, R. (1993), Habitable zones around main sequence stars. *Icarus* **101**, 108–128.

Kerr, R.A. (1994), Atmospheric scientists puzzle over high-altitude flashes. *Science* **264**, 1250–1251.

Krinov, E.L. (1966), *Giant Meteorites*. Pergamon, Oxford.

Kyte, F.T. and Wasson, J.T. (1986), Accretion rate of extraterrestrial matter: Iridium deposited 33 to 67 million years ago. *Science* **232**, 1225–1229.

Lasaga, A.C., Holland, H.D., Dwyer, M.J. (1971), Primordial oil slick. *Science* **174**, 53–55.

Lewis, J. (1974), The temperature gradient in the solar nebula. *Science* **186**, 440–443.

Lewis, J., Barshay, S.S. and Noyes, B. (1979), Primordial retention of carbon by the terrestrial planets. *Icarus* **37**, 190–206.

Love, S.G. and Brownlee, D.E. (1993), A direct measurement of the terestrial mass accretion rate of cosmic dust. *Science* **262**, 550–553.

Mason, B. (1971), *Handbook of Elemental Abundances in Meteorites*. Gordon and Breach, New York.

McKay, C.P. (1986), Exobiology and future mars missions: The search for Mars' earliest biosphere. *Adv. Space Res.* **6**, 269–285.

McKay, C.P. (1991), Urey Prize lecture: planetary evolution and the origin of life. *Icarus* **91**, 93–100.

McKay, C.P., Scattergood, T.W., Pollack, J.B., Borucki, W.J., and Van Ghyseghem, H.T. (1988), High-temperature shock formation of N_2 and organics on primordial Titan. *Nature* **332**, 520–522.

McKay, C.P., Borucki, W.R., and Kojiro, K.R., and Church, F. (1989), Shock production of organics during cometary impact. *Lunar Planet. Sci. Conf.* **20**, 671–672.

McKinnon, W.B., C.R. Chapman, and K.R. Housen (1990), Cratering of the uranian satellites. In *Uranus* (J.T. Bergstrahl, E.D. Miner, and MS. Matthews, eds.), University of Arizona Press, Tucson, pp. 629–692.

Melosh, H.J. (1989), *Impact Cratering: A Geologic Process*. Oxford University Press, New York.

Melosh, H.J. and A.M. Vickery (1989), Impact erosion of the primordial atmosphere of

Mars. *Nature* **338**, 487–489.

Miller, S.L. and Urey, H.C. (1959), Organic compound synthesis on the primitive Earth. *Science* **130**, 245–251.

Mukhin, L.M., Gerasimov, M.V., and Safonova, E.N. (1989), Origin of precursors of organic molecules during evaporation of meteorites and mafic terrestrial rocks. *Nature* **340**, 46–48.

Noyes, W.A. and Leighton, P.A. (1941), *The Photochemistry of Gases*. Reinhold, New York.

Oberbeck, V.R. and Aggarwal, H. (1992), Comet impacts and chemical evolution on the bombarded Earth. *Origins of Life* **21**, 317–338.

Oberbeck, V.R. and G. Fogleman (1989), Estimates of the maximum time required to originate life. *Origins Life* **19**, 549–560.

Oberbeck, V.R., McKay, C.P., Scattergood, T.W., Carle, G.C. and Valentin J.R. (1989), The role of cometary particle coalescence in chemical evolution. *Origins of Life* **19**, 39–55.

Oró, J. (1961), Comets and the formation of biochemical compounds on the primitive Earth. *Nature* **190**, 389–390.

Pollack, J.B., Podolak, M., Bodenheimer, P., and Christofferson, B. (1986) *Icarus* **67**, 409–443.

Prinn, R.G. and B. Fegley (1989), Solar nebula chemistry: Origin of planetary, satellite and cometary volatiles. In *Origin and Evolution of Planetary and Satellite Atmospheres* (S.K. Atreya, J.B. Pollack, and M.S. Matthews, eds.). University of Arizona Press, Tucson, pp. 78–136.

Rabinowitz, D.L. (1993), The size distribution of the Earth-approaching asteroids. *Astrophys. J.* **407**, 412–427.

Rabinowitz, D.L., Gehrels, T., Scotti, J.V., McMillan, R.S., and Perry M.L. (1993), The terrestrial asteroid belt: A new population of near-Earth asteroids. *Nature* **363**, 704–706.

Robbins, E.I. and Iberall, A.S. (1991), Mineral remains of early life on Earth? On Mars? *Geomicrobiology J.* **9**, 51–66.

Ryder, G. (1990), Lunar samples, lunar accretion and the early bombardment of the Moon. *Eos* **71**, 313, 322–323.

Ryder, G. and Wood, J.A. (1977), Serenitatis and Imbrium impact melts: Implications for large-scale layering in the lunar crust. *Proc. Lunar Sci. Conf.* **8**, 655–668.

Sagan, C. and Chyba, C. (1996), The early faint sun "paradox": organic shielding of UV-labile greenhouse gases. In preparation.

Sagan, C. and Thompson, W.R. (1984), Production and condensation of organic gases in the atmosphere of Titan. *Icarus* **59**, 133–161.

Schidlowski, M.A. (1988), 3,800-million-year isotope record of life from carbon in sedimentary rocks. *Nature* **333**, 313–318.

Schlesinger, G. and Miller, S.L. (1983a), Prebiotic synthesis in atmospheres containing CH_4, CO, and CO_2. I. Amino acids. *J. Molec. Evol.* **19**, 376–382.

Schlesinger, G. and Miller, S.L. (1983b), Prebiotic synthesis in atmospheres containing CH_4, CO, and CO_2. II. Hydrogen cyanide, formaldehyde, and ammonia. *J. Molec. Evol.* **19**, 383–390.

Schmidt, R.M. and K.R. Housen (1987), Some recent advances in the scaling of impact and explosion cratering. *Int. J. Impact Engng.* **5**, 543–560.

Schonland, B.F.J. (1928), The interchange of electricity between thunderclouds and the Earth. *Proc. Roy. Soc. A* **118**, 252–262.

Schonland, B.F.J. (1953), *Atmospheric Electricity*. Methuen, London.

Schopf, J.W. (1993), Microfossils of the early Archean apex chert: New evidence of the antiquity of life. *Science* **260**, 640–646.

Schopf, J.W. and Walter, M.R. (1983), Archean microfossils: New evidence of ancient microbes. In *Earth's Earliest Biosphere* (J.W. Schopf, ed.). Princeton University Press, Princeton, NJ, pp. 214–239.

Shoemaker, E.M. and R.F. Wolfe (1984), Evolution of the Uranus–Neptune planetesimal swarm. *Proc. Lunar Planet. Sci. Conf.* **15**, 780–781.

Sleep, N.H., K.J. Zahnle, J.F. Kasting, and H.J. Morowitz (1989. Annihilation of ecosystems by large asteroid impacts on the early Earth. *Nature* **342**, 139–142.

Stacey, F.D. (1977), *Physics of the Earth*. Wiley, New York.

Stevenson, D.J. (1983), The nature of the Earth prior to the oldest known rock record: The Hadean Earth. In *Earth's Earliest Biosphere: Its Origin and Evolution* (J.W. Schopf, ed.). Princeton University Press, Princeton, NJ, pp. 32–40.

Stevenson, D.J. (1990), Fluid dynamics of core formation. In Origin of the Earth (H.E. Newsom and J.H. Jones, eds.). Oxford University Press, New York, pp. 231–249.

Stribling, R. and Miller, S.L. (1987), Energy yields for hydrogen cyanide and formaldehyde synthesis: The HCN and amino acid concentrations in the primitive ocean. *Origins of Life* **17**, 261–273.

Strom, R.G. (1987), The Solar System cratering record: Voyager 2 results at Uranus and implications for the origin of impacting objects. *Icarus* **70**, 517–535.

Sun, S.-S. (1984), Geochemical characteristics of archaean ultramafic and mafic volcanic rocks: Implications for mantle composition and evolution. In *Archean Geochemistry* (A. Kröner, G.N. Hanson, and A.M. Goodwin, eds.). Springer-Verlag, Berlin, pp. 25–46.

Swindle, T.D., M.W. Caffee, C.M. Hohenberg, and S.R. Taylor (1986), I–Pu–Xe dating and the relative ages of the Earth and Moon. In *Origin of the Moon* (W.K. Hartmann, R.J. Phillips, G.J. Taylor, eds.). Lunar and Planetary Institute, Houston, pp. 331–357.

Tera, F., Papanastassiou, D., and Wasserburg, G. (1974), The lunar timescale and a summary of isotopic evidence for a terminal lunar cataclysm. *Lunar Planet. Sci* **5**, 792.

Tilton, G.R. (1988), Age of the Solar System. In *Meteorites and the Early Solar System* (J.F. Kerridge and M.S. Matthews, eds.). University of Arizona Press, Tucson, pp. 259–275.

Tingle, T.N, Tyburczy, J.A., Ahrens, T.J. and Becker, C.H. (1992), The fate of organic matter during planetary accretion: Preliminary studies of the organic chemistry of experimentally shocked Murchison meteorite. *Origins of Life* **21**, 385–397.

Veizer, J. (1983), Geologic evolution of the Archean-Early Proterozoic Earth. In *Earth's Earliest Biosphere* (J.W. Schopf, ed.). Princeton University Press, Princeton, NJ, pp. 240–259.

Walker, J.C.G. (1977), *Evolution of the Atmosphere*. Macmillan, New York.

Walker, J.C.G. (1986), Carbon dioxide on the early Earth. *Origins Life* **16**, 117–127.

Walter, M.R.. (1983), Archean stromatolites: evidence of the Earth's earliest benthos. In *Earth's Earliest Biosphere* (J.W. Schopf, ed.). Princeton University Press, Princeton, NJ, pp. 187–213.

Wetherill, G.W. (1977), Evolution of the earth's planetesimal swarm subsequent to the formation of the earth and moon. *Proc. Lunar. Sci. Conf.* **8**, 1–16.

Wetherill, G.W. (1990), Formation of the Earth. *Annu. Rev. Earth Planet. Sci.* **18**, 205–256.

Wilhelms, D.E. (1984), Moon. In *The Geology of the Terrestrial Planets* (M.H. Carr, ed.), pp. 107–205. NASA SP-469.

Wilhelms, D.E. (1987), The Geologic History of the Moon. U.S. Geological Survey professional paper 1348. U.S. Government Printing Office, Washington, DC.

Wilkening, L.L. (1978), Carbonaceous chondritic material in the solar system. *Naturwiss.*

66, 73–79.

Zahnle, K. (1986), Photochemistry of methane and the formation of hydrocyanic acid (HCN) in the Earth's early atmosphere. *J. Geophys. Res.* **91**, 2819–2834.

Zahnle, K. (1990), Atmospheric chemistry by large impacts. In Global Catastrophes in Earth History (V.I. Sharpton and P.D. Ward, eds.). Geological Society of America SP-247, Boulder, pp. 271–288.

Zahnle, K. and Grinspoon, D. (1990), Comet dust as a source of amino acids at the Cretaceous/Tertiary boundary. *Nature* **348**, 157–159.

Zahnle, K.J. and Walker, J.C.G. (1982), Evolution of solar ultraviolet luminosity. *Rev. Geophys. Space Phys.* **20**, 280–292.

Zhao, M. and Bada, J.L. (1989), Extraterrestrial amino acids in Cretaceous/Tertiary boundary sediments at Stevns Klint, Denmark. *Nature* **339**, 463–465.

7
Impacts and the Early Evolution of Life

K.J. Zahnle and N.H. Sleep

ABSTRACT The K/T event shows that, even today, biospheric cratering is an important process. Impacts were much larger and more frequent on the early Earth. In all likelihood impacts posed the greatest challenge to the survival of early life.

7.1 Prologue

Chiron may be the most dangerous object in the solar system. Chiron is a $\sim$ 180 km diameter object, apparently cometary (Lebofsky et al., 1984; Hartmann et al., 1990), that presently resides in an unstable Saturn-crossing orbit (Scholl, 1979; Oikawa and Everhart 1979; Hahn and Bailey 1990). Conventional wisdom has Chiron randomly walking into the inner solar system, in manner if not in mass like any short period comet (Duncan et al., 1987; 1988). Its chances of actually colliding with Earth are low, probably less than one in a million. It is far more likely that Saturn or Jupiter will eject it from the solar system. Yet the worst could happen. If Chiron were to strike Earth the impact would release on the order of 10^{34} ergs. The Earth would be enveloped by a thick rock vapor atmosphere. The upper few hundred meters of the ocean would evaporate. On land the heat pulse would sterilize something like the upper 50 m. Life would survive, but it would be thrown back into the oceans or buried in caves.

Since its formation—indeed, since before its formation—Earth has been subject to collisions with other celestial bodies. An almost inescapable inference is that Earth and the other planets accumulated by collisions of myriads of smaller objects present in the early solar system (Safronov 1972). By now the solar system is an almost empty place. Major collisions are rare. But even now there occur occasional collisions of note. The impact of a $\sim$10 km comet or asteroid into the Yucatan peninsula 65 million years ago left a 180 km diameter hole in the ground (Hildebrand et al., 1991; Swisher et al., 1992) and an even larger hole in the fossil record. This impact, which was originally inferred from the presence of extraordinary amounts of exogenous iridium in a thin global clay layer, happens to coincide with the most famous mass extinction (Alvarez et al. 1980). Although the obvious has been resisted by some, there can be little doubt that the Cretaceous/Tertiary (K/T) impact caused the death of most living things on Earth, and the extinction of a fair fraction of the species that had been living thereon, including, famously, the dinosaurs (see Silver and Schultz, 1982; Sharpton and Ward, 1990).

As one looks back toward the beginning of the solar system collisions become much more frequent and the largest among them commensurately greater. Very large events, releasing energies tens or hundreds of times as much energy as the K/T impact, were taking place on the Moon at 3.8 Ga, excavating the enormous basins we call Imbrium and Orientale. The Moon suffered at least ten such collisions between 3.8 and ~4.1 Ga (Taylor, 1986; Wilhelms, 1987). Earth would have been struck by hundreds of similar objects over the same period, and by tens of objects much larger still. To order of magnitude, the energy released by the largest would have been enough to vaporize Earth's oceans (Sleep et al., 1989).

Looking still further back, the very existence of the Moon is evidence for a truly colossal collision: a collision of a Mars-size object with the proto-Earth to form the Moon (Hartmann et al., 1986; Stevenson, 1987; Newsom and Jones, 1990). Such an impact would probably have melted the planet. If life had arisen on Earth before the Moon-forming impact, it would not have survived here. Indeed, no organic material would have survived. Biogenesis would need to start again.

Evidently life's hold on Earth would seem to be increasingly precarious at earlier times, with only the most protective niches, e.g., the mid-ocean ridge hydrothermal vents or deep aquifers, being in any sense continuously habitable (Maher and Stevenson, 1988; Oberbeck and Fogleman, 1989; Sleep et al., 1989). At some point in time there must have occurred a last collision that life could not have survived. It is unlikely that the Moon-forming impact was the last of these events. The last planet-sterilizing impact presents what is effectively a biological "event horizon" (C.P. McKay, Personal communication.), since none of the biochemical events that may have occurred before this impact can bear on the origin of life as it exists now.

7.2   Introduction

Maher and Stevenson (1988) considered the possibility that frequent impacts frustrated the origin of life, particularly for sites at the surface. They did so by comparing the time interval between dangerous impacts at a given location with the time required for biogenesis. The former can be estimated with some degree of confidence from the observed lunar impact record. The latter were arbitrary but short, with timescales assumed to range from 10^5 years to 10^7 years depending on location and luck. Maher and Stevenson (1988) concluded that most impacts, although devastating enough at the surface, would have passed unnoticed at the bottom of the ocean. They argued that the much longer time intervals available for undisturbed evolution at the mid-ocean ridges more than offset the likelihood that biogenesis would have occurred more slowly there than at the surface, and hence concluded that life probably originated at the mid-ocean ridges, and even hazarded a tentative date of 4.2–4.3 Ga, based on when the mid-ocean ridges became continuously habitable.

Sleep et al. (1989) asked a related, but inherently different question: When did the last planet-sterilizing impact occur? That is, we assumed that life had already originated in some unspecified manner, and then asked what would it have taken to wipe it out. This approach is better because the timescale for biogenesis is essentially unconstrained, whereas early ecosystems with well-defined organisms are broadly constrained by analogy to modern ones.

Oberbeck and Fogleman (1989, 1990) closely followed Maher and Stevenson (1988), but focused on the largest, globally sterilizing events. Oberbeck and Fogleman (1990) assumed that ocean vaporizing impacts reset biogenesis. From these assumptions they deduce upper limits for the time available for the origin of life that are uncomfortably short. Available time frames range from 11 million years at 3.8 Ga to 133 million years at 3.5 Ga. These results differ substantially from those given by Sleep et al. (1989), who estimated that there are no more than a few, and possibly none, or Chyba (1991), who estimates approximately four such events between 3.8 Ga and 4.4 Ga. As we will show below, the difference arises from the high impact flux used by Maher and Stevenson (1988), and inherited by Oberbeck and Fogleman (1989).

What neither Sleep et al. nor Maher and Stevenson knew when we wrote our papers was that molecular phylogeny, in the form of comparative 16-S ribosomal RNA-sequencing (Pace et al., 1986; Lake, 1988), was pointing toward a sulfur-metabolizing thermophile as the probable identity of the universal ancestor. The universal ancestor should not be confused with the first living organism. At the level of molecular biology the sulfur-metabolizing thermophile differs very little from turnips, humans, or slime molds. The gap between between a rock and a bacterium is incomparably greater. It is the enormity of this gap that defines life; it makes what is really a quantitative difference in complexity appear to be a qualitative difference. That we descend from a sulfur-metabolizing thermophile only means that the sulfur-metabolizing thermophile survived what others did not. The mid-ocean ridge hydrothermal systems are one of the habitats favored by these organisms today. We have added some discussion of this matter, which is clearly germane to the theme of natural selection by impact.

7.3   An Ocean-Vaporizing Impact

The mass of the oceans is 1.4×10^{24} g. When evaporated the oceans transform into a 270 bar steam atmosphere. Because this exceeds water's critical pressure of 220 bars, what is meant by evaporating the oceans is not entirely clear, since above the critical point there is no distinction between liquid and vapor. For specificity we will regard water's critical temperature of 647 K as the definition of total evaporation. The latent heat of vaporization of water at 273 K is 2.5×10^{10} ergs/g and the specific heat of water vapor is 1.9×10^7 ergs/g. Therefore it takes a minimum of $\sim 4.5 \times 10^{34}$ ergs to evaporate the oceans.

To illustrate the magnitude of an ocean-vaporizing impact, as a first approximation assume that roughly a quarter of the energy released by the impact is spent evaporating water, with the balance buried in the crust and mantle or radiated to space (Sleep et al., 1989). This corresponds to the impact of a 2×10^{23} g body at a typical asteroidal collision velocity of 14 km s^{-1}. This would be a 500 km diameter object—roughly the size of the large asteroids Vesta and Pallas.

At least a few impactor masses of ejecta are expected. For example, when applied to a 2×10^{23} g body striking at 14 km s^{-1}, extrapolation of a conventional crater-scaling relation (Schmidt-Housen, 1987; Eqs. (7-7) and (7-8) below) implies a (transient) crater diameter of about $\sim$1500 km, and some $\sim 5 \times 10^{23}$ g of ejecta. Much of the ejecta would be hot rock vapor created on impact; most of the rest would be melt droplets and grains destined eventually to be evaporated by later events, such as mixing with hotter vapors or atmospheric re-entry heating (Sleep et al., 1989; Zahnle, 1990; Melosh et al., 1990). The resulting $\sim$100 bars of rock vapor displaces any pre-existing thin atmosphere. Even a hypothetical thick early atmosphere offers little effective resistance to the global expansion of this much rock vapor. After a few sound crossing times—a day or three—the rock vapor atmosphere should smooth out globally and settle into hydrostatic equilibrium. It is through the rock vapor that much of the impact's energy is globally distributed.

7.3.1   Rock Vapor

Panel (a) of Figure 7.1 begins at this point. The rock vapor atmosphere radiates to space with an effective temperature of order $\sim$2000 K. The saturation vapor pressure of rock is sensitive to temperature, and therefore the cloudtop temperature is relatively insensitive to details. As an illustrative example, consider cloudtops in which 0.1% of the mass is condensed as 10μ droplets. Such a cloud becomes optically thick at 4 mbars. The saturation vapor pressure of a mixture of forsterite (Mg_2SiO_4) and silica at $T \approx 2000$ K can be approximated by (Mysen and Kushiro, 1988)

$$p_{rv} = 3.6 \times 10^{10} \exp\left(-68900/T\right) \text{ bars.} \qquad (7\text{-}1)$$

The vapor consists mostly of Mg, SiO_2, SiO, and O_2, with a mean molecular weight of $\sim$40. The scale height at 2000 K is $\sim$40 km. Equation (7-1) implies a temperature at 4 mbars of 2300 K. If the droplets were increased to 100μ, the saturation vapor pressure at optical depth unity would rise to 40 mbars and the temperature to 2500 K; if the mass fraction in condensates were increased to 1%, the saturation vapor pressure at optical depth unity would fall to 0.4 mbars and the temperature to 2150 K.

Even radiating as a 2300 K blackbody, it would take a few months to radiate away the energy ($> 10^{35}$ ergs) initially present in the rock vapor. At least as much energy is radiated down onto the oceans as is radiated directly to space, because the temperature at the bottom of the atmosphere will be higher than at the top. The high opacity of seawater to infrared radiation (Suits, 1979) concentrates radiative heating in a thin surface layer. Boiling is confined to this layer. In a sense, thermal radiation from the rock vapor ablates the surface of the ocean, leaving the ocean depths cool. The depth of the boiling layer is determined by compositional convection; the hot but extremely saline surface waters mix with enough of the relatively cool and less saline waters below to leave the surface waters at worst neutrally buoyant. In general, the density decrease caused by thermal expansion

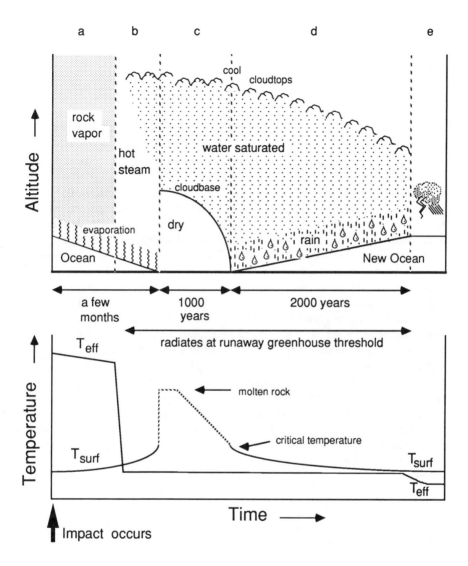

FIGURE 7.1. History of an ocean vaporizing impact. (a)–(b) An impact on this scale produces about 100 atmospheres of rock vapor. Somewhat more than half the energy initially present in the rock vapor is spent boiling water off the surface of the ocean, the rest is radiated to space at an effective temperature of order 2300 K. (c) Once the rock vapor has condensed the steam cools and forms clouds. (c)–(d) Thereafter cool cloudtops ensure that the Earth cools no faster than the runaway greenhouse threshold, with an effective radiating temperature of order 300 K. In the minimal ocean vaporizing impact the last brine pools are evaporated as the first raindrops fall. For somewhat larger impacts a transient runaway greenhouse results, with the surface temperature reaching the melting point.

outweighs the density increase from higher salinity, so that boiling is limited to surface waters. This will be discussed in more detail below in the context of global habitability.

Just above the surface there is a very thin boundary layer where the temperature rises rapidly from the relatively cool boiling waters at the surface to the much hotter rock vapors above. Since water vapor is somewhat transparent to >2000 K thermal radiation (Kasting, 1988), the oceans are not effectively shielded from the hot rock above; yet the water vapor is sufficiently opaque that it is quickly heated to the atmospheric temperature. Once heated the buoyant water vapor rises through and mixes with the denser rock vapor, stirring and homogenizing the atmosphere. The atmosphere is doubtless vigorously convective, driven at the top by radiative cooling at the cloudtops and at the bottom by rising plumes of low density, water-rich gases. The temperature profile should therefore follow the moist adiabat for rock vapor; i.e., the temperature is effectively set at all altitudes by the saturation vapor pressure of rock. For a 100 bar rock vapor atmosphere, Eq. (7-1) implies a temperature near the surface of ~ 3500 K. The radiative heat flux onto the oceans exceeds that to space by a factor of order $(3500/2300)^4 \approx 5$. This factor applies only to the oceans, which are necessarily cool. On dry land the surface quickly heats to reach radiative equilibrium. As the oceans shrink the relative importance of radiation to space grows (on Earth today the proportion of dry land would very quickly reach 40%). While the oceans remain, most of the energy in the atmosphere must go into evaporating sea water and heating water vapor. This differs from our previous estimate (Sleep et al., 1989) that the energy radiated to space and the energy spent vaporizing water were comparable. To evaporate and heat water vapor to 3500 K takes some 8.5×10^{10} ergs/g; to do this to the oceans requires $\sim 1.2 \times 10^{35}$ ergs. At the higher effective radiating temperature of 3500 K, the cooling time of the atmosphere reduces to about 1 month. The actual cooling time would be somewhere between 1 month and 1 year, depending on the relative surface areas of dry land and open seas.

Other heat sinks are probably unimportant. It takes less than 2×10^{32} ergs to heat a 1 bar N_2 atmosphere to 3500 K. Even a 60 bar CO_2 atmosphere (if all the CO_2 in all the Earth's carbonates were put into the atmosphere) requires less than 10^{34} ergs to reach 3500 K. In all likelihood the pre-existing atmosphere is quickly raised to the temperature of the rock vapor, either by radiative heating or by mixing with the rock vapor.

The heat capacity of the crust on a timescale of months is also limited. At a typical rock thermal diffusivity of $\kappa \approx 0.01$ cm^2 s^{-1}, the heat pulse is conducted to a depth of order $\sqrt{\kappa t}$, where $t \approx 10^7$ s is the approximate lifetime of the rock vapor atmosphere. This is only a few meters. With a heat capacity of 1.4×10^7 ergs g^{-1} K^{-1} and a heat of fusion of 4×10^9 ergs g$^{'}$, it would take only 3×10^{32} ergs to heat 4 m to 3500 K. On the other hand, the exposed surface would be molten, and would flow downhill, thereby exposing more rock to be melted. The speed at which mountains melt is hard to estimate because it depends on the temperature difference between the atmosphere and the rock, which in turn depends on how rapidly the rock flows away to expose cooler material below.

However, to absorb 10^{35} ergs the globally averaged depth of melting would need to be well over 1 km, which seems implausibly large; tens of meters—a few times 10^{33} ergs—would seem more likely.

Throughout this period a hot rock rain or hail falls into the ocean at a rate of meters per day. In the minimal ocean-vaporizing impact a few hundred meters of rock raindrops eventually accumulate on the ocean floor. The thermal energy in the rock rain is considerable. The rainout of a 100 bar rock vapor atmosphere delivers about 2×10^{34} ergs, or fully 20% of the energy assumed initially present in the ejecta. The raindrops are quenched near the surface, and so contribute to boiling (with a thermal diffusivity $\kappa = 0.01$ cm^{2}s^{-1} it takes about 1 s to cool a millimeter-size drop and about 1 min to cool a centimeter-size drop).

7.3.2   Steam

After a month or two the rock vapor has fully condensed and fallen from the sky. Here begins panel (b) of Fig. 7.1. There remains behind a hot steam atmosphere, and if the impact were small enough, a hot salty ocean of considerable depth. The upper regions of the steam atmosphere soon cool and condense water clouds, and the effective radiating temperature drops below 300 K. Once a cool upper atmosphere is established radiative cooling to space falls off precipitously, but the flow of energy to the oceans does not. Hence most of the thermal energy that remains in the steam is used to evaporate water until either the oceans are gone or the atmosphere reaches saturation vapor pressure equilibrium at the surface of any remaining seas. For the minimal ocean-vaporizing impact, about half the ocean remains as liquid when the rock vapor has fallen out. The rest of the ocean is afterward evaporated by excess thermal energy present in the hot steam. In the minimal ocean-vaporizing impact the last brine pools evaporate as the first rains fall on the mountains.

The properties of water vapor atmospheres in quasi-steady state have been studied in detail by Kasting (1988), Abe and Matsui (1988), and Nakajima et al. (1992). The central result of their work is that there is a maximum rate at which a water atmosphere over a liquid ocean can radiate to space. This maximum is only some $\sim$40% higher than Earth's present infrared flux. This is precisely the "runaway greenhouse" threshold often encountered in comparative planetology.[1] Higher rates of radiation correspond to atmospheres for which no liquid water is present at the surface. The threshold is reached when the infrared optical depth of the cool, moist cloudtops (the uppermost few bars of water vapor) becomes so great that no thermal radiation originating from the surface or lower regions of the atmosphere can escape to space. The planet's radiative cooling rate is then determined solely by the opacity and saturation vapor profiles of water at the cloudtops, and is effectively decoupled from conditions in the lower atmosphere or at the

---

[1] In the usual runaway greenhouse, as applied to ancient Venus or future Earth, the planet absorbs more sunlight than the atmosphere can radiate, and so the oceans evaporate.

surface. For a mass of water vapor equal to a terrestrial ocean the surface temperature must reach at least $\sim$1500 K for the planetary infrared radiative flux to significantly exceed the runaway greenhouse threshold. The runaway greenhouse threshold for Earth is 3.1×10^5 ergs cm^{-2} s^{-1}. This includes re-radiation of any incident sunlight. When the net solar irradiation four billion years ago is taken into account, the effective runaway greenhouse cooling rate would have been $\sim$1.5 $\times$ 10^5 ergs cm^{-2} s^{-1}. This assumes an albedo of 33%. With a higher albedo Earth would cool more quickly. But even with an albedo of 100% the effective cooling rate is only doubled.

The lifetime of the transient runaway greenhouse—the time it takes to cool and condense an ocean of hot steam—can be estimated by dividing the energy in the steam, plus the smaller amount of energy in hot crustal rock, by the effective runaway greenhouse cooling rate of 1.5×10^5 ergs cm^{-2} s^{-1}. It takes $\sim$4.5 $\times$ 10^{34} ergs to evaporate the ocean, and an additional 2.2 $\times$ 10^{34} ergs to raise the steam to 1500 K, at which point radiative cooling becomes more effective. It therefore takes about 900 years to cool the steam from 1500 K to the critical temperature (Fig. 7.1, panel (c)), and another $\sim$1900 years to rain out the ocean and cool it to 300 K (panel (d)). The net globally averaged rainfall is about 150 cm yr^{-1}. With allowance for the uncertain albedo, we estimate that it would have taken Earth some 2000–3000 years to return to normalcy after an ocean evaporating impact. Abe (1988) gives somewhat shorter cooling times, but he assumes a 100 bar steam atmosphere.

The steam atmosphere could be prolonged by the release of heat initially buried at the impact site. The thermal energy in melt with temperatures well above the solidus is relatively quickly convected to the surface, and so is available to prolong the lifetime of the steam atmosphere. This could constitute a considerable fraction of the buried energy, and if so, it could prolong the runaway greenhouse by a factor of a few, depending on the magnitude of the impact. But the amount of very hot melt and its temperature are greatly influenced by the extent to which cold adjacent rock slumps into and mixes with it. If the melt were to mix with enough cold rock a viscous rock slush would result. Such a slush would have little excess thermal energy available on a 3000 year timescale.

7.3.3   Salt

At minimum, a planet-sterilizing impact would have to boil away most if not all of the oceans. The depth of boiling is determined by compositional convection: the hot but extremely saline surface waters mix with enough of the relatively cool and less saline waters below to leave the surface waters at worst neutrally bouyant. As a specific model, assume that a depth h of surface water with an original salinity S has been vaporized, leaving behind a well-mixed, neutrally bouyant surface layer of depth h' with salinity $S' = (h + h')S/h$. Very approximately, the density of hot fresh water is

$$\rho(T) = 1 - 0.242 \left(\frac{T - 273}{277} \right)^{1.67} \tag{7-2}$$

and the density of saline water at 293 K is

$$\rho(S) = 1 + 0.0245 \left(\frac{S}{3.5\%}\right)^{1.05}.$$

(7-3)

Both are crude fits to data tabulated in the CRC Handbook. The effects of salinity and temperature cancel when $\rho(T) = \rho(S)$. To a first approximation the temperature of the surface water is the boiling temperature. The saturation vapor pressure (bars) of water from 273 K to 647 K is approximately

$$p_{vap} = 4.7 \times 10^5 \exp(-4870/T).$$

(7-4)

The boiling temperature is obtained by equating the vapor pressure with the total atmospheric pressure $p_{rv} + p_{atm} + hg$, where p_{rv} is the pressure of rock vapor, p_{atm} the pressure of any pre-existing atmosphere, and hg the weight of water already vaporized. Figure 7.2 shows the depth h' of the saline layer as a function of the depth of water vaporized; $p_{rv} + p_{atm}$ is given and hg calculated. The depth of boiled water remaining at the surface is determined by compositional convection. The case where the pressure is exerted by water vapor alone corresponds to the final state at the end of evaporation, when all the rock vapor has condensed and the steam atmosphere has cooled to saturation.

7.3.4   Survival?

Once the oceans are vaporized it takes little additional energy for the surface temperature to reach the melting point of rock. Yet there might remain possible refugia hidden deep below the surface, thermally insulated by several hundred meters of rock and sediment. The approximate depth of the extinction horizon can be estimated by considering the response of a semi-infinite, previously isothermal slab to a step function temperature change ΔT imposed at the surface (Turcotte and Schubert, 1982):

$$T(x, t) = T_s + \Delta T \left\{1 - \text{erf}\left(\frac{x}{2\sqrt{\kappa t}}\right)\right\},$$

(7-5)

The temperature $T(x, t)$ is the temperature at depth x at a time t after the temperature change. For simplicity, idealize the aftermath of an ocean vaporizing impact as raising the surface temperature to 1000 K for a time $\delta t \approx 3000$ yr. Assume that a temperature change of $\Delta T = 100$ K is lethal. Survival would then demand living deeper than

$$x_e \geq 2.1\sqrt{\kappa\,\delta t}.$$

(7-6)

This expression is insensitive to ΔT. The heat pulse δt probably lasts between 1000 and 5000 years, and the thermal diffusivity κ ranges approximately from 0.002 cm^2 s^{-1} for sediment to 0.01 cm^2 s^{-1} for rock. It follows that 200 m $\leq x_e \leq$ 800 m. All life above x_e is wiped out.

Deeper niches are occupied today. Oil field bacteria can live 1 km below the surface at the interface between hydrocarbons and sulfate or oxygen-rich water

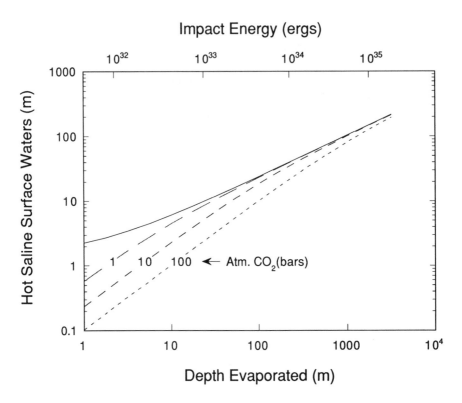

FIGURE 7.2. The depth of hot saline surface waters produced when a large impact evaporates significant amounts of the oceans. This is shown as a function of impact energy, depth of water evaporated, and the surface pressure of any preexisting CO_2 atmosphere.

(Connan, 1984). It is debatable whether similar niches, which are the biosphere's frontier outposts today, either existed or were occupied on the Hadean Earth. Owing to generally higher heat flow in the Hadean, if they existed at all such niches would have been fewer and shallower than today. The presumed absence of abundant oxygen might have posed another limitation to the habitability of these extreme environments. But even given the existence and survival of life several hundred meters below the surface, it is by no means obvious that these hypothetical subterranean creatures would have been able to recolonize an otherwise sterile planet. Life's single most important skill—photosynthesis—may well have been irretrievably lost to an ecosystem that had previously dwelt in the dark for uncounted millenia.

If photosynthetic organisms were to survive, they would need to be mixed to the bottom of the ocean (by tsunamis, say) and rapidly buried by several hundreds of meters of ejecta. They must not get crushed, eaten, or poisoned in the process. They would also need to have been buried in the midst of abundant food, chemicals both oxidized and reduced, and yet not be able to access the food so quickly that they could eat all of it while buried for millenia. They must retain the latent ability to photosynthesize in the face of competition from creatures that had long ago discarded the useless skill. They would then need to be exhumed by erosion, but not so soon that they would be scalded by hot rain as the oceans recondense. Finally, they would need to find a suitable environment where they could grow, presumably without the aid of those other creatures that had played important roles in the parent ecosystem. To meet any of these requirements would be lucky; to meet them all might be asking too much of luck.

The best strategy for surviving a truly enormous impact is to go into orbit. Impacts can lift surface rocks into orbit essentially unshocked and unheated (Melosh, 1989). This process is known to occur, most famously among the SNC meteorites, which have traveled to Earth from Mars. Rocks ejected from Earth will mostly reach Earth-like orbits, and so will spend relatively little time in space before they are swept up again, typically tens of thousands of years or less (H.J. Melosh, Personal communication). It is imaginable, if only just, that passengers could survive the journey. We also note that rocks can be exchanged between planets, although transit times are usually much longer and hence the voyage more dangerous.

7.3.5   Imbrium on Earth

The Earth must have experienced hundreds of impacts on the scale of the Imbrium and Orientale ($\sim 10^{33}$ ergs) between 3.8 Ga and 4.1 Ga, as we shall see in the next section. These impacts are about one hundred times smaller than the ocean-vaporizing event, yet dangerous nonetheless. Consider an "Imbrium-scale" 2×10^{33} erg impact that generated some 5×10^{21} g of rock vapor, enough to make a 1 bar atmosphere. The event is small enough that one can think of it as a scaled-up K/T impact. The ejecta are launched ballistically from the crater, and are thereby distributed globally (Melosh et al., 1990; Zahnle, 1990). Atmospheric re-entry ensures vaporization (Zahnle, 1990). The mixture of rock vapor and multiply shocked air that results sits on top of the original atmosphere, where owing to its high temperature it remains for some time. The layer of rock vapor, rock vapor condensates, and hot air radiates equally to space and the surface at an effective radiating temperature of order 2000–2500 K. The cooling time for an Imbrium-scale event is about 1 day.

The relative transparency of the pre-existing atmosphere to 2000 K infrared radiation would prevent it from absorbing a significant fraction of the energy. Hence the impact would boil off some 40 m of sea water, thereby adding some 4 bars of water vapor to the atmosphere. If this event occurred in a 1 bar N_2 atmosphere, there would remain at the surface a hot (150° C), highly saline (15%), boiled layer about 12 m thick (see Fig. 7.2). If the event occurred in a 10 bar CO_2 atmosphere, the surface brine layer would be 200° C and 8 m deep. In either case a small part of the extra 4 bars of water vapor in the atmosphere would rain out quickly when mixed with the cool pre-existing atmosphere. Raising a 1 bar N_2 atmosphere to 150° C uses only about 1% of the energy in the water vapor (mostly latent heat of condensation); raising a 10 bar CO_2 atmosphere to 200° C uses 10%. The rest of the injected water vapor rains out no faster than the maximum rate permitted by the runaway greenhouse threshold; i.e., 170 cm yr^{-1} (the average rainout rate is slightly higher than for the ocean-vaporizing impact because the atmosphere is cooler). It takes 24 yr to dry out the atmosphere. Because the atmospheric temperature profile would follow the moist adiabat, for the first several years the rain would be hot (> 100° C). The result would be a hot stable layer of fresh water some 40 m thick. Obviously there will be some mechanical mixing with the colder saltier water beneath, but the general impression is that normal surface water ecosystems based on photosynthesis would be annihilated. Extreme thermophiles and/or extreme halophiles would be much more likely to survive. Indeed, conditions would probably support a global bloom among the photosynthetic halophilic thermophiles. Deep water ecosystems would be relatively unperturbed. In particular, colonies living at hydrothermal vents near the mid-ocean ridges are unlikely to be greatly distressed by the events at the surface.

7.3.6 Filters

An impact ten times larger then Imbrium has qualitatively similar effects, but the major untoward effects are proportionately greater. Rather than 40 m of sea water evaporated there are 400; rather than taking 24 yr for the steam to rain out, it takes 240 yr. For a pre-existing 1 bar N_2 atmosphere the surface temperature reaches 250° C and the boiled brine layer extends 60 m deep. Over most of the next 240 yr hot rainwater floats bouyantly atop the brine. It is difficult to see how common surface ecosystems could have survived. Life in the deep oceans may proceed relatively unaffected. Photosynthesis is likeliest to survive among extreme thermophiles or extreme halophiles. Indeed, thermophiles and halophiles would have ruled the world.

It is this scale of event in particular that most strongly suggests that life on Earth should descend from organisms that either lived in the deep ocean, in hydrothermal systems, or were extremely tolerant of heat and salt. If photosynthesis survived it would have been preserved among the halophiles and thermophiles. There is no reasonable doubt that events of this scale occurred four billion years ago, with perhaps ten events scattered over the three hundred million year interval between 3.8 Ga and 4.1 Ga.

Ribosomal RNA sequencing by two competing schools (Pace et al. 1986; Lake, 1988) finds common ground in the inferrence that all extant life on Earth descends from the class of sulfur-metabolizing, thermophilic organisms that populate mid-ocean ridge hydrothermal systems today. In our opinion, the natural story is that such organisms survived because their niche was once the safest one on Earth. Inhabitants of other niches were filtered out by impact. Note well that our descent from organisms that dwelt at the mid-ocean ridges is not equivalent to life originating at the mid-ocean ridges. Life could have arisen at the surface and colonized the mid-ocean ridges. The long intervals between major events leaves plenty of time for migration. It is simply because the deep water hydrothermal systems offered sanctuary from otherwise lethal events that life survived there. The argument is strong enough that it can be inverted: because we can trace our descent to thermophiles and halophiles, we can presume that life originated before the end of the heavy bombardment.

7.4   The Lunar Record

No direct record of the earliest impacts exists on the Earth. In contrast, the lunar record is well preserved and well studied. In brief, the Moon records an early period (ca. 3.8 Ga) of high impact flux, often called the late bombardment, followed by some three and a half billion years rather like the present. The intensity of the bombardment is somewhat controversial, and the story of what may have preceeded it more controversial.

We consider four arguments that constrain the mass of material impacting the Moon between 3.8 Ga and 4.4 Ga. (1) The energies released by the more recent basin-forming impacts can be estimated without undue reliance on ambitious extrapolations of laboratory crater scaling relations. (2) Mantle samples are rare in the lunar regolith; thus impacts large enough to excavate mantle materials were rare. (3) The contamination of the lunar crust with meteoritic material provides an estimate of the total amount of meteoritic material accreted by the Moon since the lunar crust formed. (4) Impact stirring has been mild enough such that the lunar crust retains a relatively high degree of heterogeneity.

The terrestrial impact flux during this time interval can be scaled from the lunar record. We will often express the lunar impact history in terms of an equivalent thickness of impacting material. The equivalent thickness of material impacting Earth will be rather greater, for two reasons. First, Earth's gravity enhances its capture cross-section. The other reason is a consequence of small number statistics. The mass spectra of smaller solar system objects appear to obey power laws in which most of the mass is concentrated in the few largest objects (e.g., Dohnanyi,

1972; Chapman et al., 1978). Because Earth's effective collisional cross-section greatly exceeds the Moon's, in most trials Earth will accrete *all* of the ten largest objects in a population of Earth/Moon impactors (Sleep et al., 1989). As these carry most of the mass, the Moon is likely to undersample the population of impactors.

Chyba (1991) reviews the lunar record from a perspective quite similar to our own. Although there are some differences, Chyba's general conclusions are mostly agreeable with those presented here.

7.4.1   Energies of Basin-Forming Impacts

Impressive multiringed basins were formed on the Moon as recently as 3.8 BY (Wilhelms, 1987). The Imbrium basin dates to 3.83 Ga, and the spectacular Orientale basin is dated at ~3.8 Ga. The Moon experienced a half-dozen other, similar impacts dating between Nectaris and Orientale (Baldwin, 1987a, b; Wilhelms, 1987). There is some controversy about the age of Nectaris. Some authorities argue that lunar samples at 3.92 BY date to Nectaris (e.g., Tera, et al. 1974; Wilhelms, 1987; Spudis, 1993). Others argue that samples of age ~4.1 Ga were cast by Nectaris, and that the earlier date is more consistent with other evidence constraining the impact flux at still earlier times (e.g., Wetherill, 1981; Taylor, 1986; Baldwin, 1987a, b). If the young age for Nectaris is adopted, then the impact flux had a characteristic fall time of less than 70 Ma (the timescale used by Maher and Stevenson, (1988)). With this rate it almost certainly follows that the late bombardment was a short-lived "terminal cataclysm" (Tera et al., 1974), preceded by a few hundred million years of much lower impact flux. An old age for Nectaris is consistent with a monotonic decline in the impact flux. Approximately thirty basins older than Nectaris have been identified. The largest confirmed remaining basin, South Pole–Aitkin, formed early and has been nearly obliterated by later features. The older age implies a characteristic fall time greater than 144 Ma (the timescale used by Chyba (1991)).

Orientale is the youngest and best preserved of the basins. Orientale has 4 prominent rings, with diameters of 320, 480, 620, and 930 km (Wilhelms, 1987). It is by no means clear which if any is the "real" crater; each ring has its partisans. This ambiguity defeats crater-scaling as a measure of impact energy. The energy released by the Orientale impact has been estimated by examining the thermal contraction of the center of the basin (Bratt et al., 1985a). First, the fracture pattern implies that the projectile was no more than 100 km diameter. Second, the amount of buried heat can be deduced from thermal contraction. Bratt et al. (1985a) assume that one quarter of the energy was deeply buried, giving an impact energy between 4×10^{32} ergs and 3×10^{33} ergs. If it were assumed that one half of the impact energy was buried (as we have for impact on Earth; but the two are not equivalent—the lower lunar gravity allows more ejecta to escape the crater), then these energies should be divided by two.

A second, independent energy estimate is based on the mass of ejecta excavated by the impact (Bratt et al. 1985b). The ejecta mass is determined both by measuring the ejecta blanket itself and indirectly by reconstructing the hole that was later filled by an isostatically uplifted mantle. Bratt et al. (1985b) conclude that the Orientale impact expelled 2.4×10^{22} g of ejecta. Conventional crater scaling relations are much more safely inverted using the ejecta mass than using radii. According to Schmidt and Housen (1987), the ejecta mass should scale as

$$M_{ej} = 0.13 \rho E^{0.783} v^{-0.261} g^{-0.652} \sin \theta \text{ g,} \qquad (7\text{-}7)$$

where ρ is the crustal density and g is the lunar surface gravity. We have supplemented Schmidt and Housen's relation with dependence on the incidence angle θ (Melosh, 1989, p. 121). The most probable incidence angle is 45°. For a 13 km s^{-1} impact velocity, Eq. (7-7) implies an impact energy of 1.2×10^{33} ergs.

If this energy is inserted directly into Schmidt and Housens's (1987) crater scaling relation,

$$D_t = 1.4 E^{0.261} v^{-0.087} g^{-0.217} \text{ cm,} \qquad (7\text{-}8)$$

one obtains a transient crater diameter of 600 km, in accord with Bratt et al.'s (1985a) preference for the Outer Rook mountains as the true crater rim. The diameters of large craters are increased by slumping. This is a broadening and filling of the crater that takes place in the late stages of the crater's formation. Following Chyba (1991), we adopt the crater slumping prescription employed by McKinnon et al. (1990). The final (slumped) crater has diameter

$$D = 1.2 D_t^{1.13} D_c^{-0.13} \sin^{1/3} \theta, \qquad (7\text{-}9)$$

to which we have appended the empirical dependence on θ. Eq. (7-9) implies a slumped diameter for a 1.2×10^{33} erg Orientale of 950 km, in accord with Wilhems (1987, Chapter 4) preference for the Cordillera mountains as the true crater rim.

An energy of 1×10^{33} ergs corresponds to a 1.2×10^{21} g object at the average lunar impact velocity of 13 km s^{-1} (this is the average for asteroids; most comets strike at higher velocities, but the dispersion is so wide that the "average" of 25–30 km s^{-1} is more misleading than useful). With a density of 3.0 g cm^{-3}, its diameter would have been 90 km. This pushes against Bratt et al.'s (1985a) 100 km upper limit on the diameter of the body. For the same assumptions, a 3×10^{32} erg impact implies a 60 km body and a 400 km transient crater. When used with Eq. (7-8), the latter agrees with Melosh's (1989, p. 165) estimate that the transient crater was no larger than ~400 km. Melosh (1989, p. 78) notes that impacts typically excavate material from a depth equal to about 10% of the transient crater diameter. Orientale appears to have excavated only the anorthosite-rich crust, which sets an upper limit of 60 km on the depth of excavation (Spudis 1993, p. 64) (the crust is deep under Orientale). Bratt et al. (1985) estimate that the mantle under Orientale was raised 50 km; i.e., Orientale was originally some 50 km deep. Equation (7-8) then implies an impact energy of $1–2 \times 10^{33}$ ergs. We conclude that the best present estimate for Orientale's energy is 1×10^{33} ergs, uncertain by a factor of two or more.

Imbrium ejected some 3.6×10^{22} g (Spudis et al., 1988) and has a nominal outer diameter of 1160 (Wilhelms, 1987). Estimated from the relative ejecta masses, the Imbrium impact was probably no more than about twice as large as Orientale; estimated from the ratio of outer ring diameters, the Imbrium impact was $(1160/930)^{3.4} \approx 2.1$ times greater. Neither Imbrium nor Orientale dredged up mantle samples, but Imbrium did dredge up KREEP, which formed at the base of the crust (Spudis, 1993, p. 147). (Spudis, 1993, p. 169) suggests that Imbrium may have dug up a small amount of mantle material). Assuming that the crust was 50-60 km thick (Spudis, 1993, pp. 131, 162), it follows that Imbrium was no more than $(60/50)^{3.83}$ approximately twice as large as Orientale. We will assume an energy for Imbrium of $1 \times 10^{33} - 3 \times 10^{33}$ ergs, corresponding to a mass range of $1 \times 10^{21} - 4 \times 10^{21}$ g. For comparison, Baldwin (1987a, b) estimates a mass of the Imbrium impactor of $\sim 2 \times 10^{21}$ g and Wetherill (1981) estimates 4×10^{21} g.

The far side South Pole–Aitkin basin is the largest and probably oldest confirmed lunar basin (Wilhelms, 1987; Belton et al., 1992). In contrast to Imbrium and Orientale, there is spectral evidence obtained on the recent Galileo flyby of the Moon (Belton et al., 1992) that this event did dredge up some mantle material. If the crust is assumed to be 70 km thick, and it is assumed that excavation reached 10–30 km into the mantle, then the South Pole–Aitkin impactor would have been some $(80/50)^{3.83}$-$(100/50)^{3.83} \approx 6 - 15$ times greater than Orientale; i.e., roughly the impact of a 10^{22} g body releasing 10^{34} ergs. Taking the diameter of the transient crater to be ten times the excavation depth, the resulting 800–1000 km transient diameter of South Pole–Aitkin implies through Eq. (7-8) an energy of 4×10^{33}-9×10^{33} ergs, in reasonable agreement with the other estimate given.

7.4.2 Crustal Contamination by Chondrites

The mantle and crust of the Earth are highly depleted in siderophile elements, which enter the metallic iron core in preference to silicate phases. Platinum group elements like iridium are highly siderophilic. The Ir content of Earth's mantle might therefore be used to place an upper limit on the amount of material accreted by the Earth since the effective decoupling of core and mantle; i.e, since the end of core formation. This latter is something of an unknown, but it must post date the Moon-forming impact (Newsom and Taylor, 1989). There is a natural tendancy to place the Moon-forming impact at the end of accretion—to assign it pride of place—and to imagine that later impacts were of little account. This argument then implies that no more than 17 km of chondritic material accreted after the Moon-forming impact (Anders, 1989). Much of the 17 km could be left over from Moon-forming impact itself (Newsom and Taylor, 1989).

The Moon gives a much tighter limit (Sleep et al., 1989). Even more so than the Earth, the lunar mantle is highly depleted in siderophile elements (Drake, 1986). The presence of iridium in the lunar crust therefore indicates meteoritic contamination. A minimum estimate of the material which hit the Moon is obtained from the meteoritic component retained in lunar samples. Rocks exposed in the lunar highlands, which are believed to be typical of the lunar upper crust, are breccias composed of pristine rocks, impact melts, and still older breccias. It is generally agreed that Ir abundances indicate a meteoritic component of 1-2%. It is controversial whether nickel abundances indicate a comparable meteoritic component that lacks Ir (Korotev, 1987) or an internal component that was reduced to metal in the crust (Ringwood and Seifert, 1986). Studies of lunar meteorites indicate that the excess nickel is local to the nearside landing sites and that the meteoritic component is between 1–2% (Warren et al., 1989). The total thickness of lunar crust that is contaminated with siderophiles is not well constrained. Granulites believed to be deeply buried breccias as old as 4.26 Ga (Lindstrom and Lindstrom, 1986) and 4 Ga old near-surface breccias are both enriched in siderophiles, indicating that the meteoritic component is mixed in much of the upper crust, and that much or most of the meteoritic flux predates 4.26 Ga. The megaregolith thickness is one estimate of the mixing depth. Spudis and Davis (1986) give 35 km as the half thickness of the lunar crust. If the meteoritic component is between 1% and 4%, this implies a total meteoritic thickness between 0.35 km and 1.4 km, with 0.7 km our preferred estimate. For a crustal density of 3 g cm^{-3}, this is equivalent to some 8×10^{22} g. Chyba (1991) estimates a lunar veneer equivalent to 4×10^{22}-15×10^{22} g; i.e., essentially equivalent to the estimate given here.

If a magma ocean existed on the Moon, the upper crust should have frozen from the top down while being stirred by impacts (Hartmann, 1981). The date at which a mostly solid crust existed at any depth defines the start of the time interval for Ir and Ni retention in the lunar regolith. The oldest lunar rocks approach 4.5 Ga and indicate that a crust existed early in the history of the planet (Swindle, et al., 1986). The upper part of the lunar crust, ferroan anorthosite, solidified as early as 4.44 Ga (Carlson and Lugmair, 1988). The lunar crust was largely solid by 4.36 Ga (the model age of KREEP), which is believed to be final solidification of the base of the crust (Carlson and Lugmair, 1979). Much of the accretion was complete by 4.26 Ga, the age of the oldest regolith sample.

The above argument gives a lower limit on late lunar accretion, since not all material colliding with the Moon is retained. If the impact velocity is high enough a large part of the impactor escapes to space (Melosh and Vickery, 1989). For lunar impact, it must be expected that if the impactor is completely vaporized by the impact, the rock vapors will be hot enough to escape, leaving behind little or no chondritic contamination. Complete vaporization of rock impactors is expected to occur when the normal component of the impact velocity is greater than about 12 km s^{-1}ec (Melosh, 1990). For lower impact velocities vaporization is only partial. Unvaporized material does not escape as easily. Typical maximum ejecta velocities are of order one-sixth to one-tenth of the impact velocity (Melosh, 1989, p. 57). For impact velocities below 12 km s^{-1}, maximum ejection velocities should be no more than 1.2-2.0 km s^{-1}, lower than the lunar escape velocity. High-speed spalls can escape, but they are not quantitatively significant. We have assumed a typical lunar impact velocity of 13 km s^{-1}. The median impact zenith angle is 45°. Hence a typical value for the normal component of the lunar impact velocity is 10 km s^{-1}. This is small enough that Ir loss via vaporization occurs on fewer than half the asteroidal impacts. It is also useful to point out that siderophiles carried by differentiated bodies (those with iron cores) or by wholly iron bodies are more likely to be retained than are siderophiles carried by undifferentiated rocky or icy bodies. This occurs because iron cores are much denser than the lunar crust, and therefore are less severely shocked on impact than rocky bodies. In sum, we conclude with Chyba (1991) that incomplete retention of chondritic contaminants by the lunar crust does not appear likely to introduce more than a factor of two uncertainty in the asteroid impact flux, which is no worse than the uncertainty in the Ir abundance or our usage of a half-depth of 35 km for the crustal thickness.

On the other hand, the lunar crust is not likely to do a good job of integrating the comet flux. Only 4 of the 22 Earth-crossing comets would have impact velocities on the Moon below 20 km s^{-1}, and just 9 out of the 382 long-period comets (Olsson-Steel, 1987). Evidently the cometary contribution to lunar crustal iridium is small. It is therefore possible that, if the lunar basins were formed mostly by cometary impacts, lunar crustal iridium records only a small fraction of the impacting mass.

7.4.3   Impact Stirring

The lunar crust is so mildly churned that this observation by itself sets a useful upper limit on the flux of large projectiles. Large impacts were rare enough that regional heterogeneities, primary stratification, and even local igneous bodies have not been obliterated (Pieters, 1986; Davis and Spudis, 1987). Impacts similar to or larger than Imbrium appear to have been rare after the upper crust froze at 4.44 Ga, as the crustal stratification remains intact over much of the lunar surface. In particular, only the largest of the known lunar impacts, South Pole–Aitkin, may have excavated mantle. Similarly, troctolite, an expected constituent of the lower crust which was excavated by Imbrium, is quite rare (Marvin et al., 1989).

The ~600 km diameter middle ring of the Orientale basin, which Bratt et al. (1985a) identify with the excavation crater, covers about 1% of the lunar surface. The estimated mass of the Imbrium ejecta blanket, 3.6×10^{22} g, is 1% of the 35 km thick megaregolith layer (Spudis et al. 1988). In both cases the ejecta blanket covers a fair fraction of the lunar surface. Evidently the Moon could not have experienced hundreds of impacts on this scale and maintained regional hetero-geneities. A 0.7 km layer is by mass equivalent to ~40 Imbrium impactors. A much thicker layer would not seem to be consistent with this aspect of the lunar record. The total amount of material impacting since 4.44 Ga should be less than the 1.88 km equivalent layer needed to saturate the surface with Imbrium sized craters (Baldwin, 1981).

We noted above that a young age for Nectaris implies a terminal cataclysm rather than a monotonic decline in impact flux. Our reason can be summarized in the observation that only one lunar impact is known to have reached the mantle. It is not credible to posit an earlier 500 Ma years of increasingly fierce bombardment without digging deeply into the mantle, mixing the crust into the mantle, and strewing mantle samples everywhere.

7.5   Impacts on Earth

The single greatest uncertainty in scaling the lunar impact record to the early Earth lies in the mass distribution of the impactors. We will proceed on the assumption that the population of the late Earth impacting objects was a scale-invariant (power law) distribution of collisional fragments resembling the population of asteroids and comets today. If the mass distribution obeyed by the smaller objects that formed lunar craters and basins can be extrapolated to larger impactors, then the masses of the largest impactors can be estimated with some confidence. The most important residual uncertainties are the inevitable uncertainties associated with small number statistics.

7.5.1   Impactor Mass Distribution

The mass distribution of asteroids and comets is usually described by a power law of form

$$N(>m) = Cm^{1-q} \approx \left(\frac{m_1}{m}\right)^{q-1}, \tag{7-10}$$

where $N(>m)$ is the cumulative number of objects with mass greater than m, and m_1 is the largest object in the distribution. The properties of such a distribution can be visualized by binning the objects by logarithmic intervals of mass. For $q = 2$, there is equal mass in any bin. For $q = 5/3$ there is equal surface area in each bin, and most of the mass is concentrated in the largest objects. The probable origin of the power law is a self-similar fragmentation cascade. The largest objects are the source, and hence in principle could lie outside the power law. However, at the time of the late bombardment, the original source objects for the fragment cascade were probably long since gone, and even the largest objects impacting Earth would themselves have been fragments of fragments. Hence a fragmentation distribution might be expected to include even the largest remaining objects.

Popular values of q range from about 1.5 to 2. According to Safronov et al. (1986), $q \approx 1.6$ in a swarm of planetesimals dominated by coalescence; fragmentation increases q to ~ 1.8. Dohnanyi (1972) calculated that in a collisionally evolved distribution $q \rightarrow 11/6$. Turcotte (1992, pp. 20–34) argues that comminution should in general produce a fragment distribution with $q \approx 1.86$, and that this distribution explains the magnitude-frequency distribution of earthquakes. Asteroids with diameters between 100 km and 250 km have $q \approx 2$ (Hughes, 1982; Donnison and Sugden, 1984) and are very likely to be fragments; for smaller asteroids $q \rightarrow 11/6$ (Dohnanyi, 1972). Comets appear to be described by $q \approx 1.7$ (Donnison, 1986; Hughes, 1988), although this latter value is very uncertain.

The lowest values ($q \approx 1.5$) have been deduced from lunar highland craters by inverting crater-diameter scaling relations like Eq. (7-8) or Eq (7-9) ($q = 1.5$, Maher and Stevenson, 1988; $q = 1.47$, Melosh and Vickery, 1989; $q = 1.54$, Chyba, 1991). All use the newer consensus crater distribution $N(> D) \propto D^{-b}$, with $b = 1.8$ fit to craters between 10 km and 80 km diameter (Wilhelms, 1987, p. 257). (The older consensus was $b = 2$.) Because these craters date from the end of the late heavy bombardment they are directly relevant. These results do not immediately appear to be very sensitive to crater slumping. Chyba (1991) includes the latter by using Eq (7-9), he thus gets $q = 1.54 = 1 + b/3.36$ rather than the $q = 1.47 = 1 + b/3.83$ that results from application of Eq (7-8) alone. Nor is q especially sensitive to b: if Chyba had used $b = 2$, he would have gotten $q = 1.6$.

On the other hand, the same argument applied to large craters on Venus, where $b = 3$ (Phillips et al., 1990), implies that $q = 1.9$. The impactors may be different, since Venus has only sampled for 0.5–1 Ga while the Moon samples over a longer time, yet the difference is significant. Crater-scaling works when craters are self-similar. The characteristic length scale is determined by the impactor's mass and velocity, the planet's gravity, and the boundary conditions (Zel'dovich and Raizer, 1967, pp. 839-846). Larger craters are not self-similar. Breakdown of the scaling rule implies that other characteristic length scales become important for large craters. Obvious candidates include the depth of the crust ($\sim$70 km) and the porosity gradient in the regolith (several kilometers). There may be systematic effects that bias crater diameters over the relatively small span of crater diameters to which b has been fit.

Wilhelms (1987) lists 14 lunar basins from Nectaris to the present. Each of these basins would appear to require impact energies. The largest of these is Imbrium. The smallest, call it Schrödinger, is essentially a big crater; if treated as a crater its impact energy would be $\sim 3 \times 10^{31}$ ergs. For similar impact velocities Eq (7-10) implies that

$$q = 1 + \frac{\log (m_{Sc})}{\log (m_{Im})} = 1 + \frac{\log (14)}{\log (50 - 100)} = 1.57 - 1.67. \tag{7-11}$$

This q is consistent with the record for smaller craters and consistent with a cometary source. The same argument applied to the 45+ basins post-dating South Pole–Aitkin puts a lower limit of $q > 1 + (\log(45))/(\log(300 - 1000)) = 1.55 - 1.67$; this is a lower limit because many of the basins post-dating South Pole–Aitkin have since been obliterated.

The power q also affects the relation between the total mass in a distribution and the single largest body in that distribution. The total mass M_T accumulated by the Earth or the Moon is on average directly proportional to the mass of the largest impactor, although the relation between M_T and m_1 is very broad. The median value of M_T/m_1 is (Tremaine and Dones, 1993)

$$\frac{M_T}{m_1} \approx \frac{4q - q^2 - 2}{2(2 - q)(3 - q)} \ln \left(\frac{2}{q - 1} \right), \tag{7-12}$$

where $1 < q < 2$.[2] The general lack of dredged up mantle samples implies that South Pole–Aitken is the largest lunar basin post-dating the solidification of the lunar crust, and so might be identified with m_1 in Eq. (7-12). Its mass of $\sim 1 \times$

---

[2] M_T/m_1 is approximated fairly well by $M_T/m_1 \approx (3 - q)/(4 - 2q)$, an expression given by Wetherill (1975).

10^{22} g is about one-eighth of the total integrated flux required for a 0.7 km late chondritic veneer. Equation (7-12) then implies that $q \approx 1.9$. An even higher value of q would result if 0.7 km underestimated the mass of impacting matter, as would happen if much of the incident iridium escaped.

Because the power law distributions are dominated by a few large bodies, the predictive value of Eq (7-12) is necessarily limited by small number statistics: there is a 20% chance that a ratio of m_1/M_T as small as one-eighth could fall from a $q = 1.8$ distribution and a 5% chance that it could fall from a $q = 1.7$ distribution. A more likely scenario for $q < 1.8$ is to invoke a hidden, larger impact than South Pole–Aitkin. A prime candidate would be the hypothetical Procellarum basin. This impactor would be responsible for much of M_T.

A dominant cometary source is more difficult to reconcile with the above aspects of the lunar record. Like asteroids, comets make basins, but for a given basin they leave much less iridium behind. For asteroids the Moon's retention of an 8×10^{22} g chondritic veneer corresponds to an energy flux of order 10^{35} ergs; for comets the same amount of iridium corresponds to an energy flux at least an order of magnitude greater. The problem with comets is that the implicit big basins are missing. There should be a half-dozen basins as big as or bigger than South Pole–Aitkin. Mantle samples should be ubiquitous; they're not.

7.5.2   Scaling the Lunar Impact Record to Earth

The larger gravity of Earth attracts more impactors and gives them greater energy than if they had hit the Moon. The energy of an object in solar orbit hitting a planet is

$$E = \frac{m v_\infty^2}{2} \left(1 + \frac{v_{esc}^2}{v_\infty^2} \right), \tag{7-13}$$

where v_{esc} is the escape velocity and v_∞ is the approach velocity far from the planet. Similarly, the effective capture cross-section is

$$A = \pi R_p^2 \left(1 + \frac{v_{esc}^2}{v_\infty^2} \right). \tag{7-14}$$

The escape velocity from Earth is 11.2 km s^{-1} and from the Moon is 2.4 km s^{-1}. The average impact velocity on Earth of the present population of asteroids in Earth-crossing orbits is 16.1 km s^{-1}, weighted by the probability of impact (computed using Öpik's formulas (Shoemaker et al., 1982) from near-Earth asteroids tabulated by Shoemaker et al. (1990)). The average impact velocity on the Moon, also weighted by collision probability, is 13.0 km s^{-1}. The latter is effectively also the average approach velocity to Earth. With $v_\infty = 13$ km s^{-1}, the total impact probability on Earth is 23 times that on the Moon, which corresponds to a relative accretion rate per unit area by Earth 1.7 times that of the Moon. The ratios of impact velocity and impact probability don't scale precisely as predicted by (7-13) and (7-14) because gravitational focusing has its greatest effect on the slowest bodies; thus Earth's cross-sectional advantage over the Moon is greatest for the objects with the lowest encounter velocities. For comets the gravitational enhancements are unimportant.

The second factor that affects scaling the impact record from the Moon to Earth is the value of the power law exponent q. This stems from the rarity of the largest impacts in scale-invariant distributions. For a total impact probability 23 times higher on Earth than on the Moon, the Moon is hit by 4.2% of the impactors and Earth by the remaining 95.8%. Over a large number of trials the Moon will, on average, accrete 4.2% of the total mass, since in a few trials it will get hit by one of the two or three largest bodies. But in half the trials the Moon is not hit by any of the 16 largest objects, since $(0.958)^{16} > 0.5$. In most trials the Moon significantly undersamples the total mass of the impactors. The degree of undersampling is greater the more strongly the total mass of the swarm is concentrated in the largest objects; hence the dependence on q.

From a scale-invariant distribution of potential impactors the cumulative number $N_e(>m)$ of objects greater than mass m striking Earth is

$$N_e(>m) = A_e C m^{1-q}, \tag{7-15}$$

and the cumulative number of objects striking the Moon is

$$N_m(>m) = A_m C m^{1-q}, \tag{7-16}$$

where A_e and A_m are the effective cross-sectional areas of the Earth and Moon as given by Eq. (7-14). The ratio of the largest objects striking the Earth and the Moon is

$$\frac{N_e(>m_e) \approx 1}{N_m(>m_m) \approx 1} = 1 = \frac{A_e m_e^{1-q}}{A_m m_m^{1-q}}. \tag{7-17}$$

Hence the largest object hitting Earth exceeds the largest object hitting the Moon by the factor

$$\frac{m_e}{m_m} = \left(\frac{A_e}{A_m}\right)^{1/(q-1)}. \tag{7-18}$$

Depending on q, this ratio can easily reach 100. If 2×10^{33} ergs is taken as the energy of the Imbrium impact, (7-18) implies that it is quite reasonable for Earth to have suffered a more or less contemporaneous impact releasing enough energy to vaporize the oceans.

In the median case, Earth accretes $23^{1/(q-1)}$ times more material than does the Moon. For $q = 11/6$, the median case is for $\sim$40 times as much mass to hit Earth as hits the Moon. For $q = 1.5$, it becomes $\sim$500 times greater than the mass hitting the moon. The bias disappears as $q \to 2$ because most of the mass is then in a large number of small objects. When this bias is taken into account, the inferred 0.7 km blanket of accreted material on the Moon inflates to 1.2 km on Earth for $q = 2$, 2.2 km ($q = 11/6$), 6 km ($q = 5/3$), or even 30 km ($q = 1.5$). Thus, if q were small, it is possible that a 17 km late chondritic veneer supplying the excess siderophiles in the Earth's mantle could be Earth's expected complement of the same veneer that put some 700 m of chondritic material into the lunar crust (Chyba, 1991). If

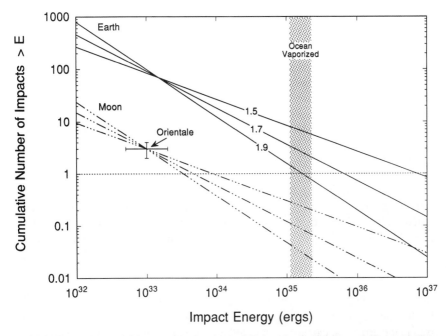

FIGURE 7.3. Terrestrial impacts at the time of Orientale, Imbrium, and Nectaris. This covers roughly the period between 3.8 Ga and 4.1 Ga. The figure shows the number of impacts on Earth with energies greater than a given energy E as scaled from the energies of the lunar Orientale and Imbrium basins. In this scaling a given object is 23 times more likely to strike the Earth than the Moon, and releases about 50% more energy on impact. The different curves are labeled by the power law index q, where the cumulative number of impactors with mass greater than m is given by $N(> m) \propto m^{1-q}$. Distributions with smaller values of q are characterized by markedly more enormous impacts.

so, most of that 17 km would necessarily have been delivered by a single object in a collision releasing $> 10^{37}$ ergs. On the other hand, it is not necessary that Earth accrete its 17 km while the Moon accreted its 700 m, since much or most of this 17 km may have predated the freezing of the lunar crust.

Figure 7.3 illustrates how the lunar basin-forming events about 4 billion years ago scale to the Earth. The time interval is approximately 3.8–4.1 Ga, corresponding to impacts after Nectaris. The point marked "Orientale" ranks Orientale third among lunar basins, the ranking given to it by Wilhelms (1987). For specificity the energy released by the Orientale impact is taken to be $1^{+1}_{-.5} \times 10^{33}$ ergs. Cumulative distributions with different plausible slopes are then drawn through the Orientale point. The error bars around "Orientale" should be understood as broadening each line into a diffuse band.

Parallel to, but offset in number and energy, are the corresponding cumulative distributions for impacts on Earth. The offset in energy is the relatively small factor of 1.5 that results from the Earth's gravity. The offset in number is Earth's higher effective cross-section, here a factor of 23. Marked also is the energy required to vaporize the oceans. Extrapolation to the largest terrestrial impacts illustrates graphically what Eq. (7-18) illustrates algebraically: the chance that one or two impacts of ocean-vaporizing scale occurred on Earth during the late heavy bombardment is considerable. There is a very high probability that during this same interval several impacts occurred that were large enough to vaporize several hundred meters of the top of the oceans; a few of these events would have occurred every hundred million years. The difficulties these would have presented for the continuous habitability of the photic zone are obvious.

The impact history of early Earth is summarized in Fig. 7.4. This figure shows the largest impacts at given times in the Earth's and Moon's history. The curve for Earth is mostly scaled from the lunar record, although Earth has independently recorded some distinctly large events. In addition to the observed basins we have added a larger, hypothetical impact predating South Pole–Aitkin that is implied by $q < 1.8$. The energies of smaller craters were estimated using Schmidt–Housen and Schmidt–Holsapple scalings with crater slumping according to Eq (7-9). The two scalings differ by about a factor of 3 for 100 km lunar craters, with those based on Schmidt–Housen scaling being the smaller. We have extended the error bars symmetrically to span an order of magnitude. The relative error is probably smaller, but the absolute error may be larger—there are many potential sources of systematic error that can affect extrapolation of laboratory cratering relationships to 200 km impact craters.

7.6   Conclusion

The famous K/T impact left a crater that is at least 180 km diameter; judged by its crater one would deduce an energy release between 1×10^{31} ergs and 3×10^{31} ergs. This is comparable to the energy of a small lunar basin, like Schrödinger, for example; nothing like it has hit the Moon in at least 3.5 billion years. This is a considerable amount of energy to release into the biosphere in a short time; a great many bad things could have happened in the aftermath and evidently many did (see Sharpton and Ward (1990) for several lists). We simply note here that ejecta from the K/T impact would have evaporated almost a meter off the oceans and left about a half meter of boiled brine on the sea surface. We also note that oceanic primary productivity was practically wiped out for thousands of years thereafter. The K/T event had many other influences of which we are all no doubt aware. The survivors were selected, naturally, but not in the usual Darwinian sense.

Impacts must have played a much bigger role in the evolution of early life. The lunar cratering record implies that hundreds of objects like those that produced the Imbrium and Orientale basins struck Earth between 3.8 Ga and 4.1 Ga. These events were typically about 30–100 times more energetic than the K/T. Impacts on this scale vaporized the uppermost tens of meters of the ocean. Surface and shallow water ecosystems would very likely have been destroyed. These impacts posed a serious recurrent hazard to life at the Earth's surface. Events on this scale

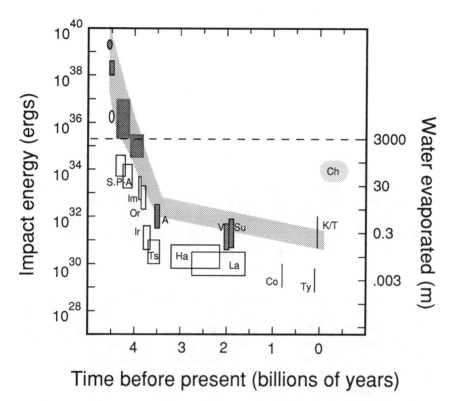

Time before present (billions of years)

FIGURE 7.4. The largest impacts before the present for the Moon and Earth. Open boxes are lunar, shaded boxes are terrestrial. The dimensions of the boxes indicates range of uncertainty. The stipled region shows the inferred largest impacts on Earth. The right-hand axis is the depth of ocean vaporized by the impact. The three earliest events (4.5 Ga) shown are the energies of the Earth and Moon formation and the moon-forming impact. Energies of basin-forming impacts (3.8–4.3 BY) are discussed in the text. The lunar craters are Tycho, Copernicus, Langrenus, Hausen, Tsiolkovskiy, and Iridum; diameters and approximate ages are adapted from Wilhelms (1987). Size and dates of terrestrial impacts at Sudbury and Vredevort are taken from Grieve (1982). The Chicxulub crater (the K/T crater, 65 Ma) is at least 180 km diameter (Hildebrand et al., 1991). The biological crater is consistent with a 10^{31} erg event (Melosh et al., 1990). Chiron remains a potential source of future troubles.

may not have entirely stopped at 3.5 Ga; one or two might be expected on Earth since, with effect dwarfing that of the K/T. Their aftermath is remembered in our genes.

More important were the tens of events during the same time period that vaporized hundreds of meters of seawater. In the aftermath some 30–100 m of hot boiled brine would float atop the oceans; hot fresh rainwater would then float atop the brine. It is this scale of event in particular that most strongly suggests that life on Earth should descend from organisms that lived in the deep ocean, in hydrothermal systems, or were extremely fond of heat and salt. If photosynthesis survived it would have been preserved among the halophiles and thermophiles.

Impacts large enough to fully vaporize the oceans must also have occurred if one looks early enough. Their number is sensitive to the mass distribution of large impactors, as illustrated by Fig. 7-3. The lunar crater record favors $q = 1.6$ (we assume a power law mass distribution of the form $n(m)\,dm \propto m^{-q}\,dm$). This low q distribution may implicate comets. The younger lunar basins also appear to favor a value of q less than 1.7, which again implicates comets. On the other hand, the apparent rarity of basins larger than Imbrium implies that the largest impactors may have obeyed an asteroidal distribution mass distribution with $q > 1.8$. Such a distribution, which originates from fragmenting collisions, is relatively deficient in large bodies. In our opinion ocean vaporizing impacts were rare after 4.4 Ga, with perhaps 2 ± 4 expected between 3.8 Ga and 4.1 Ga, and with a fair likelihood that there were none; the actual number and timing being mostly a matter of chance (or destiny). Such events, if they took place, would have strongly biased survival in favor of organisms living in mid-ocean ridge hydrothermal systems or comparably protected niches.

Molecular phylogeny indicates that "all extant life descends from sulfur-metabolizing bacteria that live at near-boiling temperatures" (Lake, 1988). Impacts on this scale provide an obvious excuse for this detail in our ancestry. The natural story is that such organisms survived because their niche was for a short time the safest one on Earth. Descent from organisms that dwelt at the mid-ocean ridges is not equivalent to life originating at the mid-ocean ridges. Life could have arisen at the surface and colonized the mid-ocean ridges. The long intervals between major events leaves plenty of time for migration. It is simply because deep water hydrothermal systems offered sanctuary from otherwise lethal events that life survived there. The argument is strong enough that it is tempting to invert it: because we can trace our descent to thermophiles and halophiles, we might presume that life originated before the end of the heavy bombardment.

In this chapter we have stressed the dark side of impacts. But not all is dark. Impacts significantly influenced the chemistry of the ocean and atmosphere, influences that remained long after the dust settled and the oceans rained out. Impacting material is, on the whole, more reduced than the crust and mantle of the differentiated Earth. Consequently, impacts tend to reduce the atmosphere and ocean. In particular, Kasting (1990) showed that before about 4 Ga the general background flux of impacting material would have been high enough to make carbon monoxide rather than carbon dioxide the dominant carbon oxide. Despite its poisonous reputation, carbon monoxide would have been more conducive to the orgin of life than CO_2.

Ejecta from impacts probably dominated ordinary sedimentation, and the deposition rate of ejecta in the open ocean was probably greater than present pelagic deposition rates. Marine ejecta, except from the largest impacts, cooled before it reached the seafloor and formed a fine-grained porous rock that was readily weathered, removing CO_2 from the atmosphere (Koster van Groos, 1988). A thick warm CO_2 atmosphere thus may not have existed on the early Earth, but rather an atmosphere with modest CO_2 pressure and temperatures like or below those of the present Earth.

Stochastic events may have been more important. Occasional large iron-rich impactors may have added enough reducing power that a transient methane-ammonia atmosphere could result. The suitability of such an atmosphere for prebiotic chemistry, even if ephemeral, is well known.

Finally, implicit in all these mass extinctions are the new radiations that followed: a biology of revolutionary rather than evolutionary change. One of the great puzzles in evolution is its slow pace over the three-plus billion years between life's origin and the Cambrian explosion. Were impacts the pacemakers of evolution? Did impacts speed the courses of evolution and so give us form?

7.7 Appendix

Oberbeck and Fogleman (1990) equate an ocean-vaporizing impact with a global sterilizing event. They then deduced upper limits for the time available for the origin of life that range from 11 million years at 3.8 Ga to 133 million years at 3.5 Ga by extrapolating back in time and energy the lunar cratering record for 10–100 km craters. Their results differ substantially from those given by Sleep et al. (1989), who estimated that there are no more than a few, and possibly no, sterilizing impacts; or those of Chyba (1991), who estimates about four such events between 3.8 and 4.4 Ga; or the estimate given in this chapter (2 ± 4 ocean-vaporizing events between 3.8 Ga and 4.1 Ga).

Maher and Stevenson (1988), Oberbeck and Fogleman (1989), and Chyba (1991) all fit the impact flux of the late heavy bombardment to an exponential function plus a small constant background that accounts for contemporary events. The functional form is

$$N(>D) = k\left(1 + \exp\left((t - t_0)/\tau\right)\right) D^{-b}. \qquad (7\text{-}19)$$

Chyba's "procrustean fit" to this equation is equivalent to $\tau = 0.144$ Ga, $t_0 = 3.25$ Ga, and $k = 5.6 \times 10^{-13}$ km^{-2} yr^{-1} for D measured in kilometers. Maher and Stevenson (1988), following Grieve (1982), use a contemporary impact flux that is about three times higher; more importantly they also take $\tau = 0.07$ Ga, which produces enormously higher impact rates earlier than 4.0 Ga. Oberbeck and Fogleman (1989) use a modern impact flux that is essentially identical to Chyba's (1991); they consider both $\tau = 150$ and $\tau = 70$.

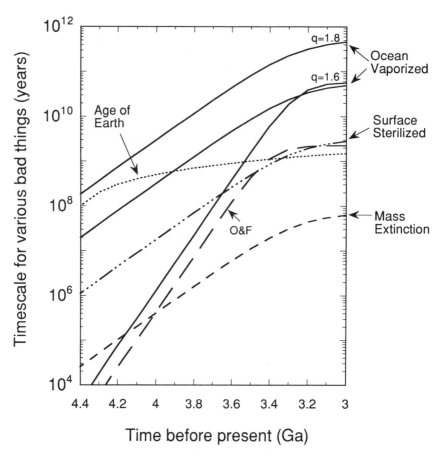

FIGURE 7.5. This figure revises Fig. 2 of Maher and Stevenson (1988). It is an ambitious extrapolation backward in time. The categories are defined in Table 7.1. The curves labeled "Ocean Vaporized" are often taken to refer to global sterilization; we leave their interpretation to the reader. We show two characteristic timescales for the decline of the heavy bombardment, $\tau = 0.144$ Ga (Chyba, 1991) and $\tau = 0.07$ Ga (Maher and Stevenson, 1988). The latter have the more agressive slope. We favor the former. The curves are prepared with $q = 1.6$, except for the uppermost curve, which with $q = 1.8$ describes a considerably safer Earth. Also shown is the corresponding curve calculated according to Oberbeck and Fogleman (1990). This curve, labeled "O&F," equates a much smaller crater with ocean vaporization; it also uses $q = 1.5$. The "Age of the Earth" assumes that the clock started at 4.5 Ga.

All three use $b = 1.8$, a consensus choice for young lunar craters between 10 km and 80 km diameter. We regard $b = 1.8$ as somewhat negotiable; the older consensus on $b = 2$ provides a fit that is at least as good.[3] For specificity we will use Chyba's impact prescription, but with $b = 2$, as a standard model in preparing Table 7.1.

We wish to scale a lunar crater count of $N (> D) = k_m D^{-b}$ km^{-2} to its counterpart $N (> D) = k_e D^{-b}$ km^{-2} on Earth. Maher and Stevenson (1988) use Schmidt and Holsapple's (1982) crater scaling; for consistency with their work we will use the same here. In this scaling the crater diameter goes like

$$D \propto v^{1/3}/g^{1/6}. \tag{7-20}$$

Because of gravitational focusing, a stray body is rather more likely to strike a given square kilometer on Earth than it is a given square kilometer on the Moon; for the same reason it will strike the Earth with a greater velocity. As in Section 7.5 above, we use a lunar impact velocity of $v_m = 13$ km s^{-1}, a terrestrial impact velocity of $v_e = 16$ km s^{-1}, and a gravitional focusing factor of 1.7. For a given impactor, the crater on Earth will be smaller than a crater on the Moon by the factor

$$D_e = D_m (v_e/v_m)^{1/3} (g_m/g_e)^{1/6} \approx 0.77 D_m. \tag{7-21}$$

Hence the number density of craters greater than a diameter D on Earth is

$$N_e(> D) \approx 1.7 (D_e/D_m)^b N_m(> D) \approx 1.06 N_m(> D). \tag{7-22}$$

For the Schmidt–Housen scaling, Eq.(7-8), the corresponding factors are 0.71 (crater diameter) and 0.92 (number). In summary, it appears that the lunar cratering record should be directly comparable to the terrestrial record.

We also need to pick new threshold "crater diameters" for various untoward events. Maher and Stevenson (1988) use thresholds of 55 km, 265 km, and 850 km diameter craters for global climate effects, surface sterilization, and global sterilization, respectively. These thresholds correspond to energies of 2×10^{30} ergs, 5×10^{32} ergs, and 4×10^{34} ergs, respectively, as determined using the formulas used by Maher and Stevenson (1988). The first two are reasonable guesses for the effects sought. For "global sterilization" we use an energy, 1.6×10^{35} ergs, that is marginally sufficient to vaporize the oceans; this is essentially the same choice made by Oberbeck and Fogleman (1990). With allowance for crater slumping, different gravities and impact velocities on the Earth and the Moon, and an average incidence angle of 45°, the three threshold energies would correspond to lunar craters of about 140 km, 900 km, and 4000 km. These diameters assume the scaling in Eqs (7-8) and (7-9). We also show the corresponding curve with $D = 1611$ km as

---

[3] The actual data for the late bombardment, shown by Wilhelms (1987) in his Figure 8.6 (pre-Nectaris) and Figure 9.22 (post-Nectaris) for this size range, are puzzling. The older surfaces are quite consistent with slope $b = 3$, but the younger surfaces fit $b = 2$. This is opposite to what one would expect of erosion and crater saturation (to which smaller craters are more vulnerable) and hints at real changes in b over time.

TABLE 7.1. Three Arbitrary Thresholds.

Effect	Energy (ergs)	Equivalent lunar crater (km)	"Last" Event (Ga)	Comments
Mass extinction				Less than K/T
	2×10^{30}	55	0.01	M&S; O&F
	2×10^{30}	144	0.1	
Photic zone				Imbrium-scale
	5×10^{32}	265	0.1	M&S; O&F
	1×10^{33}	900	3.4	
Global sterilization?				Ocean vaporized
	3×10^{34}	850	1.0	M&S; O&F
	2×10^{35}	1611	3.4	O&F
	1.6×10^{35}	4000	3.5	$\tau = 0.7$ Ga
	1.6×10^{35}	4000	3.9	
	1.6×10^{35}	4000	>4.3	$q = 1.8$

the threshold for global vaporization, the threshold used by Oberbeck and Fogleman (1990). The various thresholds are listed in Table 7.1. We assume $q = 1.6$ and $\tau = 0.144$ Ga unless stated otherwise in the table. The difference between our results and Oberbeck and Fogleman's appears to be mostly the result of different crater diameter-energy scaling.

7.8   References

Abe, Y. (1988), Conditions required for sustaining a surface magma ocean, *Proc. 21st ISAS Lun. Planet. Symp.*, 225–231 (1988).

Abe, Y. and Matsui, T. (1988), Evolution of an impact-generated H_2O-CO_2 atmosphere and formation of a hot proto-ocean on Earth, *J. Atm. Sci.* **45**, 3081–3101.

Alvarez, L.W., Alvarez, W., Asaro, F., and Michel, H.V. (1980), Extraterrestrial cause for the Cretaceous–Tertiary extinction, *Science* **208**, 1095–1108.

Anders E. (1989), Prebiotic organic matter from comets and asteroids, *Nature* **342**, 255–257.

Baldwin, R.B. (1981), On the origin of the planetesimals that produced the milti-ringed basins. In Schultz, P. and Merrill, R.B. (eds.) *Multi-ring Basins, Proc. Lun. Planet. Sci.* **12A**, Pergamon Press, New York, 19–28.

Baldwin, R.B. (1987a), On the relative and absolute ages of seven lunar front face basins I, *Icarus* **71**, 1–18.

Baldwin, R.B. (1987b), On the relative and absolute ages of seven lunar front face basins II, *Icarus* **71**, 19–29.

Belton, M.J.S., Head, J.W., Pieters, C.M., Greeley, R., McEwen, A.S., Neukem, G., Klaasen,

K. P., Anger, C.D., Carr, M.H., Chapman, C.R., Davies, M.E., Fanale, F.P., Gierasch, P.J. Greenberg, R., Ingersoll, A.P., Johnson, T., Paczkowski, B., Pilcher, C.B., and Veverka, J. (1992), Lunar impact basins and crustal heterogeneity: New western limb and far side data from Galileo, *Science* **255**, 570–576.

Bratt, S.R., Solomon, S.C., and Head, J.W. (1985a), The evolution of impact basins: Cooling, subsidence, and thermal stress, *J. Geophys. Res.* **90**, 12415–12433.

Bratt, S.R., Solomon, S.C., Head, J.W. and Thuber, C.H. (1985b), The deep structure of lunar basins: Implications for basin formation and modification, *J. Geophys. Res.* **90**, 3049–3064.

Carlson, R.W. and Lugmair, G.W. (1979), *Earth Planet Sci. Lett.* **45**, 123–132.

Carlson, R.W. and Lugmair, G.W. (1988), *Earth Planet Sci. Lett.* **90**, 119–130.

Chapman, C.R., Williams, J.G., and Hartmann, W.K. (1978), The Asteroids, *Ann Rev. Astron. Astrophys.* **16**, 33–75.

Chyba, C. (1991), Terrestrial mantle siderophiles and the lunar impact record, *Icarus* **92**, 217–233.

Connan, J. (1984), *Adv. Petroleum Geochem.* **1**, 299–335.

Davis, P.A. and Spudis, P. (1987), *J. Geophys. Res.* **92**, E387–E395.

Drake, M.J. (1986), Is lunar bulk material similar to Earth's mantle? In W.K. Hartmann, R.J. Phillips, and G.J. Taylor (eds.), *Origin of the Moon*, (Lunar and Planetary Institute, Houston), pp. 105–143.

Donnison, J.R. (1986), The distribution of cometary magnetudes, *Astron. Astrophys.* **167**, 359–363.

Donnison, J.R. and Sugden, R.A., The distribution of asteroidal diameters (1984), *Mon. Not. Roy. Astron. Soc.* **210**, 673–682.

Dohnanyi, J.S. (1972), Interplanetary objects in review: statistics of their masses and dynamics, *Icarus* **17**, 1–48.

Duncan, M., Quinn, T., and S. Tremaine (1987), The formation and extent of the solar system comet cloud, *Astron. J.* **94**, 1330–1338.

Duncan, M., Quinn, T., and S. Tremaine (1988), The origin of short period comets, *Astrophys. J.* **328**, L69–L73.

Grieve, R.A.F. (1982), The record of impact on Earth: Implications for a major Cretaceous/Tertiary impact event. In Silver, L.T., and Schultz, P.H., eds., *Geological Implications of Impacts of Large Asteroids and Comets on the Earth*, Geological Society of America Special Paper 190, 25–37.

Hahn, G. and Bailey, M.E. (1990), Rapid dynamical evolution of giant comet Chiron, *Nature* **348**, 132–136.

Hartmann, W.K. (1981). In *Proceedings of the Conference on the Lunar Highlands Crust*, 155-173 (Pergamon Press, New York).

Hartmann, W.K., Phillips R.J. and Taylor, G.J., eds. (1986), *Origin of the Moon*, Lunar and Planetary Institute, Houston.

Hartmann, W.K., D.J. Tholen, K.J. Meech, and D.P. Criukshank (1990), 2060 Chiron: Colorimetry and possible cometary behavior, *Icarus* **83**, 1–15.

Hildebrand, A.R., Penfield, G.T., King, D.A., Pilkington, M., Camargo Z.,A., Jacobsen, S.B. and Boynton, W.V. (1991), Chicxulub crater: a possible Cretaceous/Tertiary boundary impact crater on the Yucatan Peninsula, Mexico, *Geology* **19**, 867–871.

Hughes, D.W. (1982), Asteroidal size distribution, *Mon. Not. Roy. Astron. Soc.* **199**, 1149–1157.

Hughes D.W. (1988), Cometary distribution and the ratio between the numbers of long- and short-period comets, *Icarus***73**, 149–162.

Kasting, J.F. (1988), Runaway and moist greenhouse atmospheres and the evolution of Earth and Venus, *Icarus* **74**, 472–494.

Kasting, J.F. (1990), Bolide impacts and the oxidation state of carbon in the Earth's early atmosphere, *Orig. Life* **20**, 199–231.

Korotev, R.L. (1987), *J. Geophys. Res.* **92**, E447–E461.

Koster van Groos, A.F. (1988), Weathering, the carbon cycle, and the differentiation of the continental crust and mantle, *J. Geophys. Res.* **93**, 8952–58.

Lake, J.A. (1988), Origin of the eukaryotic nucleus determined by rate-invariant analysis of rRNA sequences, *Nature* **331**, 184–186.

Lebofsky, L.A., Tholen, D.J., Rieke, G.H., and Lebofsky, M.J. (1984), 2060 Chiron: Visual and thermal infrared observations, *Icarus* **60**, 532–537.

Lindstrom, M.M. and Lindstrom, D.J. (1986), *J. Geophys. Res.* **91**, D263–D276.

Maher, K.A. and Stevenson, D.J. (1988), Impact Frustration of the origin of life, *Nature* **331**, 612–614.

Marvin, U.B., Carey, J.W., and Lindstrom, M.M. (1989), *Science* **243**, 925–931.

McKinnon, W.B., Chapman, C.R., and Housen, K.R. (1991), Cratering of the Uranian satellites. In J.T. Bergstrahl, E.D. Miner, and M.S. Matthews (eds.), *Uranus* (University of Arizona Press, Tucson), pp. 1177–1252.

Melosh, H.J. (1989), *Impact Cratering: A Geological Process*, Oxford University Press, New York.

Melosh, H.J. (1990), Giant impacts and the thermal state of the Earth. In Newsom, H.E. and Jones, J.H., eds., *Origin of the Earth*, Oxford University Press, pp. 69–84.

Melosh, H.J., and Vickery, A.M. (1989), Impact erosion of the primitive atmosphere of Mars, *Nature* **338**, 487–490.

Melosh, H.J., Schneider, N., Zahnle, K., and Latham, D. (1990), Ignition of global wildfires at the Cretaceous/Tertiary boundary, *Nature* **343**, 251–254.

Mysen, B.O. and Kushiro, I. (1988), Condensation, evaporation, melting, and crystallization in the primitive solar nebula, *Am. Min.* **73**, 1–19.

Nakajima, S., Hayashi, Y.-Y., and Abe, Y. (1992), A study of the "runaway greenhouse effect" with a one-dimensional radiative-convective equilibrium model, *J. Atm. Sci.* **49**, 2256–2266.

Newsom, H.E. and Jones, J.H. (1990), *Origin of the Earth*, Oxford University Press.

Newsom, H.E. and Taylor, S.R. (1989), Geochemical implications of the formation of the Moon by a single giant impact *Nature* **338**, 29–34.

Oberbeck, V. and Fogleman, G. (1989). Impacts and the origin of life, *Nature* **339**, 434.

Oberbeck, V. and Fogleman, G (1990). Estimates of the maximum time required for the origin of life, *Orig. of Life* **340**.

Oikawa S. and Everhart, E., Past and future orbit of 1977 UB, object Chiron (1979), *Astron. J.* **84**, 134–139.

Olsson-Steel, D. (1987) Collisions in the solar system. IV. Cometary impacts upon the planets *Mon. Not. Roy. Astron. Soc.* **227**, 501–524.

Pace, N., Olsen, G.J., and Woese, C.R. (1986), Ribosomal RNA phylogeny and the primary lines of evolutionary descent *Cell* **45**, 325–326.

Pieters, C.M. (1986), Composition of the lunar highland crust from near-infrared spectroscopy. *Rev. Geophys.* **24**, 557–578.

Ringwood, A.E. and Seifert, S. (1986) *in* Hartmann, W.K., Phillips R.J. and Taylor, G.J., eds., *Origin of the Moon*, Lunar and Planetary Institute, Houston, 331–358.

Rivera, M.C. and Lake, J.A. (1992), Evidence that eukaryotes and eocyte prokaryotes are immediate relatives, *Science* **257**, 74–76.

Safronov, V.S. (1972), *Evolution of the Protoplanetary Cloud and Formation of the Earth and the Planets*. NASA TT F-677.

Safronov, V.S., G.V. Pechernikova, E.I. Ruskol, and A.V. Vitjazev (1986), Protosatellite swarms. In Burns, J. and Matthews, M.S., eds. *Satellites*, The University of Arizona Press, Tucson, pp. 89–116.

Schmidt, R.M., and Holsapple, K.A. (1982), Estimates of crater size for large body impact. In Silver, L.T., and Schultz, P.H., eds., *Geological Implications of Impacts of Large Asteroids and Comets on the Earth*, Geological Society of America Special Paper 190, 93–102.

Schmidt, R.M., and Housen, K.R. (1987), Some recent advances in the scaling of impact and explosion cratering, *Int. J. Impact Mech.* **5**, 543–560.

Scholl, H. (1979), History and evolution of Chiron's orbit, *Icarus* **40**, 345–349.

Sharpton, V., and Ward, P., eds. (1990), *Global Catastrophes in Earth History*, Geological Society of America Special Paper 247.

Shoemaker, E.M., R.F. Wolfe, and C.S. Shoemaker (1982), Cratering timescales for the Galilean satellites. In Morrison, D., ed., *Satellites of Jupiter*, The University of Arizona Press, Tucson, pp. 277–339.

Shoemaker, E.M., Wolfe, R.F., and Shoemaker, C.S. (1990), Asteroid and comet flux in the neighborhood of Earth. In V.L. Sharpton and P.D. Ward, eds., *Global Catastrophes in Earth History*. Geol. Soc. of Am. Special Paper 247, pp. 155–180.

Silver, L.T., and Schultz, P.H., eds. (1982), *Geological Implications of Impacts of Large Asteroids and Comets on the Earth*. Geological Society of America Special Paper 190.

Sleep, N.S., Zahnle. K., Kasting, J.F., and Morowitz, H. (1989), Annihilation of ecosystems by large asteroid impacts on the early Earth, *Nature* **342**, 139–142.

Spudis, P.D. (1993), *The Geology of Multi-Ring Impact Basins*, Cambridge University Press.

Spudis, P.D. and Davis, P.A. (1986) *J. Geophys. Res.* **91**, E84–E90.

Spudis, P.D., Hawke, B.R., and Lucey, P.G. (1988), *Proc. Lunar Planet. Sci. Conf.* **18**, 155–168.

Stevenson, D.J. (1987), Origin of the Moon – the Collision Hypothesis, *Ann. Rev. Earth Planet. Sci.* **15**, 271–315.

Suits, G.W. (1979), Natural Sources, in Wolfe, W. and Zissis, G., *The Infrared Handbook*, Office of Naval Research, Washington, DC, 3-1–3-154.

Swindle, T.D., Caffee, M.W., Hohenberg, C.M., and Taylor, S.R. (1986), I-Pu-Xe dating and the relative ages of the Earth and Moon. In W.K. Hartmann, R.J. Phillips, and G.J. Taylor (eds.), *Origin of the Moon*, (Lunar and Planetary Institute, Houston), pp. 331–358

Swisher, C., Grajales-Nishimura, J., Montanari, A., Margolis, S., Claeys, P., Alvarez, W., Renne, P., Cedillo-Pardo, E., Maurrasse, F., Curtis, G., Smit, J., and McWilliams, M. (1992), Coeval ^{40}Ar/^{39}Ar ages of 65.0 million years ago from Chicxulub crater melt rock and Cretaceous-Tertiary boundary tektites, *Science* **257**, 954–958.

Taylor, S.R. (1986), *Planetary Science: A Lunar Perspective*, Lunar and Planetary Institute, Houston.

Tera, F., Papanastassiou, D.A., and Wasserburg, G.J. (1974), Isotopic evidence for a terminal lunar cataclysm, *Earth Planet. Sci. Lett.* **22**, 1–21.

Tremaine, S. and Dones, L. (1993), On the statistical distribution of massive impactors, *Icarus* **106**, 335–341.

Turcotte, D.L. (1992), *Fractals and Chaos in Geology and Geophysics*, Oxford University Press.

Turcotte, D.L., and Schubert G. (1982), *Geodynamics*, Wiley.

Warren, P.H., Jerde, E.A., and Kallemeyn, G.W. (1989), *Earth Planet. Sci. Lett.* **91**, 245–260.

Wetherill, G.W. (1975), Late heavy bombardment of the moon and terrestrial planets, *Proc. Lunar Sci. Conf.* **6**, 1539–1561.

Wetherill, G.W. (1981), Nature and origin of basin-forming projectiles. In Schultz, P. and Merrill, R.B., *Multi-ring Basins, Proc. Lun. Planet. Sci. 12A*, Pergamon Press, New York, 1–18.

Wilhelms, D.E. (1987), *The Geologic History of the Moon*, U.S.G.S. Professional Paper 1348.

Vickery, A.M., and Melosh, H.J. (1990), Atmospheric erosion and impactor retention in large impacts, with application to mass extinctions. In Sharpton, V.L., and Ward, P.D., eds., *Global Catastrophes in Earth History*, Geological Society of America Special Paper 247, 289–300.

Zahnle, K. (1990), Atmospheric chemistry by large impacts. In Sharpton, V.L., and Ward, P.D., eds., *Global Catastrophes in Earth History*, Geological Society of America Special Paper 247, 271–288.

Zel'dovich, Ia.B., and Raizer, Yu.P. (1967), *Physics of Shock Waves and High Temperature Hydrodynamic Phenomena*, Academic.

8

Cometary Impacts on the Biosphere

D. Steel

ABSTRACT   There is now a wealth of evidence to link impacts by comets and asteroids with catastrophic disruptions of the biosphere and mass extinctions. Such evidence includes impact craters formed close to boundary events, iridium and other siderophile element/isotope anomalies, microtektites, and shocked quartz. Mass extinctions, crater excavations, and various geologic upheavals all seem to follow a common periodicity of ~30 Myr. Another cycle of ~250 Myr supports a link with the solar motion about the galaxy, this being its orbital period, whilst 26 Myr–32 Myr is its half-period for oscillations about the galactic disk. Oort cloud disturbances and therefore comet showers produced as the solar system passes through the disk seem the most likely explanation for the observed phenomena, although there is currently a problem with understanding the phase of the showers. This may be reconciled if the cyclicity is, in fact, ~15 Myr. An alternative (which does not explain the 250 Myr cycle) is the solar companion star or "Nemesis" theory, invoking Oort cloud disturbances when this hypothetical star passes perihelion every 30 Myr. The fact that the biosphere may be significantly affected by comets through means other than actual impacts (e.g., dust veiling of the atmosphere) is emphasized, such considerations leading to an understanding of the structure of boundary events through prolonged periods of influx to the Earth of meteoroids and dust. The possible contemporary modulation of the biosphere by cometary decay products is mentioned. Finally, the role of large impacts in panspermia—the spreading of life from one planet to another—is briefly discussed.

8.1   Introduction

This chapter is concerned with an interdisciplinary area of science, in which researchers from a wide variety of disciplines—geology, geophysics, geochemistry, biology, botany, evolutionary studies, paleontology, physics, astronomy, atmospheric physics and chemistry, to list but a few—have been involved over the past 15 years since the publication of the famous paper by Alvarez et al. (1980). That is not to say that the field is only 15 years old, as will be pointed out below, but many major advances have been made over that time. The field is starting to attain some maturity, and confidence is growing that in broad sweep the ways in which comets may affect the biosphere are understood, even if the details are lacking.

One of the reasons that a list of the scientific areas involved was given at the start of this section was to accentuate the fact that any review of this subject is bound to be focused upon the parts of the general topic that are within the range of the writer's specific background. This chapter has a predominantly astronomical perspective, but also a sufficiently wide-ranging view so as to contain pertinent material for everyone concerned with this topic. Nevertheless, since the main focus is astronomical, some specific discussion of the astronomical background is given (in particular, in Section 8.2), whereas the reader wanting details at a similar level of the geological or paleontological context will need to look elsewhere, suitable references being given. In this connection it is noted that recent dedicated conferences on global catastrophes—impact-induced or otherwise—have led to much-needed cross-disciplinary discussions, the constraints applied by disparate areas of science being recognized (e.g., Silver and Schultz, 1982; Sharpton and Ward, 1990; Johnson, 1993). Alvarez (1990) has discussed how the need for interdisciplinary studies has led to much progress, although many maintain an isolationist stance. Apart from the isolationists there are many who oppose the idea of a link between impacts and mass extinctions, favoring other mechanisms (e.g., Officer et al., 1987; Courtillot et al., 1988; Crowley and North, 1988, 1991). However, given the many correlations between impacts, geological boundary events and mass extinctions discussed in this chapter, and their common periodicities, it is difficult to see how a denial of the impacts at least being the triggers of global catastrophes can be maintained. For general discussions of mass extinctions, see Jablonski and Raup (1986), Elliott (1986), Raup (1986a), Stanley (1987), or Donovan (1989).

Mainstream geology was founded upon the uniformitarian principles developed in the late eighteenth century and the first half of the nineteenth, predominantly by British geologists, at the expense of the French school of Georges Cuvier, who had a catastrophist outlook. The crux of the uniformitarian argument was that the face of the planet has been shaped solely by processes which are seen in action in the present epoch, and it is the long-term action of these that have resulted in the features that we witness now. These ideas are now outmoded. For a discussion of what modern geological uniformitarianism correctly entails, see Shea (1982); for a discussion of how the old concepts of uniformitarianism (what Gould (1965) terms "substantive uniformitarianism") have limited the progress of geology, see Alvarez et al. (1989); for a defence of the geological profession from the accusation that adherence to the old-style uniformitarian philosophy has limited its progress, see Hallam (1989).

A biological corollary to that old concept of uniformitarianism or gradualism is Darwinian evolutionary theory, whereby gradual adaptation of species leads to the fittest surviving. The catastrophist philosophy is to the contrary: in geology, catastrophism says that extraordinary (indeed, often extraterrestrial) events are dominant, with the large impacts upon the Earth playing a major role in its history. In biology, instead of evolutionary change occurring through gradual alterations over very long time periods, many adaptation and speciation events are viewed as occurring rapidly: this is what has been called "punctuated equilibrium" (Gould and Eldridge, 1993), although the scenario whereby the punctuations are largely due to impacts has been alternatively termed "punctured equilibrium". Indeed, the understanding developed whereby mass extinctions occur due to massive impacts (or related events brought on by comets, such as dust veiling of the atmosphere causing rapid climatic deterioration) does not allow for gradual adaptational change in the flora or fauna, with previous evolutionary pressures being of no great consequence with regard to whether certain species or families survive or not, since the extremal conditions suddenly imposed have not previously been met. That is, a species succumbs in a mass extinction due to bad luck rather than bad genes (Raup, 1991), and a corollary of that idea is that survival is due to good luck in that a species has been equipped with the features necessary for survival by accident rather than by evolutionary pressure. This is an extreme (but perhaps fundamentally correct) application of the concept of catastrophism to evolutionary biology.

It was mentioned above that, whilst there has been an explosion in work in this area over the past 15 years, that time-span is certainly not the full extent of the history of research on the role of impacts in shaping the face of the Earth, the influence of such impacts upon geological epochs, and the connection between impacts and the evolution of life. When Edmund Halley recognized that the comet that bears his name would return periodically, determining its heliocentric orbit and earning himself perpetual fame, it also occurred to him that it was possible that the comet in question, or bodies like it, could and would strike the Earth from time to time with devastating effect. Halley gave a talk to the Royal Society of London on the topic in December 1694, suggesting that various large structures on the Earth, such as the Caspian Sea, had been formed by cometary impacts (Jones, 1988). Halley's suggestion was soon followed by William Whiston, in 1696 (Marvin, 1990). The possibility of catastrophic impact by comets resurfaced from time to time before the modern era, for example, in the writings of Thomas Wright of Durham (in 1755), and Laplace (in 1816); for background information see Clube and Napier (1986, 1990) and Bailey et al. (1990).

If the above suggestions of cometary impacts causing calamitous terrestrial upheavals are to be taken seriously, one must ask whether the likelihood of their occurring is significant. Even today only about 30 comets are known which repetitively come back on periodic orbits, and might strike the Earth on any one; P/Halley is an example, other well-known comets in this category being P/Encke and P/Swift-Tuttle. In addition mankind has observed about 450 comets which have Earth-crossing orbits but do not return frequently to the inner solar system. It is straightforward to calculate proper collision probabilities with the Earth for each of these, but for present purposes it is sufficient to note that the cross-sectional area of the Earth is about one part in two billion of the area of a sphere of radius 1 AU, so that for a randomly oriented comet path which crosses that sphere twice in each orbit the Earth-collision probability is of order one in a billion; in fact, it is higher than this by a factor of a few since some parts of that sphere cannot be crossed by most comets. Even if the $\sim$10 Earth-crossing comets spotted each year comprised only 10% of the total population, this would still imply that comet impacts occur on a once-in-ten-million-years timescale, to first order, dependent upon the size of impactor in question. Whilst such infrequent impacts might be significant with regard to geological epoch changes, they could hardly explain the density of craters on the Moon, nor be of immediate concern with regard to the future of mankind. (Having written that, the reader may note that the calculations were done from a uniformitarian perspective in that it was assumed that present-day observations of the cometary population are indicative of the long-term population. As will be seen later, this seems to be an incorrect assumption. It may also be noted that most evaluations of the contemporary hazard to the Earth/mankind are based on an assumption that the long-term impact rate is a determinant of the present rate; e.g., Chapman and Morrison 1994.)

However, a big change occurred in the 1930s with regard to the state of astronomical knowledge of possible Earth impactors. The first asteroid (or minor planet) had been discovered on the first day of the nineteenth century, and from then through to 1932 many more had been charted in roughly circular orbits between Mars and Jupiter. Since these did not have orbits coming close to our planet, they were of no concern with regard to terrestrial impacts. Whilst the origin of meteorites continued to be a subject of debate (e.g., see Marvin, 1990), the idea of huge meteorites—in fact, asteroid impacts—was not considered until a few such bodies on Earth-crossing orbits were discovered in the 1930s. Since these are very difficult to find, being much darker than comets which are relatively bright due to their extensive comae and tails, these discoveries implied that there were many such asteroids which could impact our planet. Watson (1941) and Baldwin (1949) recognized that not only did impacts by such objects provide an explanation for the pock-marked face of the Moon, but also the consequences of a latter-day impact would be horrific for the human race. Despite this, however, it was not until after the first lunar samples were returned that a majority of geologists accepted the implications of the discovery of Earth-crossing (Apollo-type) asteroids with regard to lunar cratering.

In the case of the Earth, acceptance of impact cratering has been even longer coming. Earlier this century a major dispute occurred concerning the origin of the best-known terrestrial impact crater, Meteor or Barringer Crater in Arizona, with the Odessa crater in Texas being the first to be accepted as having an impact origin due to the association with it of indisputable meteoritic material (Hoyt, 1987). One complicating factor was that Earth science in general was in the throes of taking on board the concept of plate tectonics during the 1960s and into the 1970s (Marvin, 1990); how could the significance of the results of lunar studies be appreciated whilst the community was preoccupied with continental drift? Present figures show that an impact by an asteroid at least 1 km in size, producing a crater 10 km–20 km in diameter, occurs about once every $(1 - 3) \times 10^5$ yr on the Earth (Olsson-Steel, 1987; Chapman and Morrison, 1994; Gehrels, 1994), so that impact cratering must be considered as being a major factor in shaping the features of our planet, as it is for all solid bodies in the solar system.

The reason for spelling out this change in perception of the importance of impacts in geology, which change is continuing, is in order to give a context for the resistance encountered by the impact hypothesis when it is invoked as a factor affecting the biosphere. Some account of the earlier suggestions of asteroidal (as opposed to cometary) impacts and their effect upon the biosphere would be useful, so as to dispel the notion that the idea was invented by Alvarez et al. (1980). In 1942 the American meteoriticist Harvey Nininger published a short account of the implications of the discovery of three Apollo-type asteroids (1862 Apollo, 2101 Adonis and 1937 UB Hermes) in the previous decade. In a far-sighted piece which went beyond a mere description of what these discoveries implied for the origin of lunar and terrestrial craters, Nininger pointed out that Earth scientists had to then been unable to explain the various phenomena which occurred at the boundaries between geological epochs, a plausible explanation coming from the catastrophes that would be wreaked by such energetic impacts. Further, he noted that evolutionary biologists had been unable to explain the apparent rapid changes in flora and fauna close to geological boundaries, and the subsequent flourishing of new species, for which impacts would also provide an answer. Nininger mentioned the widespread volcanism, rises and falls of sea level, continental flooding, atmospheric and climatic changes, which would be expected to occur when large impacts occurred, and which are evidenced in the terrestrial record. Nininger also correctly pointed out the Earth would be expected to be depleted in crater density compared to the Moon since erosion would have removed the obvious evidence aeons ago.

The first person to specifically suggest that the dinosaur extinction at the Cretaceous/Tertiary (K/T) boundary 65 million years ago was due to an asteroid impact appears to have been de Laubenfels (1956). He considered the Tunguska event, an $\sim$10 megaton explosion which occurred over Siberia in 1908 (Hills and Goda, 1993; Chyba et al., 1993), and suggested that since far larger impacts had occurred on Earth over geologic time, they should be examined as possible mass extinction mechanisms.

Öpik (1958) considered the general case of massive asteroid impacts causing geological boundary events, and is often erroneously quoted as being the originator of the idea; Nininger was clearly ahead of him. Urey (1973) seems to have been the first to have introduced physical evidence to link impacts with geological boundaries, showing that tektites from certain strewn fields have ages which agree (to within measurement uncertainties) with the boundaries of the last four major geological periods.

It is in the above context that Alvarez et al. (1980) suggested an impact by an $\sim$10 km asteroid as the root cause of the K/T boundary extinction. Alvarez et al. identified an anomalous amount of iridium at the K/T boundary, and proposed that this was delivered to Earth in an asteroid impact, being spread globally as a consequence of the huge explosion; iridium is rare in the Earth's crust, but more common in many meteorites (Kyte and Wasson, 1986; but see Courtillot, 1990). Although some Earth scientists and paleontologists still adhere to a gradualist/uniformitarian explanation for changing geological epochs and mass extinctions (e.g., see Alvarez et al., 1989; Marvin, 1990), it now seems clear that such impacts by asteroids and comets have played a very significant role in not only the geological history of this planet, but also its biological history. The evidence for the major influence of such impacts and related phenomena is discussed in this chapter.

One of the paleontological problems in linking mass extinctions to impact events is the apparent selectivity of the fauna which died out. For example, the well-known division in surviving land vertebrates at body masses of 25 kg–30 kg. This

TABLE 8.1. Environmental Stresses Produced by Large Impacts such as That at the K/T Boundary (after Gilmour et al. 1989).

Stress induced	Time-scale
Darkness: loss of photosynthesis	Months
Cold: impact winter	Months
Winds: 500 km hr^{-1} plus	Hours
Tsunamis	Hours/Days
Greenhouse: H_2O	Months
Greenhouse: CO_2	10^4–10^5 yr ([†])
Fires	Months
Pyrotoxins	Years
Acid rain	Years
Ozone layer destruction	Decades
Volcanism triggered by impacts	Millennia
Mutagens	Millennia

[†] From O'Keefe and Ahrens, 1989.

selectivity may be explained by the large number of environmental stresses applied by the impacts and their aftermath (Table 8.1). There is a long history of studies aimed at understanding the causes of the selectivity (e.g. Cowles, 1939; Emiliani et al., 1981; Rhodes and Thayer, 1991). In a wider context the response of the lifeforms of Earth to global changes is discussed in Walliser (1986), but what is critical for any forward progress is that the gross environmental changes produced in a large impact be understood (e.g., the formation of "Strangelove" oceans (Hsü and McKenzie, 1990) and global cooling in an "impact winter" (Covey et al., 1990); see also Toon et al., 1994), and then factored into studies of survival selectivity (Sheehan and Russell, 1994; Smit, 1994).

8.2   Astronomical Considerations

8.2.1   The Nature of the Impactors

Since this volume is concerned with comets as opposed to asteroids, it is appropriate to note here that recent astronomical work has led to a blurring in the distinction drawn between these two types of object. There is growing evidence that a sizeable fraction of objects termed "asteroids," especially those on Earth-crossing orbits, are in fact extinct or moribund comets (e.g. Wetherill, 1991). Throughout this chapter it will be assumed that there is no fundamental difference between these classifications of objects, the distinction being etymological rather than physical. There are, however, certainly differences between the impact characteristics of comets and asteroids. For example, the former have most-likely impact speeds of order 55 km s^{-1}, whereas the latter are closer to 15–20 km s^{-1}, but this merely represents the fact that it is the low-inclination comets which are more likely to be captured into short-period orbits, maybe producing Apollos when they are devolatilized. Characteristic speeds, compositions, and densities are important, however, in determining the effects of particular impacts. Modeling has been done by several researchers, with quite different results being obtained. For example, Wetherill (1989) finds that only about one-third of terrestrial craters larger than 10 km in size being formed in the present epoch may be due to classical asteroids (i.e. bodies from the main belt), whereas Bailey (1991) finds that the majority of such craters are produced by asteroids. As a rule of thumb these are formed by ~1 km impactors (Schmidt and Holsapple, 1982). However, when Wetherill includes extinct comets as plausible Apollo asteroids, he finds that it may be such asteroids that produce the majority of craters, with only ~30% being due to active comets. Olsson-Steel (1987) found an even smaller fraction of large craters would be expected to be due to comets, either short- or long-period. Other discussions of the cratering influx include those by Shoemaker (1983) and by Weissman (1990).

The nature of the comet nucleus will control: (i) models of the production of asteroids from comets after volatile depletion, since the silicate/volatile ratio will govern the remnant mass; and (ii) for small comets, whether the nucleus penetrates through the atmosphere, since low-density projectiles are much more susceptible to atmospheric disruption. The data from the spacecraft that flew by P/Halley indicated that at least that comet has large volatile component with a significant fraction of organics (Jessberger et al., 1988), and extending by inference to other comets we might believe a "generic comet" to be composed of something like 60–70% water ice, ~10% other inorganic volatiles, ~10% organics, and ~10% silicates, although estimates for the silicate/volatile (or dust/gas) ratios of comets have ranged from 0.1–3.0, and more (i.e., the silicate fraction could be over 90%, especially for a largely devolatilized comet). The dominant view of the nature of the cometary nucleus remains that of Whipple (1950), an icy conglomerate model of a "dirty snowball"; a variant upon this is the rubble pile idea of Weissman (1986). However, Sykes and Walker (1992a) point out that the IRAS dust trails (not *tails*) found in the wakes of comets (Sykes and Walker, 1992b) are indicative of a refractory-to-volatile ratio of the order of 3, leading to the suggestion that cometary nuclei are more like "icy mudballs" than "dirty snowballs".

The largest Apollo asteroid observed in the present epoch is about 8 km in size, whilst the periodic comets P/Halley and P/Swift-Tuttle are 10 km–20 km in dimension. Occasionally much larger comets of order 100 km or more are seen, for example, the great comet of 1729 (Comat Sarabat). There are several outer solar system objects known which are thought to be cometary in nature. One example is the ~250 km object known as 2060 Chiron. This was originally classified as an asteroid, but it has later shown cometary activity; it is on an unstable orbit which may lead to an Earth-crossing path on a timescale of ~1 Myr (Hahn and Bailey, 1990). There are several similar objects known, such as 5145 Pholus and 1993 HA_2, as well as (at the time of writing) seven trans-Neptunian asteroids/comets which may be large (100 km–300 km) Kuiper belt objects on unstable orbits. The influence of such objects upon the terrestrial environment is discussed by Bailey et al. (1994).

The very largest impact events are therefore to be expected to be due to comets, because of their sizes and impact speeds. If asteroidal densities are similar to those of meteorites (i.e., ~2.5–3.0 g cm^{-3} for chondritic meteorites, ~8 g cm^{-3} for nickel-irons), then their masses for similar sizes will be higher, since cometary densities appear to be of the order of 0.5 g cm^{-3} (Rickman et al., 1987), with the density of even P/Halley being uncertain by a factor of 3–4 (Sagdeev et al., 1988; Peale, 1989; Rickman, 1989). Despite this the observed populations of objects would indicate that the very highest energy events (10 km plus impactors, > 10– 100 million megatons of energy released) are due to comets. Our major barrier to understanding the frequency of cometary impacts of greater than stipulated energy limits is our ignorance of their masses, which could be wrong by a factor of 10 (Bailey, 1990); the problem is in converting from an observed brightness/magnitude to the mass.

Since comets have relatively short physical lifetimes, they clearly have a replenishing source, as discussed by Bailey (1992). However, the Apollo asteroids must also have some source since their collisional lifetimes with the planets (10^7– 10^8 yr), representing their maximal lifetimes (they may physically decay on shorter timescales), are much less than the age of the solar system. It has already been mentioned that the asteroid belt and cometary cores are viewed as plausible sources

(Wetherill, 1991). Whilst the asteroid belt could supply Earth-crossers at a more-or-less constant rate, if a substantial proportion of such asteroids are in fact dormant/moribund/ extinct cometary cores, then the supply from *that* source would be variable in time, reflecting perhaps the recent arrival of a cometary wave from the Oort cloud, or the break-up of a single giant comet. That is, the Apollo population in the present epoch is not necessarily a measure of the long-term average population.

An important question is the nature of the impactors arriving on the Earth at geological boundaries/mass extinctions, and indeed at other times in which the biosphere is affected. Alvarez et al. (1980) suggested an asteroid, and yet later investigations have pointed to a wave of comets or some similar event. The above discussion concerning asteroids and comets shows that trying to differentiate between these classifications of objects may be futile, from the perspective of a discussion of how comets affect the biosphere. For example, the suggestion by Sykes and Walker (1992a) of a high ratio of silicates to volatiles in comets may explain why there is an "asteroidal" remnant signature found in some craters believed to have been formed at the times of boundary events (Weissman, 1985).

8.2.2   Meteoroids and Dust

Asteroid and comet impacts do not occur frequently, whereas cometary/ asteroidal material rains down upon the biosphere in a continual stream. The long-term averaged annual influx of solid particles with individual masses ranging from picograms to 10^{12} tons is of the order of 170,000 tons (Ceplecha, 1992), with $\sim$80% of that influx being due to occasional massive objects, submillimeter particles contributing (40 $\pm$20),000 tons per year on a continual basis (Love and Brownlee, 1993). The dominant mass influx on the year-in, year-out basis is in the mass range from 10^{-6} g to 10^{-4} g, although recent data from the the the *Spacewatch* telescope (Rabinowitz, 1993) may imply that the influx at $\sim$$10^9$ g rivals the dust influx (Ceplecha, 1992); indeed over long timescales the $\sim$$10^9$ g meteoroids may rival the large asteroids and comets as the predominant source of material influx to the Earth. The mass distribution of the terrestrial influx averaged over long periods (i.e. periods longer than the time between the most massive impacts) is shown in Fig. 8.1.

Particles smaller than about a microgram (equivalent to a size of about 100 μm) are decelerated in the atmosphere and gradually settle out to the surface. Such particles are termed "dust," and they are observable in space through satellite impact experiments, and through the sunlight that they scatter so as to produce the zodiacal light. Comets are observed to have distinct ion and dust tails, these tails providing another way in which dust may be observed, soon after its production. Although heating in atmospheric deceleration will lead to depletion of the volatile components of dust, their refractory constituents are largely unchanged. Such surviving interplanetary dust may be sampled in a number of ways, including collection at high altitude or in arctic/antarctic ice cores. About 5–10% of the mass of small particles survives entry (see Love and Brownlee, 1993).

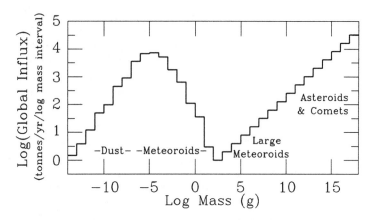

FIGURE 8.1. Distribution of the mass influx to the Earth, after Kyte and Wasson (1986) and Anders (1989). The unit is tons per logarithmic mass interval per year. The more recent spacecraft data of Love and Brownlee (1993) indicate that the mass influx at 10^{-6}–10^{-4} g may be twice as high as that shown here. Ceplecha (1992) finds subsidiary peaks at $\sim$1 g and $\sim$$10^9$ g. The actual mean mass influx from the very largest objects is very uncertain, but clearly the mass arriving on the Earth as small particles (meteoroids and dust) may be at least 20–30% of the total. Comets may therefore have pronounced effects upon the biosphere through mechanisms other than giant impacts.

Solid particles larger than about 100 μm and smaller than about 10 m are termed "meteoroids." Upon arriving in the upper atmosphere, at altitudes of about 140 km down to 80 km deceleration and heating occur, with the temperature rising to above 1000° C. Meteoroids will be totally vaporized, whereas smaller dust grains may survive since they undergo lesser heating due to their higher surface area to mass ratio. Larger meteoroids penetrate deeper into the atmosphere, most eventually detonating (Hills and Goda, 1993; Chyba, 1993).

The largest particles commonly observed as bright meteors (or fireballs) are only a few meters in diameter, whereas until recent years the smallest objects observed in space by means of telescopes were 100 m or more in size (small comets, near-Earth asteroids). That is, there was no overlap between observed meteoroids and observed asteroids. However, the development of the *Spacewatch* telescope at the University of Arizona has made it possible to detect a sample of small asteroids/large meteoroids in space, with sizes down to 5–10 m (Rabinowitz, 1993). Therefore a definition is now needed as to where meteoroids stop, and where asteroids begin. Meteoroids are here defined to be those objects smaller than 10 m in size, and asteroids to be those larger.

Comets (i.e., solid objects with substantial volatile components producing observable outgassing effects) of only 10 m sizes would have short physical lifetimes. Typically, comets lose volatiles equivalent to a meter or more of the surfaces of their nuclei in each perihelion passage, so that cometary objects smaller than $\sim$100 m would soon become totally devolatilized, and would then be anticipated to appear asteroidal in nature. There is evidence that the smaller asteroidal objects observed by *Spacewatch* are in fact the decay products of comet break-up (Rabinowitz, 1993). A claim made during the 1980s that the Earth is being bombarded by a huge number of "mini-comets" has been argued against on many grounds (see Dessler, 1991).

The ~40,000 ton annual terrestrial influx of small particles is not invariant from day to day. Whilst dust particles are quickly perturbed from the orbits of their parent objects, due to high ejection speeds and subsequent radiation pressure effects, meteoroids follow broadly similar orbits to their parent comets since they have lower ejection speeds and the radiative forces, which vary as the reciprocal of the particle size, are not so important. These meteoroids therefore remain in coherent streams until such time as the streams are disrupted in some way. Simulations have shown that close approaches to the planets, in particular Jupiter, are the most important factors in determining how long a stream will last, and how long it takes for the meteoroids to be thrown out into the semirandom "sporadic background;" the physical lifetimes of individual meteoroids, however, are limited by catastrophic collisions with smaller zodiacal dust particles (Olsson-Steel, 1986). Various surveys have put the fraction of the meteoroidal influx to the Earth which arrives in showers (which occur when the planet passes through a stream) as opposed to sporadics at between 20% and 60%. Each year there are various recurrent streams which put up the visual meteor count rate from the normal background of 5–10 per hour to 50–100 per hour. Every decade or so there occurs a "meteor storm," which is due to the Earth passing close by a comet in its orbit, whereupon we run into the higher density of meteoroids near the parent object as opposed to being spread around the complete orbit (producing annual showers). At such times the count rates may be 100,000 per hour or even higher.

Meteoroid ablation in the upper atmosphere (80–140 km) deposits metallic species at these altitudes which have significant effects upon the aeronomy. Probably of more importance, however, would be dust ladening of the atmosphere when the influx to the Earth alters from the long-term average. Three distinct reasons for this could be invoked: (i) a near passage by a comet; (ii) a period of enhanced cometary/asteroidal flux increasing the mass held in the sporadic swarm and zodiacal dust cloud; and (iii) encounters with interstellar dust clouds due to the motion of the solar system about the galaxy.

Mechanism (i) was suggested by Hoyle and Wickramasinghe (1978) as being the possible cause of biological and geological catastrophes, including that at the K/T boundary. In connection with (ii), Wickramasinghe et al. (1989) pointed out that changes in the dust influx can alter the climate significantly, and such a consideration is of import with regard to present-day studies of the greenhouse effect. In fact, the total mass of dust in the zodiacal dust cloud is only equivalent to a solid body about 5 km in diameter, so that clearly a big comet breaking up would lead to a large enhancement in the terrestrial dust influx, and a concomitant alteration of the climate. This enhancement of the dust influx would continue for ~10^5 yr, this being the timescale over which such a complex of material would disperse (Olsson-Steel, 1986; Steel, 1992). This corresponds to the approximate time over which a layer of amino acids were apparently deposited at the K/T boundary (Zhao and Bada, 1989; note that the amino acid anomaly was anticipated: Sack 1988). The suggestion has been made that these were brought to Earth through cometary dust rather than a singular impact (Zahnle and Grinspoon, 1990). Volatiles in dust (rather than larger particles) are more likely to survive atmospheric entry for a number of reasons (Steel, 1992). Indeed Rampino (1982) has even cautioned that the iridium anomaly at the K/T reported by Alvarez et al. (1980) could have been produced through gradual meteoroidal influx rather than through one giant impact, and MacLeod and Keller (1991) believe that the iridium peaks identified may be due to incomplete stratigraphic records. McLaren and Goodfellow (1990) have discussed the significance of the occurrence of enhanced siderophile elements like iridium at geological boundaries.

Mechanism (iii) above deserves special attention since this is an idea with a long history. Even at the turn of the century the idea was in place that there were distinct cycles of geologic events (Williams, 1981; Shaw, 1987). For example, Holmes (1927) suggested that there were $\sim$30 Myr cycles of orogeny correlated with sea-level changes, superimposed on a cycle of $\sim$250 Myr (see Clube and Napier, 1986; Rampino and Stothers, 1986). The longer timescale led Shapley (1921) to suggest that, this being close to the orbital period of the solar system about the galaxy, the cause was due to passages through interstellar clouds of hydrogen. This basic idea was also taken up by Hoyle and Lyttleton (1939), McCrea (1975, 1981), Begelman and Rees (1976), and Talbot and Newman (1977). Since variations in the flux of micrometeorites to the Moon have a similar period, Lindsay and Srnka (1975) suggested that the root cause might be that the solar system meets dust lanes in the galactic spiral arms with that frequency. Yabushita and Allen (1989) considered the effects of both the hydrogen and dust met by the solar system as it passes through a giant molecular cloud, showing that the combined environmental perturbations could lead to the onset of ice epochs and geological boundaries; in particular, they suggested the the K/T event was due to passage through the core of such a cloud.

An alternative concept, still based upon a galactic pacemaker, was championed by Napier and Clube (1979), Clube and Napier (1982a, 1984a, 1986), and Clube (1989, 1992). This saw the terrestrial disruptions as being due not so much to a direct influx of interstellar dust, but rather the capture into inner solar system orbits of large comets from galactic spiral arms, or perturbations of comets from the Oort cloud during discrete episodes (e.g., encounters with giant molecular clouds). These comets would then undergo fragmentation events, producing smaller comets, asteroids, meteoroids and dust, each of which would affect the terrestrial environment in different ways. The two periodicities (250 Myr and 30 Myr) would arise from discrete events occurring in episodes spaced according to the Sun's orbital period about the galactic center, producing passages through spiral arms every $\sim$250 Myr, and also the Sun's oscillation up and down through the galactic plane, with a half-period of $\sim$30 Myr (e.g., Innanen et al., 1978; Bahcall and Bahcall, 1985).

The point of the above discussion is to indicate that there is a plethora of mechanisms whereby comets may affect the biosphere quite apart from catastrophic impacts upon the Earth. Whilst these are the most obvious manifestations on the Earth of the trans-terrestrial cometary/asteroidal flux, through the craters that are evacuated and the associated stratigraphic deposits, it does not necessarily follow that the impacts are the sole (or even major) perturbers of the terrestrial environment. For example, Jansa et al. (1990) studied the effects of impacts by objects below 3 km in size on biological diversity and found that there is a threshold below which mass extinctions are not directly caused, pointing toward other effects (e.g., associated dust influx) as the direct cause of the environmental degradation producing the extinctions. The decay products of a massive comet arriving in near-Earth space will impose long-term ($\sim 10^5$ yr) effects upon the planet (e.g., Clube, 1989, 1992; Steel, 1992).

8.3   The Impact of Impacts

There is no doubt that the Earth has been subject to a bombardment from space in the past, and that this will continue in the future (Chapman and Morrison, 1994; Gehrels, 1994). Evidence for the latter includes astronomical observations of numerous asteroids and comets on Earth-crossing orbits, and from time to time a near miss is picked up: the record at the time of writing is 1993 KA_2 which passed within 150,000 km of the Earth in May of that year, just a shade closer than 1991 BA two years before. However, these are only ~ 10 m objects; for kilometer-size asteroids the record is about 650,000 km by 4581 Asclepius in 1989. Small asteroids in fact pass closer than the Moon on a daily basis, but in the main remain undetected.

The evidence for past impacts includes the numerous impact craters found around the planet. There are a number of excellent reviews available on this topic, including Shoemaker (1983), Mark (1987), Grieve (1987, 1990a, b, 1991), Melosh (1988a), and Pilkington and Grieve (1992). To differing extents these discuss the impactor flux, the actual craters known, the importance of impacts as a geologic agent (see also Marvin, 1990), and the form and characteristics of impact craters. With the scientific community now sensitized to the existence of impact craters, these are being found/recognized at an increasing rate (Grieve, 1993). The identification of heavily eroded impact structures through the unambiguous evidence of the mineral stishovite (a form of quartz only produced in the exceptional transient pressures of a massive impact) has meant that many impact events can now be demonstrated unequivocally both at the impact site (Grieve, 1990a) or from the strata far away (Bohor et al., 1987).

Given that impacts by relatively small (kilometer-size) objects can help to shape the face of the Earth by excavating craters which are tens of kilometers wide but also several kilometers deep, so that they punch through some fraction of the depth of the crust, it is natural to ask whether the less frequent giant impacts might be the boundary producing events in Earth history, with the concomitant mass extinctions. Or, if there are from time to time enhancements in the flux of comets and asteroids, do the many smaller impacts that occur produce epoch changes from their combined effects? Stothers (1992) shows that the lunar cratering record from the past 3.8 billion years demonstrates six major episodes of bombardment, and that as far as the accuracy of the dating goes these coincide with the six main episodes of orogenic tectonism on Earth. That is, it appears that the impacts have influenced plate tectonic motion over virtually the whole history of our planet since the crust became stable. It is known that impacts occurred at a higher rate in the first billion years of the Earth's history (Baldwin, 1985; Sleep et al., 1989; Barlow, 1990), and evidence is known on Earth for individual impacts dating back over at least 3.4 billion years (Lowe et al., 1989).

There is a number of reasons that linking craters with geological boundary events is fraught with difficulty. The first is that craters, although numerous, are hard to find; they may have been eroded or overlain since they formed on land, and the majority will be oceanic in any case. Second, separate dating of craters and geological boundaries, both of which dates will be uncertain to some degree, makes age coincidence very hard to demonstrate (e.g., see Hodych and Dunning, 1992), so that stratigraphic evidence is more secure (e.g., iridium anomalies, microtektites, stishovite fragments). Third, and of pertinence here, the numerical majority of impactors (although not necessarily the majority of the incoming mass; see Fig. 8.1) will not form craters at all, even though they may contribute to the iridium layers. Cometary/stony objects of dimensions up to $\sim$100 m detonate in the atmosphere at altitudes above 5 km (Hills and Goda, 1993). It is therefore a rare event that forms a small (less than $\sim$1 km) crater on Earth, when compared to the actual influx of bodies which would form such a scar in the absence of any atmosphere. In fact, almost all craters smaller than $\sim$2 km (produced by bolides 100–200 m in size) appear to have been produced by iron or stony-iron impactors (Grieve, 1987), even though such compositions comprise only a minor fraction of the population of meteoroids/asteroids/comets in space. In the case of Venus, craters less than $\sim$8 km in size are absent due to the shielding properties of its atmosphere, which is about 90 times as massive as that of our own planet (Ivanov, 1991; Zahnle, 1992).

In Subsection 8.2.2 the evidence for $\sim$30 Myr and $\sim$250 Myr periodicities in the geologic record (e.g., volcanism, oregeny) was briefly mentioned. This is not a new realization, the existence of such cycles being recognized for many decades (Clube and Napier, 1986). Such regular cycles appear to require an astronomical pacemaker, although an endogenic cause could not be rejected ab initio. The realization that extinctions also follow a $\sim$30 Myr cyclicity (Raup and Sepkoski, 1984) linked the geological with the biological cycles. The following realization that craters are produced periodically on a similar cycle (Alvarez and Muller, 1984) confirmed that the cause must surely be exogenic, since the impacts might cause the geologic changes, whereas the volcanoes and mountains could not produce the asteroids and comets. These periodicities, and their similarities, are discussed in more detail below.

As a final introductory paragraph to this section, we may note a rival has arisen to the "Gaia hypothesis" (which holds that the Earth may be thought of as an organism which regulates the conditions in the biosphere so as to perpetuate the existence of life). This is the "Shiva hypothesis", which regards the control of the biosphere to be largely extraterrestrial in nature, with periodic or episodic catastrophes caused by impacts leading to the extinction of many species, and the subsequent proliferation of others (Rampino, 1992). Shiva is the Hindu god of destruction and rebirth.

8.3.1   Impacts: Links to Boundary Events and Environmental Effects

Urey (1973) showed that the ages of various tektite strewn fields are similar to those of geological boundary events of the last 40 Myr, thereby linking the boundaries to massive impacts. Napier and Clube (1979) and Clube and Napier (1984a, 1986) suggested that the planetary region might be subject to periodic comet/planetesimal showers as the Sun passes through the spiral arms of our galaxy, and invoked the resulting impacts as being causes of the breaks in geologic periods within the last 400 Myr, arguing that there were craters linked in time to the geological boundaries.

With the interest provoked by the Alvarez et al. (1980) paper, this form of argument has been taken up enthusiastically by many others, with it being recognized that extraterrestrial factors must be involved if impacts are correlated with boundary/extinction events. If such events occur randomly in time, then the "stray asteroid/comet" concept of Alvarez et al. (1980) would be acceptable. However, if there is a periodicity in the events, then these may be linked to galactic dynamics or other astrophysical phenomena (e.g., see the review by Torbett, 1989). The evidence for periodicities of $\sim$30 Myr in the geological and biological records is discussed in the following subsections. In this subsection the links between impacts and boundaries is the topic.

The existence of an iridium anomaly at the K/T has previously been discussed, and its implication that a large impact occurred at that time. Iridium anomalies have also been identified at several other boundaries (Orth et al., 1990). However, as Orth et al. (1990) point out, not all impactors might be expected to produce iridium anomalies, so that small tektites (microspherules) and shocked quartz (stishovite) needs also to be searched for at geological boundaries in order to provide an unequivocal indicator that an impact had occurred. Such evidence is present at the K/T (Bohor et al., 1984, 1987; Cisowski, 1990) and its global distribution has been interpreted as implying more than one impact around the time of the boundary (Robin et al., 1993). There is also evidence for multiple impacts through the identification of plant growth between ejecta layers at the K/T boundary, implying that at least one growing season (and likely many seasons) passed between impacts (Shoemaker and Izett, 1992). The occurrence of extraterrestrial amino acids at the K/T has already been mentioned in Subsection 8.2.2, their depositional history implying an extended influx (Zahnle and Grinspoon, 1990).

Only a few years ago Grieve (1990b) was able to write that the only large ($\sim$100 km) impact crater known which could be linked to a boundary event was the Popigai crater in Siberia, whose age coincides with the Eocene–Oligocene mass extinction at $\sim$36 Myr to within the dating uncertainties (and the boundary also has an associated iridium anomaly: Alvarez et al., 1982). This is no longer the case in that a very large crater has now been associated with the K/T boundary on the Yucatan Peninsula of Mexico (Pope et al., 1991; McKinnon, 1992). There is now a wealth of geophysical and geochemical data which support a link between this crater—the Chicxulub structure—and the K/T boundary producing event (e.g., Izett 1991; Izett et al., 1991; Alvarez et al., 1992; Sharpton et al., 1992; Blum et al., 1993). Initial estimates of an $\sim$180 km diameter for Chicxulub have since been expanded to closer to 300 km (Sharpton et al., 1993). Recent work has unequivocally tied the ejected material found in Colorado and Haiti to the Chicxulub crater (Krogh et al., 1993), the geology of the impact area being of importance with regard to the sort of environmental stresses produced (e.g., sulfate-bearing rocks producing sulfuric acid and hence atmospheric acidification; carbonate rocks liberating CO_2 and hence prolonged greenhouse warming; see O'Keefe and Ahrens (1989), and Blum (1993) and references cited therein).

Especially, if an impact occurs within or close to an ocean, gigantic tsunami generation is expected. The Chicxulub impact at the K/T gave rise to tsunami which deposited rubble in Haiti, and across the region now occupied by Texas (e.g., Hildebrand and Boynton, 1990; Maurrasse and Sen, 1991; Alvarez et al., 1992; but see also Keller et al., 1993).

Whilst the K/T is the best-known and the best-studied geologic boundary/extinction episode, in fact more extensive mass extinctions (in terms of percentage loss of species, families and genera) have happened, for example, at the Frasnian-/Famennian (or Late Devonian) boundary, when $\sim$70% of all species were extinguished. That boundary/mass extinction has also been linked to a large impact through the identification of microtektites in the stratigraphic record, with two candidate craters being known (Claeys et al., 1992). McLaren and Goodfellow (1990) review the other extinctions/boundary events in an impact context. Other boundaries still await study. In particular, the great Permo-Triassic extinction $\sim$250 Myr ago when over 90% of oceanic species and 70% of terrestrial vertebrate families died still requires an explanation (Erwin, 1993, 1994). The periodicity argument described below may be taken to indicate that a series of impacts was responsible, but physical evidence is required. Some early reports of an iridium anomaly at this boundary have since been dismissed as inconclusive (Sarjeant, 1990; Erwin, 1994; but see also Rampino and Haggerty, 1994).

A 10 km impactor will excavate $\sim$$10^5$ cubic kilometers of crustal rock from its crater, clearly a major perturbation of the environment. The energy released is enough to produce an earthquake of magnitude 12.4 (Roddy et al., 1987). O'Keefe and Ahrens (1982), Jones and Kodis (1982), Melosh et al. (1990), Zahnle (1990), and Melosh (1991) discuss what would be expected to happen in the few hours after an impact. A 10 km projectile would scatter perhaps 1% of this huge mass of target rock onto ballistic trajectories, passing out above atmosphere, whilst a large fraction will be deposited locally; for example, ejecta from the Lake Acraman impact in South Australia 600 Myr ago has been found in the Flinders Ranges, $\sim$300 km away (Gostin et al., 1986; Williams, 1986). Some ejecta will have speeds in excess of the Earth's escape velocity, making capture into heliocentric orbits possible, and hence transfer to the Moon, Mars, or other planets (see Section 8.4 for the implications of this). However, most will have speeds insufficient to escape the Earth, and these will re-enter at typical speeds of $5-10\,\mathrm{km\ s^{-1}}$. This will produce a radiation flux at the Earth's surface equivalent to about ten solar constants (Zahnle, 1990), persisting for an hour or two, which would quickly incinerate the majority of the biomass. In fact, a pervasive soot and charcoal layer has been found, coincidental with the layer of iridium and other heavy metals, the mass of which indicates that upward of 90% of the biomass was incinerated at that time in global wildfires (Wolbach et al., 1985, 1988, 1990). Fires local to the impact site (i.e., on a continental scale) would be expected from the expanding fireball produced by the impact (Hassig et al., 1987), but not globally, implicating the radiation flux from re-entering ejecta as the major ignitor.

However, not all material will ablate to atomic dimensions in the re-entry. Interplanetary dust is decelerated in the atmosphere and does not burn up, for sizes below about 100 μm (Subsection 8.2.2). Pulverized rock from impacts has a greater chance of surviving intact since it has a lower entry speed ($5-10\,\mathrm{km\ s^{-1}}$) than incoming interplanetary dust (above $11\,\mathrm{km\ s^{-1}}$: see Steel (1992)), and much will be have oblique entry angles. Dust from the impact plume would also be expected to be injected above the troposphere, taking some months to settle out (Pollack et al., 1983; Toon et al., 1994), and similarly any enhancement in the dust density in interplanetary space will also feed into the terrestrial environment. The effect of increased dust density in the upper atmosphere will be a cooling of the global climate, and thus an "impact winter" (Wolfe, 1991). Although this name derives from the idea of a nuclear winter produced by the soot liberated in a major nuclear war (Turco et al., 1991), in fact the original concept of nuclear winter stemmed from studies of the climatic aftermath of the Tunguska detonation of 1908, and also modeling of the K/T event.

There are numerous other effects upon the atmosphere that might be expected to result from large impacts, in particular, the deposition of nitrogen oxides which would lead to atmospheric acidification (Zahnle, 1990); these are produced through direct interaction by the bolide with the atmosphere as opposed to liberation of target rock constituents as mentioned above. Prinn and Fegley (1987) and Toon et al. (1994) discuss how the biosphere would be severely stressed under such circumstances.

The above has mainly been concerned with the links between impact events and environmental changes as delineated by the stratigraphic/fossil record. One of the problems faced by the impact hypothesis for mass extinctions is that various other massive geological upheavals are also known to have occurred at the times of boundary events, for example widespread volcanism at the K/T (e.g., see Alvarez and Asaro, 1990; Courtillot, 1990). As aforesaid this seems to require a causal link between the impacts and the volcanism, rather than being an alternative to an impact trigger (i.e., although it may well be that the environmental degradation leading to the extinctions was largely due to the volcanic eruptions, these were themselves triggered by the impacts). Clube and Napier (1982b) address this question of the link between impacts and geophysics. There are many facets of geophysical behavior which have been linked to impacts. That geomagnetic reversals might be impact-induced has been a long-term suggestion (see Creer and Pal (1989) for a review), and tektite deposition has been linked temporally to field reversals (Glass et al., 1979). Throughout the Cenozoic there appears to be a relationship between impact events and tectonic episodes (Burek and Wanke, 1988), which supports the suggestion of Seyfert and Sirkin (1979) that very energetic impacts might be responsible for the break-up of continental plates, and the stimulation of mantle plumes. However, it is the common periodicities of extinctions, geological events, and impacts that provide the strongest evidence of connections between these phenomena.

8.3.2   Extinction/Impact/Geologic Upheaval Periodicity

The study of the link between impacts and mass extinctions received a massive impetus in 1984 when Raup and Sepkoski (1984) published their first paper, although this was essentially a rediscovery of a trend identified by Fischer and Arthur (1977). Raup and Sepkoski used marine microfossil data from the past 260 Myr and found that there was strong evidence to support a 26 Myr periodicity in the extinction rates. Figure 8.2 shows the nine major peaks of extinctions, whilst Fig. 8.3 plots their ages against cycle number along with a straight line showing the trend expected for a 26 Myr periodicity. Later investigations have tended to confirm this result (Sepkoski and Raup, 1986; Raup and Sepkoski, 1986, 1988), although the exact periodicity is still debatable in view of the various difficulties in interpreting the paleontological record (Stigler and Wagner, 1987; Rampino, 1992). Study of the various geologic timescales and their uncertainties shows that the actual periodicity, if regular, is between 24 Myr and 33 Myr (Stothers 1989); the possibility exists that the period is not regular but shows some jitter, as might be expected for astronomical pacemakers discussed below. This may be the reason that Yabushita (1992), in applying stringent statistical tests to the extinction data, finds that the 26 Myr is apparent but nevertheless does not pass those tests for significance.

Sepkoski (1990) has shown that the taxonomic spread of the extinctions are indicative of some external environmental perturbation rather than internal dynamics of faunal evolution. For general descriptions of the history of the periodic extinction idea, and its astronomical interpretations, see Raup (1986b, 1991), Goldsmith (1986), and Muller (1988)

As aforementioned, Alvarez and Muller (1984) showed that impact craters on the Earth appear to have a periodicity near 26 Myr (see also Seyfert and Sirkin, 1979). Although there have been some disputes, this suggestion has been backed up by later studies which also have a bearing upon the source and nature of the impactors, and the way in which they are perturbed from their previous orbits (e.g., Stothers, 1988; Yabushita, 1992). Single sporadic impacts like that suggested by Alvarez et al. (1980) would lead to a stochastic impact record rather than a periodicity, meaning that multiple impacts are expected at each boundary. Mass extinctions are therefore seen to be due to comet showers originating in the outer solar system, beyond the planetary region (Hut et al., 1987; Perlmutter and Muller,

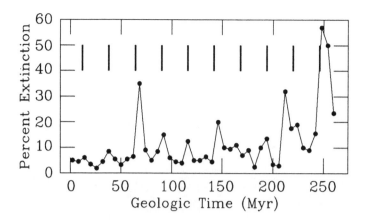

FIGURE 8.2. Percentage extinction record for marine animal genera over the past 260 Myr. The lines at top show the best-fit 26 Myr cycle, these correlating in nine cases out of the ten with an extinction event of some magnitude. After Sepkoski (1990).

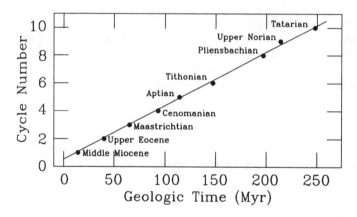

FIGURE 8.3. The dates of the nine mass extinctions in Fig. 8.2 plotted with a straight line of slope 26 Myr. A mass extinction might be expected in the Middle Jurassic (between the Pliensbachian and Tithonian extinctions) but is not clearly defined in the presently available data as plotted in Fig. 8.2.

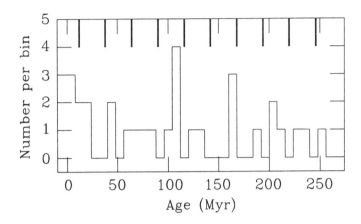

FIGURE 8.4. Histogram of the ages of craters younger than 260 Myr and larger than 5 km in size, in 8 Myr bins, from the tabulation of Grieve (1990b). Whilst the craters, many of which are poorly dated, appear to have a periodic formation record (see the various citations in the text) there is a problem with the phasing of these compared to the extinction record. The superposed 26 Myr cycle is the same as that plotted in Figs. 8.2 and 8.3; in the case of the crater ages, the cycle appears to be out of phase to some extent, as discussed in the text.

1988) rather than occasional large asteroids arriving from the main belt between Mars and Jupiter. Mass extinctions, from the paleontological record, do not occur instantaneously, but rather appear to take place over a series of steps stretching through one to three million years. Oort cloud comets have orbits with periods of a few million years, so that any disturbance of the cloud would lead to an enhancement in the near-parabolic flux continuing through such timespans. A few of these (perhaps massive) comets might strike the Earth and cause severe disruption of the biosphere, a few also being captured into short-period orbits and disintegrating so as to leave an enhanced population of planet-crossing asteroids, meteoroids, and dust. The structure of the environmental degradation at a boundary event is therefore expected to be quite different to that originally visualized by Alvarez et al. (1980), with many continuing and recurring perturbations being imposed.

Returning to the dated impact craters, the period generally obtained is about 30 Myr. Figure 8.4 shows the ages of craters of diameter greater than 5 km and ages of less than ~260 Myr.

Common periodicities for mass extinctions and astronomical phenomena could have an origin other than the obvious effects of impacts/dust influx. For example, if the extinctions were to occur as the solar system passes through the spiral arms of the galaxy, these being regions where star formation is occurring and massive young stars rapidly evolve and explode as supernovae, then it would be possible that proximal supernovae might cause the extinctions and boundaries through short-lived gross alterations of the terrestrial environment. Van den Berg (1989) discusses this possibility, and finds that the probability of a supernova within about sixteen light years of the Sun in the last billion years is only about 6%, so that multiple mass extinctions within the past 600 Myr are very unlikely. Other astrophysical phenomena which might conceivably affect the Earth have been reviewed by Torbett (1989). Even if supernovae could be argued to be more frequent within lethal distances, this would not explain the cratering and stratigraphic records.

The common periodicity in mass extinctions and large impact craters was quickly realized to correspond to the half-period with which the Sun oscillates through the galactic plane (Rampino and Stothers, 1984a, b, 1986; Bahcall and Bahcall, 1985; Stothers, 1985; Torbett, 1989), suggesting a plausible source for a wave of comets in a disturbance of the Oort cloud either through stellar encounters or passages through giant molecular clouds (see Hills, 1981; Thaddeus and Chanan, 1985; Stothers, 1988; Bailey, 1992; Clube, 1992).

Whilst the motion of the Sun about the galaxy provided a plausible source for the periodicity, two other ad hoc suggestions were made for the origin of cometary waves that could produce the impact craters and mass extinctions. These are the "Nemesis" and "Planet X" theories.

The Nemesis theory holds that there is a distant companion star of the Sun, with a period near 30 Myr and an eccentric orbit which leads to disturbances of the Oort cloud when the star passes perihelion. This idea was independently suggested by Davis et al. (1984) and by Whitmire and Jackson (1984). There are a number of objections to this theory on the basis of the stability of the star's orbit under galactic perturbations, and also the stability of the planetary system under perturbations caused by the star (see Torbett and Smoluchowski, 1984; Torbett, 1989). Nevertheless, the Nemesis theory does have some advantages over the rival theories, given the present state of our knowledge providing restrictions on the mechanisms involved (e.g., durations of comet showers). See Perlmutter et al. (1990) and Vandervoort and Sather (1993) for discussions.

The second ad hoc theory is that involving a hypothetical "Planet X". (Note that the idea of a tenth planet causing perturbations of the orbits of Uranus and Neptune has now been negated in light of improved mass determinations for the outer planets (Standish, 1993).) The idea under consideration here is that suggested by Whitmire and Matese (1985); see also Matese and Whitmire (1986), with the supposed planet in an orbit of period 60 Myr. Every 30 Myr orbital precession would bring its perihelion and aphelion around so as to perturb comets at the edges of a gap in the condensed inner Oort cloud (Bailey, 1992) which the planet was assumed to have swept out earlier in its history. There are various arguments against the Planet X hypothesis, as discussed by Torbett (1989).

This leaves the Nemesis and galactic motion ideas. There being strong arguments for the latter (Clube and Napier, 1984b) apart from the former invoking an unknown phenomenon (Nemesis itself). Torbett (1989) discusses the pros and cons of each idea. Bailey et al. (1987) show that known stars and molecular clouds cannot produce the 30 Myr comet waves unless the mass density of such objects is higher than presently believed. An alternative (as suggested by Bailey et al., 1987) might be the "dark matter" believed by some to comprise 90% of the mass of the Universe,

TABLE 8.2. Extinction and Geologic Event Periodicities, the Periods Being Ordered According to Their Powers in the Spectral Analysis of Rampino and Caldeira (1992, 1993). Crater Periodicity from Yabushita (1991, 1992).

Event	Period(s) (Myr)
Extinctions	26.0
Sequence boundaries	28.2
Flood basalts	23.1, 15.4, 26.2
Orogenic events	30.6
Sea-floor spreading	18.4, 26.7
Ocean anoxic events	39.7, 26.1
Evaporite (salt) deposits	28.2
Craters	16.5, 30–32

if this is in the form of brown dwarf stars or Jupiter-mass objects. The recent detection of the possible trace of such an object (Alcock et al., 1993) lends weight to the hypothesis that these might be responsible, although it has been assumed that these have a galactic halo distribution which would *not* produce the required comet wave modulation.

So far the discussion in Subsection 8.3.2 has been limited to the common periodicities of mass extinctions and crater-forming impacts on the Earth. However, there are a number of geological phenomena which have occurred over the past 600 Myr and which are also known to have the common ~30 Myr and ~250 Myr periodicities, as pointed out by Rampino and Stothers (1984b) and associated by them with the half-period of the motion of the Sun vertically about the galactic plane and also the "Cosmic Year" of 250±50 Myr, the period of the Sun's orbit about the galactic center. One of the better-studied types of event is the occurrence of flood basalt eruptions, for which Rampino and Stothers (1988) examined the 11 distinct episodes over the past 250 Myr and found a periodicity near 32 Myr. Later studies covering extinctions, major sea-level changes, ocean anoxic events, black shale events, evaporite deposits, discontinuities in sea-floor spreading, orogenic episodes, flood-basalt eruptions, atmospheric composition variations, tektite formation, crater production, geomagnetic reversals, and geochemical anomalies have confirmed the periods of ~30 Myr and ~250 Myr (Rampino and Caldeira, 1992, 1993; Liritzis, 1993). For example, see Table 8.2. The ability of the galactic motion theory to explain both periodicities, whilst the Nemesis hypothesis only produces a 30 Myr cycle, is another argument in favor of the former.

Another periodicity of ~15 Myr is also claimed as being evidenced by impact crater ages, geomagnetic reversals, and global volcanism (see Mazaud et al., 1983; Napier, 1989; Creer and Pal, 1989; Clube, 1992; Yabushita, 1991, 1992). In particular, Napier (1989) argues that paucity of data and a lack of objectivity in some studies has led to the 30 Myr cycle being adopted with interpulse events (and hence a 15 Myr cycle) being neglected.

8.3.3   Phase Relation Between Biological/Geological Upheavals and the Solar Motion

Apart from the common period, another feature of the apparent interrelationship between extinctions/geological upheavals and the solar oscillation through the galactic plane is the phase relation. If comet waves are produced in some way as the solar system passes through the galactic plane, then the terrestrial events and impacts should occur in-phase with those plane crossings. If these events are found to be in-phase then this gives additional weight to the galactic hypothesis as opposed to the Nemesis hypothesis, since if the common periods of the galactic cycle and the terrestrial events/comet waves are by chance, then the phase would be expected to be random.

The bounds on the galactic oscillation half-period were found to be 26–37 Myr by Bahcall and Bahcall (1985), although the astrophysical models which rendered the longer possible periods have since been countermanded, producing limits closer to 26–32 Myr (see Napier, 1989). The mass extinction periodicity seems to be in the range 24–33 Myr, depending on the timescale used (Stothers 1989), so that there is excellent agreement between these. Bahcall and Bahcall (1985) find that it is less than 3 Myr since the solar system last passed through the galactic plane. Since the infall time from the Oort cloud is 1–3 Myr, if the comet waves are produced during plane crossings then we are currently in, or almost in, an episode of bombardment. That is, the phase would be very close to zero at the present time. Yabushita (1992), studying crater ages, finds that the current phase is between 1 Myr and 5 Myr (i.e., we are now in a comet shower, or have recently experienced one). In an early study, Rampino and Stothers (1984b) found that this low current phase value is not contradicted by the terrestrial evidence, mainly because the phase relations obtained are dominated by the uncertainties in each as produced by the periodicity fitting. However, an analysis of more extensive data sets (Rampino and Caldeira, 1992, 1993) indicates that with a best-fit periodicity of 26.6 Myr the present phase is 8–9 Myr, which is at variance with the concept of cometary waves being produced at crossings of the galactic plane. Clube (1992) notes that this would imply that the waves are induced at close to the times of the Sun's maximum deviation from the galactic plane. Napier (1989) studied the phase/time-lag question and found additional evidence to support the ~15 Myr rather than ~30 Myr periodicity, on the basis that the phases then agree with a recent passage through the plane and thus a recent impact episode. The source of the Oort cloud perturbations at different points in the solar orbit, and hence an explanation for the phase found by Rampino and Caldeira (1992, 1993) in terms of a ~15 Myr periodicity, is given by Clube

(1992).

8.4   The Contemporary Environment

If the inner solar system has recently suffered an enhancement in the comet-ary/planetesimal flux (Section 8.3.3 above), a natural question is whether the global environment has been subject to significant perturbations of this origin; for example, could the most recent ice age, or its cessation, have been due to changes in the small body flux over the past 10,000–20,000 yr (Asher and Clube, 1993)? This was part of the reason for emphasizing the role of the meteoroid/dust influx in Section 8.2; macroscopic impacts are not necessary for the perturbation of the climate. Glaciation events such as that of the Pleistocene are most often explained as being predominantly due to the Milankovich cycle (orbital/spin axis variations of the Earth), but there are objections to that idea (Clube, 1992). An alternative could be that the inner solar system environment is currently subject to the substantial control of the products of the break-up of a giant comet within the past $\sim$20,000 yr (Steel et al., 1994). There is evidence from historical records for such an idea, including periods of greatly enhanced seasonal fireball fluxes (Bailey et al., 1990; Clube and Napier, 1990; Hasegawa, 1992). Indeed, many of the problems associated in explaining the short-period comet population might be explicable in terms of a break-up of a giant comet within the past $\sim$10^4 yr (Bailey, 1992).

The various concepts discussed in this chapter therefore pertain not only to mass extinctions and environmental changes produced in the past; they also apply to investigations of the contemporary environment.

8.5   Panspermia

Any review of impacts by comets and asteroids into the biosphere should mention how such impacts might be causal agents in spreading life from Earth to other planets (or perhaps other viable life-supporting environments), or indeed the reverse: transfer of life forms from other locales to our planet. For example, Hoyle and Wickramasinghe (1993, and many earlier publications) have suggested that microbial life is continually raining down upon the Earth from passing comets.

The main focus of this brief section is the possibility that lifeforms might be blasted off the surface of a planet, enclosed in the ejecta from a large impact. The recent discovery that microbial life is pervasive deep within the terrestrial crust, with a total mass which may be as large as that of surface lifeforms (see Gold, 1992; Stetter et al., 1993) suggests that the spreading of such colonies to other planets might be possible, since such life-bearing rock could be ejected from the Earth onto heliocentric orbits in an impact. Such a mechanism for spreading life (panspermia) was discussed by Melosh (1988b). He noted that there are classes of meteorite that are believed to have originated as ejecta from the Moon and Mars, which makes the reverse process seem plausible, and he discussed the various problems involved. One of these is whether sufficiently large rocks might be blasted off of the planet in a largely unshocked state, a question that has recently been answered in the affirmative by Gratz et al., (1993). A question yet to be answered is whether the microbial life could exist long enough to survive the voyage through interplanetary space before arriving at, say, Mars. Although the collisional lifetimes of planet-crossing objects like this are generally 10^7 yr or more, this does not preclude a few rocks arriving at Mars in a much shorter time. Thus, for example, meteorites have space exposure ages (derived from cosmic ray track densities) which are typically $\sim 10^7$ yr (Perlmutter and Muller, 1988), as expected from the theoretical collisional lifetimes, but nevertheless the youngest meteorite known (in terms of this space exposure age) has an age of only 10^4 yr (Heymann and Anders, 1967). In principle an ejected rock could arrive at Mars within a handful of years, and in practice this has almost certainly occurred. Whether viable microbial life could survive in space for that period, and whether it could then survive the impact on Mars, subsequently flourishing, are questions that need to be addressed.

Acknowledgments: Support from the Australian Research Council and the Department of Employment, Education and Training is gratefully acknowledged.

8.6   References

Alcock, C., Akerlof, C.W., Allsman, R.A., Axelrod, T.S., Bennett, D.P., Chan, S., Cook, K.H., Freeman, K.C., Griest, K., Marshall, S.L., Park, H-S., Perlmutter, S., Peterson, B.A., Pratt, M.R., Quinn, P.J., Rodgers, A.W., Stubbs, C.W., and Sutherland, W. (1993), Possible gravitational microlensing of a star in the Large Magellanic Cloud, *Nature*, **365**, 621–623.

Alvarez, W. (1990), Interdisciplinary aspects of research on impacts and mass extinctions; A personal view, *Geol. Soc. Amer., Spec. Pap.*, **247**, 93–98.

Alvarez, W. and Asaro, F. (1990), An Extraterrestrial Impact, *Sci. Amer.*, **263**, 44–52.

Alvarez, W. and Muller, R.A. (1984), Evidence from crater ages for periodic impacts on the Earth, *Nature*, **308**, 718–720.

Alvarez, L.W., Alvarez, W., Asaro, F., and Michel, H.V. (1980), Extraterrestrial Cause for the Cretaceous-Tertiary Extinction, *Science*, **208**, 1095–1108.

Alvarez, W., Asaro, F., Michel, H.V., and Alvarez, L.W. (1982), Iridium anomaly approximately synchronous with terminal Eocene extinctions, *Science*, **216**, 886–888.

Alvarez, W., Hansen, T., Hut, P., Kauffman, E.G., and Shoemaker, E.M. (1989), Uniformitarianism and the response of Earth scientists to the theory of impact crises, pp. 13–24 in Clube (1989).

Alvarez, W., Smit, J., Lowrie, W., Asaro, F., Margolis, S.V., Claeys, P., Kastner, M., and Hildebrand, A.R. (1992), Proximal impact deposits at the Cretaceous-Tertiary boundary in the Gulf of Mexico: A restudy of DSDP Leg 77 Sites 536 and 540, *Geology*, **20**, 697–700.

Anders, E. (1989), Pre-biotic organic matter from comets and asteroids, *Nature*, **342**, 255–

257.

Asher, D.J. and Clube, S.V.M. (1993), An extraterrestrial influence during the current glacial-interglacial, *Ql. J. Roy. Astron. Soc.*, **34**, 481–511.

Bahcall, J.N. and Bahcall, S. (1985), The Sun's motion perpendicular to the galactic plane, *Nature*, **316**, 706–708.

Bailey, M.E. (1990) Cometary masses, pp. 7–35 in *Baryonic Dark Matter*, eds. D. Lynden-Bell and G. Gilmore, Kluwer, Dordrecht, Holland.

Bailey, M.E. (1991), Comet craters versus asteroid craters, *Adv. Space Res.*, **11(6)**, 43–60.

Bailey, M.E. (1992), Origin of short-period comets, *Cel. Mech. Dyn. Astron.*, **54**, 49–61.

Bailey, M.E., Wilkinson, D.A., and Wolfendale, A.W. (1987), Can episodic comet showers explain the 30-Myr cyclicity in the terrestrial record?, *Mon. Not. Roy. Astron. Soc.*, **227**, 863–885.

Bailey, M.E., Clube, S.V.M., and Napier, W.M. (1990), The Origin of Comets, Pergamon Press, Oxford.

Bailey, M.E., Clube, S.V.M., Hahn, G., Napier, W.M., and Valsecchi, G.B. (1994), Hazards due to giant comets: Climate and short-term catastrophism, in Gehrels (1994).

Baldwin, R. (1949), *The Face of the Moon*, University of Chicago Press, Chicago.

Baldwin, R.B. (1985), Relative and absolute ages of individual craters and the rate of infalls in the Moon in the post-Imbrium period, *Icarus*, **61**, 63–91.

Barlow, N. (1990), Application of the inner Solar System cratering record to the Earth, *Geol. Soc. Amer., Spec. Pap.*, **247**, 181–187.

Begelman, M.C. and Rees, M.J. (1976), Can cosmic clouds cause climatic catastrophes?, *Nature*, **261**, 298–299.

Blum, J.D. (1993), Zircon can take the heat, *Nature*, **366**, 718.

Blum, J.D., Chamberlain, C.P., Hingston, M.P., Koeberl, C., Marin, L.E., Schuraytz, B.C., and Sharpton, V.L. (1993), Isotopic comparison of K/T boundary impact glass with melt rock from the Chicxulub and Manson impact structures, *Nature*, **364**, 325–327.

Bohor, B.F., Foord, E.E., Modreski, P.J., and Triplehorn, D.M. (1984), Mineralogic evidence for an impact event at the Cretaceous-Tertiary boundary, *Science*, **224**, 867–869.

Bohor, B.F., Modreski, P.J. and Foord, E.E. (1987), Shocked quartz in the Cretaceous-Tertiary boundary clays: Evidence for a global distribution, *Science*, **236**, 705–709.

Burek, P.J. and Wanke, H. (1988), Impacts and glacio-eustasy, plate-tectonic episodes, and geomagnetic reversals: A concept to facilitate detection of impact events, *Phys. Earth Planet. Int.*, **50**, 183–194.

Ceplecha, Z. (1992), Influx of interplanetary bodies onto Earth, *Astron. Astrophys.*, **263**, 361–366.

Chapman, C.R. and Morrison, D. (1994), Impacts on the Earth by asteroids and comets: Assessing the hazard, *Nature*, **367**, 33–40.

Chyba, C. (1993), Explosions of small Spacewatch objects in the Earth's atmosphere, *Nature*, **363**, 701–703.

Chyba, C., Thomas, P., and Zahnle, K. (1993), The 1908 Tunguska explosion: Atmospheric disruption of a stony asteroid, *Nature*, **361**, 40–44.

Cisowski, S.M. (1990). A critical review of the case for, and against, extraterrestrial impact at the K/T boundary, *Surveys Geophys.*, **11**, 55–131.

Claeys, P., Casier, J-G., and Margolis, S.V. (1992), Microtektites and Mass Extinctions: Evidence for a Late Devonian Asteroid Impact, *Science*, **257**, 1102–1104.

Clube, S.V.M. (1989), The catastrophic role of giant comets, pp. 81–112 in *Catastrophes and Evolution: Astronomical Foundations*, ed. S.V.M. Clube, Cambridge University Press, Cambridge, UK.

Clube, S.V.M. (1992), The fundamental role of giant comets in Earth history, *Cel. Mech. Dyn. Astron.*, **54**, 49–61.

Clube, S.V.M. and Napier, W.M. (1982a), Spiral arms, comets, and terrestrial catastrophism, *Ql. J. Roy. Astron. Soc.*, **23**, 45–66.

Clube, S.V.M. and Napier, W.M. (1982b), The role of episodic bombardment in geophysics, *Earth Planet. Sci. Lett.*, **57**, 251–262.

Clube, S.V.M. and Napier, W.M. (1984a), The microstructure of terrestrial catastrophism, *Mon. Not. Roy. Astron. Soc.*, **211**, 953–968.

Clube, S.V.M. and Napier, W.M. (1984b), Terrestrial catastrophism: Nemesis or galaxy?, *Nature*, **311**, 635–636.

Clube, S.V.M. and Napier, W.M. (1986), Giant comets and the galaxy: implications of the terrestrial record, pp. 260–285 in *The Galaxy and the Solar System*, eds. R. Smoluchowski, J.N. Bahcall, and M.S. Matthews, University of Arizona Press, Tucson.

Clube, S.V.M. and Napier, W.M. (1990), The Cosmic Winter, Blackwell, Oxford.

Courtillot, V.E. (1990), A Volcanic Eruption, *Sci. Amer.*, **263**, 53–60.

Courtillot, V.E., Féraud, G., Maluski, H., Vandamme, D., Moreau, M.G. and Besse, J. (1988), The Deccan flood basalts and the Cretaceous/Tertiary boundary, *Nature*, **333**, 843–846.

Covey, C., Ghan, S.J., Walton, J.J. and Weissman, P.R. (1990), Global environmental effects of impact-generated aerosols; Results from a general circulation model, *Geol. Soc. Amer., Spec. Pap.*, **247**, 263–270.

Cowles, R.B. (1939), Possible implications of reptilian thermal tolerance, *Science*, **90**, 465–466.

Creer, K.M. and Pal, P.C. (1989), On the frequency of reversals of the geomagnetic dipole, pp. 113–132 in Clube (1989).

Crowley, T.J. and North, G.R. (1988), Abrupt climate change and extinction events in earth history, *Science*, **240**, 996–1002.

Crowley, T.J. and North, G.R. (1991), *Paleoclimatology*, Oxford University Press, New York.

Davis, M., Hut, P., and Muller, R.A. (1984), Extinction of species by periodic comet showers, *Nature*, **308**, 715–717.

De Laubenfels, M.W. (1956), Dinosaur extinction: one more hypothesis, *J. Paleont.*, **30**, 207–212.

Dessler, A. (1991), The small comet hypothesis, *Rev. Geophys.*, **29**, 355–382 and 609–610.

Donovan, S.K., ed. (1989), *Mass Extinctions: Processes and Evidence*, Belkhaven, London.

Elliott, D.K., ed. (1986), *Dynamics of Extinction*, Wiley, New York.

Emiliani, C., Kraus, E.B., and Shoemaker, E.M. (1981), Sudden death at the end of the Mesozoic, *Earth Planet. Sci. Lett.*, **55**, 317–334.

Erwin, D.H. (1993), *The Great Paleozoic Crisis: Life and Death in the Permian*, Columbia, New York.

Erwin, D.H. (1994), The Permo-Triassic extinction, *Nature*, **367**, 231–236.

Fischer, A.G. and Arthur, M.A. (1977), Secular variations in the pelagic realm, *Soc. Econ. Paleont. Mineral Spec. Publ.*, **25**, 19–50.

Gehrels, T., ed. (1994), *The Hazard due to Comets and Asteroids*, University of Arizona Press, Tucson.

Gilmour, I., Wolbach, W.S., and Anders, E. (1989), Major wildfires at the Cretaceous–Tertiary boundary, pp. 195–213 in Clube (1989).

Glass, B.P., Swinki, M.B. and Zwart, P.A. (1979), Australiasian, Ivory Coast and North American tektite strewn fields: Size, mass and correlation with geomagnetic reversals and other Earth events, *Proc. Lunar Planet. Sci. Conf.*, **10**, 25–37.

Gold, T. (1992), The deep, hot biosphere, *Proc. Natl. Acad. Sci. (USA)*, **89**, 6045–6049.

Goldsmith, D. (1986), *Nemesis: The death star and other theories of mass extinction*, Walker, New York.

Gostin, V.A., Haines, P.W., Jenkins, R.J.F., Compston, W., and Williams, I.S. (1986), Impact ejecta horizon within late Precambrian shales, Adelaide geosyncline, South Australia, *Science*, **233**, 198–203.

Gould, S.J. (1965), Is uniformitarianism necessary?, *Amer. J. Sci.*, **263**, 223–228.

Gould, S.J. and Eldridge, N. (1993), Punctuated equilibrium comes of age, *Nature*, **366**, 223–227. But see also Nelson, G. (1994), Older than that, *Nature*, **367**, 108.

Gratz, A.J., Nellis, W.J., and Hinsey, N.A. (1993), Observations of high-velocity, weakly shocked ejecta from experimental impacts, *Nature*, **363**, 522–524.

Grieve, R.A.F. (1987), Terrestrial impact structures, *Ann. Rev. Earth Planet. Sci.*, **15**, 245–270.

Grieve, R.A.F. (1990a), Impact cratering on the Earth, *Sci. Amer.*, **262**, 44–51.

Grieve, R.A.F. (1990b), The record of impact on Earth: Implications for a major Cretaceous/Tertiary impact event *Geol. Soc. Amer., Spec. Pap.*, **247**, 25–37.

Grieve, R.A.F. (1991), Terrestrial impact: The record in the rocks, *Meteoritics*, **26**, 175–194.

Grieve, R.A.F. (1993), When will enough be enough? *Nature*, **363**, 670–671.

Hahn, G. and Bailey, M.E. (1990), Rapid dynamical evolution of giant comet Chiron, *Nature*, **348**, 132–136.

Hallam, A. (1989), Catastrophism in geology, pp. 25–55 in Clube (1989).

Hasegawa, I. (1992), Historical variation in the meteor flux as found in Chinese and Japanese chronicles, *Cel. Mech. Dyn. Astron.*, **54**, 129–142.

Hassig, P.J., Rosenblatt, M., and Roddy, D.J. (1987), Analytical simulation of a 10-km-diameter asteroid impact into a terrestrial ocean: Part 2—Atmospheric response, *Lunar Planet. Sci. Conf., Abstracts*, **17**, 321–322.

Heymann, D. and Anders, E. (1967), Meteorites with short cosmic ray exposure ages, as determined from their Al^{26} content, *Geochim. Cosmochim. Acta*, **31**, 1793–1809.

Hildebrand, A.R. and Boynton, W.V. (1990), Proximal Cretaceous-Tertiary boundary impact deposits in the Caribbean, *Science*, **248**, 843–847.

Hills, J.G. (1981), Comet showers and the steady-state infall of comets from the Oort cloud, *Astron. J.*, **86**, 1730–1740.

Hills, J.G. and Goda, M.P. (1993), The fragmentation of small asteroids in the atmosphere, *Astron. J.*, **105**, 1114–1144.

Hodych, J.P. and Dunning, G.R. (1992), Did the Manicouagan impact trigger end-of-Triassic mass extinction?, *Geology*, **20**, 51–54.

Holmes, A. (1927), *The Age of the Earth: An Introduction to Geological Ideas*, Benn, London.

Hoyle, F. and Lyttleton, R.A. (1939), The effect of interstellar matter on climatic variation, *Proc. Cambridge Phil. Soc. Math. Phys. Sci.*, **35**, 405–415.

Hoyle, F. and Wickramasinghe, N.C. (1978), Comets, ice ages and ecological catastrophes, *Astrophys. Space Sci.*, **53**, 523–526.

Hoyle, F. and Wickramasinghe, N.C. (1993), *Our Place in the Cosmos: The Unfinished Revolution*, Dent, London.

Hoyt, W.G. (1987), *Coon Mountain Controversies*, University of Arizona Press, Tucson.

Hsü, K.J. and McKenzie, J.A. (1990), Carbon-isotope anomalies at era boundaries; Global catastrophes and their ultimate cause, *Geol. Soc. Amer., Spec. Pap.*, **247**, 61–70.

Hut, P., Alvarez, W., Elder, W.P., Hansen, T., Kauffman, E.G., Keller, G., Shoemaker, E.M., and Weissman, P.R. (1987), Comet showers as a cause of mass extinctions, *Nature*, **329**, 118–126.

Innanen, K.A., Patrick, A.T., and Duley, W W. (1978), The interaction of the spiral density wave and the Sun's galactic orbit, *Astrophys. Space Sci.*, **57**, 511–515.

Ivanov, B.A. (1991), Impact crater processes, *Adv. Space Res.*, **11(6)**, 67–75.

Izett, G.A. (1991), Tektites in Cretaceous-Tertiary boundary rocks on Haiti and their bearing on the Alvarez impact extinction hypothesis, *J. Geophys. Res.*, **96**, 20879–20905.

Izett, G.A., Dalrymple, G.B. and Snee, L.W. (1991), $^{40}Ar/^{39}Ar$ age of the Cretaceous-Tertiary boundary tektites from Haiti, *Science*, **252**, 1539–1542.

Jablonski, D. and Raup, D.M., eds. (1986), *Patterns and Processes in the History of Life*, Springer-Verlag, New York.

Jansa, L.F., Aubry, M.-P., and Gradstein, F.M. (1990), Comets and extinctions; Cause and effect?, *Geol. Soc. Amer., Spec. Pap.*, **247**, 223–232.

Jessberger, E.K., Christoforidis, A., and Kissel, J. (1988), Aspects of the major element composition of Halley's dust, *Nature*, **332**, 691–695.

Johnson, K.R. (1993), Extinctions at the antipodes, *Nature*, **366**, 511–512.

Jones, H.D. (1988), Halley and comet impacts, *J. Brit. Astron. Assoc.*, **98**, 339.

Jones, E.M. and Kodis, J.W. (1982), Atmospheric effects of large body impacts: The first few minutes, *Geol. Soc. Amer., Spec. Pap. 190*, 175–186.

Keller, G., MacLeod, N., Lyons, J.B., and Officer, C.B. (1993), Is there evidence for Cretaceous-Tertiary boundary-age deep-water deposits in the Caribbean and Gulf of Mexico?, *Geology*, **21**, 776–780.

Korobejnikov, V.P., Chushkin, P.I. and Shurshalov, L.V. (1991), Combined simulation of the flight and explosion of a meteoroid in the atmosphere, *Solar System Res.*, **25**, 242–255.

Krogh, T.E., Kamo, S.L., Sharpton, V.L., Marin, L.E., and Hildebrand, A.R. (1993), U–Pb ages of single shocked zircons linking distal K/T ejecta to the Chicxulub crater, *Nature*, **366**, 731–734.

Kyte, F.T. and Wasson, J.T. (1986), Accretion rate of extraterrestrial material: Iridium deposited 33 to 67 million years ago, *Science*, **232**, 1225–1229.

Lindsay, J.F. and Srnka, L.J. (1975), Galactic dust lanes and lunar soil, *Nature*, **257**, 776–777.

Liritzis, I. (1993), Cyclicity in terrestrial upheavals during the Phanerozoic Eon, *Ql. J. Roy. Astron. Soc.*, **34**, 251–260.

Love, S.G. and Brownlee, D.E. (1993), A direct measurement of the terrestrial mass accretion rate of cosmic dust, *Science*, **262**, 550–553.

Lowe, D.R., Byerly, G.R., Asaro, F., and Kyte, F.J. (1989), Geological and geochemical record of 3400-million-year-old terrestrial meteorite impacts, *Science*, **245**, 959–962.

MacLeod, N. and Keller, G. (1991), Hiatus distributions and mass extinctions at the Cretaceous/Tertiary boundary, *Geology*, **19**, 497–501.

Mark, K. (1987), *Meteorite Craters*, University of Arizona Press, Tucson.

Marvin, U.B. (1990), Impact and its revolutionary implications for geology, *Geol. Soc. Amer., Spec. Pap.*, **247**, 147–154.

Matese, J.J. and Whitmire, D.P. (1986), Planet X as the source of the periodic and steady-state flux of short period comets, pp. 297–309 in *The Galaxy and the Solar System*, eds. R. Smoluchowski, J.N. Bahcall and M.S. Matthews, University of Arizona Press, Tucson.

Maurrasse, F.J.-M.R. and Sen, G. (1991), Impacts, tsunamis, and the Haitian Cretaceous-Tertiary boundary layer, *Science*, **252**, 1690–1693.

Mazaud, A., Laj, C., de Seze, L., and Verosub, K.L. (1983), Long time-scale fluctuations in the evolution of the Earth, *Nature*, **304**, 328–330; see also **311**, 396 (1984).

McCrea, W.H. (1975), Ice ages and the galaxy, *Nature*, **255**, 607–609.

McCrea, W.H. (1981), Long time-scale fluctuations in the evolution of the Earth, *Proc. Roy. Soc. A*, **375**, 1–41.

McKinnon, W.B. (1992), Killer acid at the K/T boundary, *Nature*, **357**, 15–16.

McLaren, D.J. and Goodfellow, W.D. (1990), Geological and biological consequences of giant impacts, *Ann. Rev. Earth Planet. Sci.*, **18**, 123–171.

Melosh, H.J. (1988a), *Impact Cratering: A Geologic Process*, Oxford University Press, Oxford and New York.

Melosh, H.J. (1988b), The rocky road to panspermia, *Nature*, **332**, 687–688.

Melosh, H.J. (1991), Atmospheric impact processes, *Adv. Space Res.*, **11(6)**, 87–93.

Melosh, H.J., Schnieider, N., Zahnle, K., and Latham, D. (1990), Ignition of global wildfires at the Cretaceous-Tertiary boundary, *Nature*, **343**, 251–254.

Muller, R. (1988), *Nemesis: The Death Star*, Weidenfeld and Nicolson, New York.

Napier, W.M. (1989), Terrestrial catastrophism and galactic cycles, pp. 133–167 in Clube (1989).

Napier, W.M. and Clube, S.V.M. (1979), A theory of terrestrial catastrophism, *Nature*, **282**, 455–459.

Nininger, H.H. (1942), Cataclysm and evolution, *Pop. Astron.*, **50**, 270–272.

Officer, C.B., Hallam, A., Drake, C.L., and Devine, J.D. (1987), Late Cretaceous and paroxysmal Cretaceous/Tertiary extinctions, *Nature*, **326**, 143–149.

O'Keefe, J.D. and Ahrens, T.J. (1982), The interaction of the Cretaceous/Tertiary extinction bolide with the atmosphere, ocean and solid Earth, *Geol. Soc. Amer., Spec. Pap. 190*, 103–120.

O'Keefe, J.D. and Ahrens, T.J. (1989), Impact production of CO_2 by the Cretaceous/Tertiary extinction bolide and the resultant heating of the Earth, *Nature*, **338**, 247–249.

Olsson-Steel, D. (1986), The origin of the sporadic meteoroid component, *Mon. Not. Roy. Astron. Soc.*, **219**, 47–73.

Olsson-Steel, D. (1987), Collisions in the solar system–IV. Cometary impacts upon the planets, *Mon. Not. Roy. Astron. Soc.*, **227**, 501–524.

Öpik, E.J. (1958), On the catastrophic effects of collisions with celestial bodies, *Irish Astron. J.*, **5**, 34–36.

Orth, C.J., Attrep, M., Jr., and Quintana, L.R. (1990), Iridium abundance patterns across bio-event horizons in the fossil record, *Geol. Soc. Amer., Spec. Pap.*, **247**, 45–59.

Peale, S.J. (1989), On the density of Halley's comet, *Icarus*, **82**, 36–49.

Perlmutter, S. and Muller, R.A. (1988), Evidence for comet storms in meteorite ages, *Icarus*, **74**, 369–373.

Perlmutter, S., Muller, R.A., Pennypacker, C.R., Smith, C.K., Wang, L.P., White, S., and Yang, H.S. (1990), A search for Nemesis: Current status and review, *Geol. Soc. Amer., Spec. Pap.*, **247**, 87–91.

Pilkington, M. and Grieve, R.A.F. (1992), The geological signature of terrestrial impact craters, *Rev. Geophys.*, **30**, 161–181.

Pollack, J.B., Toon, O.B., Ackerman, T.P., McKay, C.P., and Turco, R.P. (1983), Environmental effects of an impact-generated dust cloud: Implications for the Cretaceous-Tertiary extinctions, *Science*, **219**, 287–289.

Pope, K.O., Ocampo, A.C., and Duller, C.E. (1991), Mexican site for K/T impact crater?, *Nature*, **351**, 105.

Prinn, R.G. and Fegley, B. (1987), Bolide impacts, acid rain, and biospheric traumas at the Cretaceous-Tertiary boundary, *Earth Planet. Sci. Lett.*, **83**, 1–15.

Rabinowitz, D. (1993), The size distribution of the Earth-approaching asteroids, *Astrophys. J.*, **407**, 412–427.

Rampino, M.R. (1982), A non-catastrophist explanation for the iridium anomaly at the Cretaceous/Tertiary boundary, *Geol. Soc. Amer., Spec. Pap. 190*, 455–460.

Rampino, M.R. (1992), Gaia versus Shiva: Cosmic effects on the long-term evolution of the terrestrial biosphere, pp. 382–390 in *Scientists on Gaia*, eds. S.H. Schneider and P.J. Boston, MIT Press, Cambridge, MA and London.

Rampino, M.R. and Caldeira, K. (1992), Episodes of terrestrial geologic activity during the past 260 million years: a quantitative approach, *Cel. Mech. Dyn. Astron.*, **54**, 143–159.

Rampino, M.R. and Caldeira, K. (1993), Major episodes of geologic change: correlations, time structure and possible causes, *Earth Planet. Sci. Lett.*, **114**, 215–227.

Rampino, M.R. and Haggerty, B.M. (1994), Impacts and mass extinctions, in Gehrels (1994).

Rampino, M.R. and Stothers, R.B. (1984a), Terrestrial mass extinctions, cometary impacts, and the Sun's motion perpendicular to the galactic plane, *Nature*, **308**, 709–712.

Rampino, M.R. and Stothers, R.B. (1984b), Geologic rhythms and cometary impacts, *Science*, **226**, 1427–1431.

Rampino, M.R. and Stothers, R.B. (1986), Geologic periodicity and the galaxy, pp. 241–259 in *The Galaxy and the Solar System*, eds. R. Smoluchowski, J.N. Bahcall, and M.S. Matthews, University of Arizona Press, Tucson.

Rampino, M.R. and Stothers, R.B. (1988), Flood basalt volcanism during the past 250 million years, *Science*, **241**, 663–668.

Raup, D.M. (1986a), Biological extinction in Earth history, *Science*, **231**, 1528–1533.

Raup, D.M. (1986b), *The Nemesis Affair: A Story of the Death of the Dinosaurs and the Ways of Science*, Norton, New York.

Raup, D.M. (1991), *Extinction: Bad Genes or Bad Luck?*, Norton, New York.

Raup, D.M. and Sepkoski, J.J. (1984), Periodicity of extinctions in the geologic past, *Proc. Natl. Acad. Sci. (USA)*, **81**, 801–805.

Raup, D.M. and Sepkoski, J.J. (1986), Periodic extinction of families and genera, *Science*, **231**, 833–836.

Raup, D.M. and Sepkoski, J.J. (1988), Testing for periodicity of extinction, *Science*, **241**, 94–96.

Rhodes, M.C. and Thayer, C.W. (1991), Mass extinctions: Ecological selectivity and primary production, *Geology*, **19**, 877–880.

Rickman, H. (1989), The nucleus of comet Halley: Surface structure, mean density, gas and dust production, *Adv. Space Res.*, **9(3)**, 59–71.

Rickman, H., Kamél, L., Festou, M., and Froeschlé, Cl. (1987), Estimates of masses, volumes and densities of short-period comet nuclei, pp. 471–481 in *Symposium on the Diversity and Similarity of Comets*, European Space Agency, Paris, **SP–278**.

Robin, E., Froget, L., Jéhano, C., and Rocchia, R. (1993), Evidence for a K/T impact in the Pacific Ocean, *Nature*, **363**, 615–617.

Roddy, D.J., Schuster, S.H., Rosenblatt, M., Grant, L.B., Hassig, P.J., and Kreyenhagen, K.N. (1987), Analytical simulation of a 10 km diameter asteroid impact into a terrestrial ocean: Part 1—Summary, *Lunar Planet. Sci. Conf., Abstracts*, **17**, 720–721.

Sack, N.J. (1988), Organic-chemical clues to the theory of impacts as a cause of mass extinctions, *Earth, Moon and Planets*, **43**, 131–143.

Sagdeev, R.Z., Elyasberg, P.E. and Moroz, V.I. (1988), Is the nucleus of comet Halley a low density body?, *Nature*, **331**, 240–242.

Sarjeant, W.A.S. (1990), Astrogeological events and mass extinctions: global crises or scientific chimaerae?, *Modern Geology*, **15**, 101–112.

Schmidt, R.M. and Holsapple, K.A. (1982), Estimates of crater size for large-body impact: Gravity-scaling results, *Geol. Soc. Amer., Spec. Pap.* **190**, 93–102.

Sepkoski, J.J. (1990), The taxonomic structure of periodic extinctions, *Geol. Soc. Amer., Spec. Pap.*, **247**, 33–44.

Sepkoski, J.J. and Raup, D.M. (1986), Periodicities in marine extinction events, pp. 3–36 in *Dynamics of Extinctions*, ed. D.K. Elliot, Wiley-Interscience, New York.

Seyfert, C.K. and Sirkin, L.A. (1979), Earth History and Plate Tectonics, Harper and Row, New York.

Shapley, H. (1921), Note on a possible factor in changes of geological climate, *J. Geol.*, **29**, 502–504.

Sharpton, V.L. and Ward, P.D., eds. (1990), Global catastrophes in Earth history: An interdisciplinary conference on impacts, volcanism, and mass mortality, *Geol. Soc. Amer., Spec. Pap.*, **247**.

Sharpton, V.L., Dalrymple, G.B., Marin, L.E., Ryder, G., Schuraytz, B.C., and Urrutia-Fucugauchi, J. (1992), New links between the Chicxulub impact structure and the Cretaceous/Tertiary boundary, *Nature*, **359**, 819–821.

Sharpton, V.L., Burke, K., Caamargo-Zanoguera, A., Hall, S.A., Lee, D.S., Marin, S.E., Suárez-Reynoso, G., Quezada-Muñeton, J.M., Spudis, P.D., and Urrutia-Fucugauchi, J. (1993), Chicxulub multiring impact basin: Size and other characteristics derived from gravity analysis, *Science*, **261**, 1564–1567.

Shaw, H.R. (1987), The periodic structure of the natural record, and nonlinear dynamics, *Eos*, **68**, 1651–1665.

Shea, J.H. (1982), Twelve fallacies of uniformitarianism, *Geology*, **10**, 455–460.

Sheehan, P.M. and Russell, D.A. (1994), Faunal changes following the Cretaceous-Tertiary impact: Using paleontological data to assess the hazards of impacts, in Gehrels (1994).

Shoemaker, E.M. (1983), Asteroid and comet bombardment of the Earth, *Ann. Rev. Earth Planet. Sci.*, **11**, 464–494.

Shoemaker, E.M. and Izett, G.A. (1992), Stratigraphic evidence from Western North America for multiple impacts at the K/T boundary, *Lunar and Planet. Sci. Conference, Abstracts*, **23**, 1293–1294.

Silver, L.T. and Schultz, P.H., eds. (1982), Geological Consequences of Impacts of Large Asteroids and Comets on the Earth, *Geol. Soc. Amer., Spec. Pap.* **190**.

Sleep, N.H., Zahnle, K.J., Kasting, J.F., and Morowitz, H.J. (1989), Annihilation of ecosystems by large asteroid impacts on the early Earth, *Nature*, **342**, 139–142.

Smit, J. (1994), Extinctions at the Cretaceous-Tertiary boundary: The link to the Chicxulub crater, in Gehrels (1994).

Standish, E.M. (1993), Planet X: No dynamical evidence in the optical observations, *Astron. J.*, **105**, 2000–2006.

Stanley, S.M. (1987), *Extinctions*, Scientific American Books, New York.

Steel, D. (1992), Cometary supply of terrestrial organics: Lessons from the K/T and the present epoch, *Origins Life Evol. Biosphere*, **21**, 339–357.

Steel, D.I., Asher, D.J., Napier, W.M., and Clube, S.V.M. (1992), Are impacts correlated in time?, in Gehrels (1994).

Stetter, K.O., Huber, R., Blöchl, E., Kurr, M., Eden, R.D., Fielder, M., Cash, H., and Vance, I. (1993), Hyperthermophilic archaea are thriving in deep North Sea and Alaskan oil reservoirs, *Nature*, **365**, 743–745.

Stigler, S.M. and Wagner, M.J. (1987), A substantial bias in nonparametric tests for periodicity in geophysical data, *Science*, **238**, 940–945.

Stothers, R.B. (1985), Terrestrial record of the Solar System's oscillation about the galactic plane, *Nature*, **317**, 338–341.

Stothers, R.B. (1988), Structure of Oort's comet cloud inferred from terrestrial impact craters, *The Observatory*, **108**, 1–9.

Stothers, R.B. (1989), Structure and dating errors in the geologic time scale and periodicity in mass extinctions, *Geophys. Res. Lett.*, **16**, 119–122.

Stothers, R.B. (1992), Impacts and tectonism in Earth and Moon history of the past 3800 million years, *Earth, Moon & Planets*, **58**, 145–152.

Sykes, M.V. and Walker, R.G. (1992a), The Nature of Comet Nuclei, pp. 587–591 in *Asteroids, Comets, Meteors 1991*, eds. A.W. Harris and E. Bowell, Lunar and Planetary Institute, Houston, TX.

Sykes, M.V. and Walker, R.G. (1992b), Cometary dust trails. I. Survey, *Icarus*, **95**, 180–210.

Talbot, R.J. and Newman, M.J. (1977), Encounters between stars and dense interstellar clouds, *Astrophys. J. Suppl.*, **34**, 295–308.

Thaddeus, P. and Chanan, G.A. (1985), Cometary impacts, molecular clouds, and the motion of the Sun perpendicular to the galactic plane, *Nature*, **314**, 73–75.

Toon, O.B., Zahnle, K., Turco, R.P., and Covey, C. (1994), Environmental perturbations cause by asteroid impacts, in Gehrels (1994).

Torbett, M.V. (1989), Solar system and galactic influences on the stability of the Earth, *Palaeogeography, Palaeoclimatology, Palaeoecology*, **75**, 3–33.

Torbett, M.V. and Smoluchowski, R. (1984), Orbital stability of the unseen solar companion linked to periodic extinction events, *Nature*, **311**, 641–642.

Turco, R.P., Toon, O.B., Ackerman, T.P., Pollack, J.B., and Sagan, C. (1991), Nuclear winter: Physics and physical mechanisms, *Ann. Rev. Earth Planet. Sci.*, **19**, 383–422.

Urey, H.C. (1973), Cometary collisions and geological periods, *Nature*, **242**, 32–33.

Van den Bergh, S. (1989), Life and death in the inner solar system, *Publ. Astron. Soc. Pacific*, **101**, 500–509.

Vandervoort, P.O. and Sather, E.A. (1993), On the resonant orbit of a solar companion star in the gravitational field of the galaxy, *Icarus*, **105**, 26–47.

Walliser, O.H., ed. (1986), *Global Bio-Events*, Springer-Verlag, Berlin.

Watson, F. (1941), *Between the Planets*, Blakiston, Philadelphia.

Weissman, P.R. (1985), Terrestrial impactors at geological boundary events: Comets or asteroids?, *Nature*, **314**, 517–518.

Weissman, P.R. (1986), Are cometary nuclei primordial rubble piles?, *Nature*, **320**, 242–244.

Weissman, P.R. (1990), The cometary impactor flux at the Earth, *Geol. Soc. Amer., Spec. Pap.*, **247**, 171–180.

Wetherill, G.W. (1989), Cratering of the terrestrial planets by Apollo objects, *Meteoritics*, **24**, 15–22.

Wetherill, G.W. (1991) End products of cometary evolution: Cometary origin of Earth-crossing bodies of asteroidal appearance, pp. 537–556 in *Comets in the Post-Halley Era*, eds. R.L. Newburn, Jr., M. Neugebauer, and J. Rahe, Kluwer, Dordrecht.

Whipple, F.L. (1950), A comet model I. The acceleration of comet Encke, *Astrophys. J.*, **111**, 375–394.

Whitmire, D.P. and Jackson, A.A. (1984), Are periodic mass extinctions driven by a distant solar companion?, *Nature*, **308**, 713–715.

Whitmire, D.P. and Matese, J.J. (1985), Periodic comet showers and planet X, *Nature*, **313**, 36–38.

Wickramasinghe, N.C., Hoyle, F., and Rabilizirov, R. (1989), Greenhouse dust, *Nature*,

341, 28.

Williams, G.E., ed. (1981), *Megacycles*, Hutchinson Ross, Stroudsburg, Pennsylvania.

Williams, G.E. (1986), The Acraman impact structure: Source of ejecta in late Precambrian shales, South Australia, *Science*, **233**, 200–203.

Wolbach, W.S., Lewis, R.S., and Anders, E. (1985), Cretaceous extinctions: Evidence for wildfires and search for meteoric material, *Science*, **230**, 167–170.

Wolbach, W.S., Gilmour, I., Anders, E. Orth, C.J., and Brooks, R.R. (1988), Global fire at the Cretaceous–Tertiary boundary, *Nature*, **334**, 665–669.

Wolbach, W.S., Gilmour, I., and Anders, E. (1990), Major wildfires at the Cretaceous–Tertiary boundary, *Geol. Soc. Amer., Spec. Pap.*, **247**, 391–400.

Wolfe, J.A. (1991), Palaeobotanical evidence for a June impact winter at the Cretaceous/Tertiary boundary, *Nature*, **352**, 420–423. But see also *Nature*, **356**, 295–296.

Yabushita, S. (1991), A statistical test for periodicity hypothesis in the crater formation rate, *Mon. Not. Roy. Astron. Soc.*, **250**, 481–485.

Yabushita, S. (1992), Periodicity in the crater formation rate and implications for astronomical modelling, *Cel. Mech. Dyn. Astron.*, **54**, 161–178.

Yabushita, S. and Allen, A.J. (1989), On the effect of accreted interstellar matter on the terrestrial environment, *Mon. Not. Roy. Astron. Soc.*, **238**, 1465–1478.

Zahnle, K. (1990), Atmospheric chemistry by large impacts, *Geol. Soc. Amer., Spec. Pap.*, **247**, 271–288.

Zahnle, K. (1992), Airburst Origin of Dark Shadows on Venus, *J. Geophys. Res.*, **97**, 10243–10255.

Zahnle, K. and Grinspoon, D. (1990), Comet dust as a source of amino acids at the Cretaceous/Tertiary boundary, *Nature*, **348**, 157–160.

Zhao, M. and Bada, J. (1989), Extraterrestrial amino acids in Cretaceous/Tertiary boundary sediments at Stevns Klint, Denmark, *Nature*, **339**, 463–465.

9
The Contemporary Hazard of Cometary Impacts

D. Morrison

ABSTRACT Cosmic impacts pose a continuing hazard of loss of human life and property. Significant contemporary risk is associated with projectiles in the energy range from about 10 megatons of TNT up to the size of the K/T impactor. The lower threshold for damage is defined by the atmosphere of the Earth, which effectively shields us from smaller projectiles. Up to energies of about a gigaton of TNT, the effects are local or regional for impacts on the land, or coastal for ocean impacts, which can generate large tsunamis. A greater risk is associated with still larger impacts, which are capable of causing global ecological catastrophe, possibly leading to mass mortality from starvation and epidemics. If such an impact took place anywhere on Earth during our lifetimes, we would each be in danger, independent of where the projectile struck. Statistical estimates indicate that each human on this planet runs a risk of roughly 1 in 20,000 of dying from this cause. Prudence suggests that we should be concerned about such impacts and seek ways of avoiding them or mitigating their consequences. The primary objective of any program to deal with this hazard is to determine whether or not such a near-term impact is likely. The best approach for the asteroidal component is a comprehensive telescopic survey, which can discover all Earth-crossing asteroids larger than 1 km in diameter and provide decades of warning in which to plan ways to deflect or destroy a threatening object. Long-period comets, however, pose a much greater challenge, since they cannot be discovered long in advance of a possible impact, their orbits are harder to predict, and they are significantly more difficult to deflect or destroy.

9.1 Introduction

Recognition of a contemporary impact hazard is a result of changing perceptions of catastrophic influences on evolutionary history. Although a few prescient planetary scientists (e.g., Baldwin, 1949) had recognized that the impact origin for the lunar craters implied a comparable cratering history for the Earth, the majority of scientists first became aware of the importance of impacts in the history of life from the identification of a cosmic impactor as the probable cause of the K/T extinction (Alvarez et al., 1980). The Alvarez paper not only established the presence of an extraterrestrial component in the K/T boundary layer, but also suggested a mechanism–short-term global climate change caused by stratospheric dust–whereby an impact could influence the entire ecosphere. Thus an impact of negligible energy or momentum on a geophysical scale could have profound influence on the history of life (e.g., Steel, this volume).

It is a relatively short step from recognition of the role of impacts in evolutionary history to the realization that impacts also pose a contemporary hazard. As long as this hazard is limited to the fall of meteorites (which generally strike the ground at terminal velocity for an object falling through air), there is not a major hazard from impacts; indeed, there is no authenticated record of any human having been killed by a meteorite. Larger impacting objects may not be ablated and decelerated to harmless speeds, however, but can penetrate to the lower atmosphere or surface at nearly their initial velocity. The only historic record of such a destructive hypervelocity impact occurred on June 30, 1908, when the atmospheric disruption of an approximately 60 m stony object released 10–20 megatons of explosive energy over a wilderness region near the Tunguska River in Siberia (Krinov, 1963; Chyba et al., 1993). There were at most two deaths. The best-known impact crater on the Earth, Meteor Crater in Arizona, also represented an explosive energy of 10–20 megatons, but in this case the projectile was metallic rather than stony. Here there were certainly no casualties, since the impact occurred 50,000 years ago, before the occupation of North America by humans. This historical record, while limited, confirms our intuitive expectation that large hypervelocity impacts are much rarer than comparably destructive volcanic eruptions or earthquakes, and thus that the risk they pose is less than that of many other natural hazards.

In contrast, the hazard associated with very large (but exceedingly rare) impacts can be considerable, as a consequence of their global influence. Unlike other natural disasters, which are limited in size by various physical constraints (e.g., an earthquake will relieve crustal stress whenever it approaches some maximum value), there is no practical limit to the magnitude of an impact (Chapman and Morrison, 1989). Comets and asteroids exist today that are substantially larger than the K/T impactor. It is not impossible that an object as large as Chiron (~200 km diameter) could eventually strike the Earth, possibly bringing the history of terrestrial life to an abrupt conclusion, as discussed in detail by Zahnle and Sleep (this volume).

Even at smaller scales than the K/T event, an impact could blanket the Earth's atmosphere with dust and precipitate a severe, if brief, environmental catastrophe. This potential for globally catastrophic impacts was recognized in the first conference on the impact hazard, held in 1981 at Snowmass Colorado under NASA sponsorship, with Gene Shoemaker as chairperson. Although the report of that Snowmass Conference was never published (see Chapman and Morrison, 1989), it has influenced all subsequent thinking about the contemporary impact hazard. The public has also been exposed to vivid accounts of such environmental disasters through the popular science fiction novel *Lucifer's Hammer* (Niven and Pournelle, 1977), as well as by press accounts of the nuclear winter debate of the 1980s.

In addition to its much larger scale, the impact hazard differs from other natural calamities in that we are capable, at least in principle, of avoiding major impacts. We cannot suppress an earthquake or tame a hurricane, but we probably could deflect a cosmic projectile before impact. These possibilities were first evaluated in 1968 by a group of MIT students who were given the assignment of devising a system to protect the planet from a hypothetical impact by the asteroid Icarus.

Even 25 years later, their *Project Icarus Report* (Kleiman, 1968) retains its value. One reason we are interested in the impact hazard is that, unlike the weather, we can do something about it.

In this chapter I discuss the nature of the impact hazard and estimate its magnitude relative to other natural disasters, following the previous analyses published by Chapman and Morrison (1994) and Morrison et al. (1994). I then summarize possible mitigation schemes, with particular attention to the special problems posed by comets, which are more difficult to deal with than asteroids. The chapter concludes with a brief discussion of public policy issues raised by the impact danger.

9.2   Impactor Population

The first step in analyzing cosmic impacts is to determine the flux of comets and asteroids striking the Earth. The average total impact flux over the past 3 Gyr can be found from the crater density on the lunar maria (Hartmann et al., 1986). Given the absence of significant erosion or geological activity since the end of widespread volcanism on the Moon, these lava plains are a scorecard for the integrated flux of cosmic debris in near-Earth space. With adjustments for the Earth's greater gravity, the average flux at the top of the Earth's atmosphere can be derived from the lunar flux (Zahnle and Sleep, this volume). We can also estimate the contemporary impact rate on the Earth from a census of existing Earth-crossing asteroids and comets together with estimates of their dynamical lifetimes (Shoemaker, 1983; Shoemaker et al., 1990; Steel, this volume).

The Earth-crossing asteroids are primarily rocky or metallic fragments of main-belt asteroids or extinct comet nuclei (Wetherill, 1988; 1989). In addition, there are impacts from active comets. The bulk properties of the projectiles thus range from metallic (like iron–nickel meteorites) to stony (like chondritic meteorites) to cometary (low-density silicates, organics, and volatiles). Because the lunar cratering history, which is our primary resource for determining the impact flux, does not distinguish among stony, metallic, or volatile-rich projectiles, I consider all of these Earth–crossing asteroids and comets together in evaluating the current impact hazard. We will distinguish cometary from asteroidal impacts, however, when we later discuss ways to deflect or destroy incoming projectiles.

By the end of 1992, 163 Earth-crossing asteroids had been cataloged. The largest are 1627 Ivar and 1580 Betulia, with diameters of 6-8 km. From the asteroid discovery statistics and the lunar crater size distribution, Shoemaker (1983) derived the now-standard population and size-distribution for Earth-crossing asteroids larger than a given diameter (see also Morrison, 1992). A small additional source of impactors are long-period and short-period active comets; they amount to about 5% of the asteroid flux. However, comets average significantly greater impact velocities than asteroids and, therefore, constitute a larger share (perhaps as much as 25%) of the impact hazard.

In his chapter, Steel discusses evidence for possibly significant variations in the impact flux, noting that comet "showers" could lead to major short-term increases in the impact hazard (see also Steel et al., 1994). However, the current impact rate as derived from the known population of Earth-crossing comets and asteroids is at least approximately equal to the long-term average flux recorded by the lunar craters, suggesting that we do not live in a special time with respect to possible large-scale variations. For the purposes of this evaluation of the current impact hazard, it is adequate to treat impacts as occurring at their long-term average rate.

Fig. 9.1 summarizes the average terrestrial impact flux as a function of projectile kinetic energy, measured in megatons (1 MT = 4.2×10^{15} J). Also shown for comparison is the associated diameter for a stony asteroid striking the Earth with a velocity of 20 km s^{-1}. These flux values, derived by Shoemaker more than a decade ago (Shoemaker, 1983), still represent our best estimate of the average impact rate for energies ranging from about 1 megaton to millions of megatons (Rabonowitz et al., 1994). Note the steep power-law dependence of size on energy, a relationship equally apparent when we look at the size distribution of small lunar craters.

9.3   Nature of the Hazard

Based on the average flux of comets and asteroids striking the Earth, we can evaluate the danger posed by impacts of different magnitudes. Of particular interest are the threshold for penetration though the atmosphere, and the threshold at which impacts have major global as well as local effects.

9.3.1   Penetration Through the Atmosphere

The atmosphere protects us from smaller projectiles. Fig. 9.1 indicates that an impact event with the energy of the Hiroshima nuclear bomb occurs roughly annually, while a megaton event is expected at least once per century. Obviously, however, such relatively common events are not destroying cities or killing people. Even at megaton energies, most bodies break up and are consumed before they reach the lower atmosphere.

Unless they are composed of iron, meteors as large as a few tens of meters are subject to aerodynamic stresses during atmospheric entry that cause fragmentation and dispersal at high altitude (Chyba et al., 1993; Hills and Goda, 1993). The height of fragmentation depends primarily on the meteoroid's physical strength, with loose cometary aggregates and carbonaceous meteorites fragmenting at altitudes above 30 km, as observed for most meteors. Stronger stony objects can penetrate to perhaps 20 km, but still cause little or no damage when they explode. Such atmospheric explosions are routinely detected by military surveillance satellites; the largest reported bolide exploded 21 km above the western Pacific Ocean on 1 February 1994 with an estimated energy of 50–100 kilotons (Tagliaferri, 1994; and Private Communication).

If the projectile is large enough and strong enough to penetrate below about

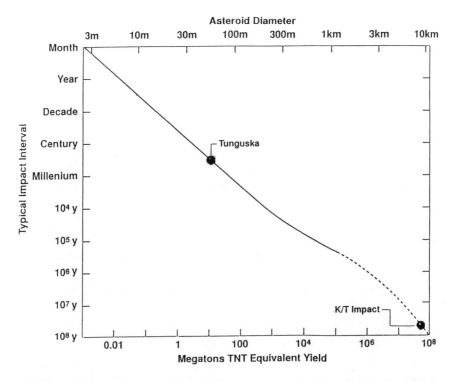

FIGURE 9.1. Typical intervals between impacts equal to or larger than the specified yields, based on Shoemaker (1983), extended to an estimate of the K/T impact. Equivalent asteroid diameters are shown, assuming 20 km s^{-1} impact velocity and 3 g cm^{-3} density.

15 km altitude before it explodes, the resulting airburst can be highly destructive. Numerical models of atmospheric fragmentation and dispersal show that rocky objects >50 m in diameter (10 megatons) and cometary objects >100 m (100 megatons) penetrate deep enough to pose significant hazards (Chyba et al., 1993; Hills and Goda, 1993; Chyba, 1994). The area of devastation scales approximately as the explosive yield to the two-thirds power and is somewhat greater for an airburst than for a groundburst explosion (Glasstone and Dolan, 1977).

Tunguska provides a calibration, with a shock wave sufficient to fell trees over an area of 10^5 hectares. The yield of the Tunguska blast has been estimated at 10–20 MT from microbarograph measurements in Europe and comparison of the blast damage with later nuclear airbursts. If we assume that the radius of forest devastation would apply also to destruction of many buildings, the area of damage is given by

$$A = 10^4 Y^{2/3} \tag{9-1}$$

where Y is the yield in megatons and A is in hectares (Chyba et al., 1993; Hills and Goda, 1993; Chapman and Morrison, 1994). For example, this formula indicates that at 100 MT, the radius of destruction is 20 km and the fraction of the Earth's surface that is effected is 0.001%. At 100,000 MT, the radius is 200 km and the area is about 0.1%.

9.3.2   Globally Catastrophic Impacts

At sufficiently great energies, an impact has global consequences. An obvious, if extreme, example is the K/T impact 65 million years ago, which disrupted the global ecosystem and led to a major mass extinction (see also the chapter by Steel). This impact of a 10–15 km object released approximately 10^8 MT of energy and excavated a crater (Chicxulub in Mexico) >200 km in diameter. Among the environmental consequences were devastating wildfires and changes in atmospheric and oceanic chemistry as well as a dramatic short-term perturbation in climate produced by some 10^{16} kg of submicrometer dust injected into the stratosphere (Alvarez et al., 1980; Sharpton and Ward, 1990; Toon et al., 1982, 1994).

Chapman and Morrison (1994) and Morrison et al. (1994) have argued that even much smaller projectiles can perturb the global climate by injecting dust into the stratosphere, producing climate changes sufficient to reduce crop yields and precipitate mass human starvation (but not a mass extinction). They define a globally catastrophic impact as one that would lead to the death of more than a quarter of the world's population. The primary cause of such death would probably be starvation and disease caused by crop failures, but such an unprecedented catastrophe might also threaten the stability and future of modern civilization.

To appreciate the scale of the global catastrophe thus defined, we must be clear what it is not. Chapman and Morrison (1994) have defined a catastrophe far larger than the effects of the great World Wars but far smaller than the K/T impact. Although such a catastrophe could destabilize modern civilization, it would not threaten the survival of the human species. Nor would such a catastrophe leave any fossil record, although this fact would provide scant comfort to those people who experienced such an impact.

9.3.3   Threshold for a Globally Catastrophic Climate Perturbation

A drop of a few degrees Centigrade in surface temperature over many months is sufficient to reduce crop yields dramatically (Harwell and Hutchinson, 1989; Covey et al., 1990). The energy threshold for a globally catastrophic impact is therefore determined by the explosive yield required to loft sufficient submicrometer dust into the stratosphere to lower the surface temperature by this amount. Models originally developed for nuclear winter (Turco et al., 1991) suggest that a global stratospheric dust layer with an optical depth >2 would depress land temperatures by as much as $10°$ C, and lead to the climatic effects associated with nuclear winter. In the most comprehensive discussion of these effects, Toon et al. (1994) find that major climate effects would almost certainly follow from an impact with a yield of 10^6 MT, and perhaps from one as much as an order of magnitude less. For a stony object striking at 20 km s^{-1}, a million megatons corresponds to a diameter of about 2 km.

Following the analysis by Toon et al. (1994), Morrison et al. (1994) associated a global catastrophe (with agricultural collapse and the death of more than a quarter of the world population) with a threshold yield between 10^5 MT and 10^6 MT. Their nominal threshold energy of 3×10^5 MT, corresponding to a diameter for an asteroid of about 1.4 km, yields an average frequency (from Fig. 9.1) of about once per 300,000 years. Given the many uncertainties in determining this threshold energy, the frequency of such global catastrophes can only be estimated to lie between 10^5 years and 10^6 years. Particularly uncertain in this estimate is the response of society to such an "impact winter."

9.4   Hazard Analysis

We now address the scale of destruction and the numerical hazard associated with impacts of various magnitude, still following the analyses published by Chapman and Morrison (1994) and Morrison et al. (1994). By numerical hazard is meant the probability of death for an individual due to this event.

If the average zone of mortality from an impact explosion can be identified roughly with the area of devastation defined earlier from the Tunguska example, then the average number of fatalities per impact can be calculated (using the average world population density of 0.1 person per hectare) as

$$N = 10^3 Y^{2/3}. \tag{9-2}$$

Of course, most of the world's population is concentrated in a tiny fraction of the surface area, so most such impacts strike uninhabited parts of the globe and kill few people or none at all, just as in the 1908 example of Tunguska. Indeed, the average interval between a Tunguska-scale impact on an urban zone is about 10,000 years for the current population distribution.

At energies above about 1000 MT, the danger from tsunami damage for ocean impacts dominates the mortality calculations (Pike, 1993; Hills et al., 1994; Morrison et al., 1994). Because tsunamis decline in amplitude in proportion to a $1/r$ power law rather than the roughly $1/r^2$ dependence of the direct blast wave, their effects can extend thousands of kilometers beyond the point of impact. For ocean impacts, therefore, the distinction between local and global scales is blurred. Preliminary calculations suggest that the average mortality N exceeds the values calculated from the formula above by a factor of several when the role of tsunamis is accounted for.

Above the threshold for global catastrophe, the number of fatalities is (by definition) more than one fourth of the Earth's total population. Just above the nominal threshold of 3×10^5 MT, the expected average local casualties from the direct blast are tens of millions, while indirect casualties are (by definition of the global threshold) one and a half billion. This difference reflects the different areas affected: less than 1% of the Earth's surface for the direct blast, but the entire surface for the indirect effects.

Is the greatest risk from the smaller, more frequent impacts that cause local or regional destruction, from larger impacts near the global catastrophic threshold, or from the even rarer mass extinction events? Chapman and Morrison (1994) have integrated the cumulative damage over segments of the impactor size-frequency distribution to yield some quantitative estimates, as summarized in Table 9.1.

The smaller, frequent events larger than the 10 MT atmospheric cut-off yield equivalent annual fatality rates of tens of deaths per year for the current world population, up to perhaps a few hundred per year if the tsunami risk is included. In reality, of course, centuries to tens of thousands of years pass with practically no fatalities, followed by an impact that would kill huge numbers if it struck a major urban area or generated a large tsunami. Such risks, while not insignificant, are substantially lower than the mortality associated with many smaller and more frequent natural hazards, such as earthquakes, hurricanes, volcanic eruptions, floods, or mud slides. Thus the hazard from Tunguska-like impacts does not inspire special concern or justify heroic efforts either to predict such events or to attempt to avert them.

At the opposite, high-yield extreme, a mass-extinction event would probably kill almost the entire world population, but these impacts are so infrequent that the annual fatality rate is only a few hundred (the world's present population of five billion people killed every 10–30 million years). We are relatively certain that no asteroid large enough (diameter >5 km) to cause a disaster of this magnitude exists today in an Earth-crossing orbit. However, we cannot exclude the possibility of a large comet appearing at any time and dealing the Earth such a devastating blow—a blow that might lead to human extinction. I will return to this issue later when I discuss mitigation strategy.

The greatest risk is associated with impacts near the threshold for global catastrophe, which we have taken as 3×10^5 MT. If one and a half billion people are

TABLE 9.1. Fatality Rates (Equivalent Annual Average Deaths) as a Function of Impact Energy.

Type of Event	Energy (MT)	Diameter	World deaths/yr
High atmosphere break-up	<10	<50 m	<1
Tunguska-like events	$10 - 2 \times 10^3$	50 m - 300 m	55
Sub-global land impacts	$2 \times 10^3 - 5 \times 10^5$	300 m - 2 km	30
Sub-global ocean impacts	$2 \times 10^3 - 5 \times 10^5$	300 m - 2 km	300(?)
Threshold global catastrophes	$10^5 - 10^6$	1 km - 2 km	3000
Mass extinction events	$>10^7$	>4 km	<300

killed by events that occur every 300,000 years or so, then the fatality rate would be of the order several thousand per year. The annual risk per individual is thus near one in a million, with a corresponding lifetime risk of about 1:20,000. This risk is still small compared to many other causes of death, both natural and accidental, but it approaches that associated with other natural disasters, such as earthquakes or severe storms. Further, there is the qualitative difference noted previously between the globally catastrophic impact and all other natural dangers, in that only the impact has the potential to kill billions and destabilize civilization.

Human reaction to such risk estimates varies greatly, especially since the impact hazard represents such an extreme combination of low probability together with high consequence (Slovic, 1987; Morrison et al., 1994). Since no one has been killed by an impact in all of recorded history, it is easy to dismiss the risk as negligible and to regard those who express concern as alarmist. Further, the calculated annual risk of about one in a million is near the level at which many persons consider risks to be effectively zero. On the other hand, modern industrial societies spend large sums to protect people from even less likely hazards, ranging from hurricanes to terrorist attacks to trace quantities of possibly carcinogenic toxins in food and water.

9.5   Risk Reduction and Mitigation

For most natural hazards, risk reduction or mitigation strategies can only deal with the consequences of the disaster. Thus, for example, we cannot stop an earthquake or even reduce its force, but we can mandate higher standards in building construction and develop plans to treat casualties and restore public services after the event. If impacts could be predicted weeks or months in advance, similar approaches could be taken, including evacuation of the populace from the target area. In addition, however, the possibility exists of avoiding the impact entirely by deflecting or destroying the projectile before it hits.

Prediction is the key to any strategy of risk reduction. Because impacts are such rare events, it makes no sense to maintain standing defenses against asteroids and comets or even to have civil-defense plans in place to deal with an impact if one takes place. By any standards of cost-effectiveness, such efforts are nonproductive or even counterproductive, in that the risks inherent in the defense preparations or the opportunity costs represented by preparing to deal with events that are unlikely to happen for centuries or millennia far exceed any positive effects of such efforts. (See also discussions by Sagan and Ostro (1994a, b) and Harris et al. (1994) of additional risks that may be associated with development of asteroid defenses.) Any approach to this problem must first consider the search for potentially hazardous asteroids and comets.

9.5.1   Impact Prediction

The source of impacts that is easiest to deal with is the Earth-crossing asteroids and their cousins, the short-period comets; fortunately, this population also represents the primary impact risk. It is possible with current technology to undertake a complete census (down to some size limit) of the asteroids and comets in near-Earth space in order to determine if any will strike the Earth in the near future. Such an approach, called the Spaceguard Survey, has been outlined in a NASA study undertaken at the request of the U.S. Congress (Morrison, 1992, 1993).

The strategy recommended in the *Spaceguard Survey Report* is to use ground-based astronomical telescopes to survey the sky for asteroids and short-period comets, which can be identified by their motion with respect to the stellar background. Modeling of the problem (Bowell and Muinonen, 1994) suggests that to achieve a reasonably complete census within a nominal period of two decades requires that the telescopes be capable of detecting asteroids at a range of 200 million km down to the size for which a complete survey is desired. For completeness to a 1 km diameter, it is required to reach visual magnitude 22, implying telescopes of about 2 m aperture. The *Spaceguard Survey Report* specifically recommends an international network of six telescopes of this size, scanning a total of 6,000 square degrees of sky each month. The scale of the task requires large format CCD detectors and automated detection schemes similar to those pioneered by the Arizona Spacewatch Camera (Gehrels, 1991). Such a system would discover several thousand Earth-crossing asteroids and comets each year and achieve 90%

completeness to the 1 km level.

The most likely result of the Spaceguard Survey or its functional equivalent would be to find no asteroids or comets larger than 1 km on Earth-impact trajectories. Indeed, we expect that the survey would need to achieve a high level of completeness down to 100 m before there is a substantial probability of finding an impactor within a 100 year time horizon. However, if any asteroid exists that might be capable of precipitating a global catastrophe within our lifetimes or the lifetimes of our children, there is a high probability that the Spaceguard Survey would find it with several decades of advanced warning. A measured international response could therefore be undertaken, as is discussed below. This is not true for the long-period comets, however, a subject we return to later.

9.5.2   Deflection or Destruction

In the 1968 MIT study of a postulated impact threat from Icarus (Kleiman, 1968), it was decided that there was insufficient time to deflect the asteroid, so the only solution was to destroy it. If, however, a specific impact threat can be identified several years (or more) in advance, then deflection becomes the strategy of choice.

The most straightforward approach to deflection for an object in a short-period orbit is to apply an impulse that changes the orbital period. If such an impulse is applied several orbits before the threatened collision, only a very small velocity change (a few cm s^{-1}) is required (Ahrens and Harris, 1992). Based on discussions at several workshops that have examined this problem (e.g., Canavan et al., 1994), the optimum way to impart such an impulse without risking accidental disruption of the body would be a stand-off neutron bomb explosion. Bombs of the appropriate yield exist within current nuclear arsenals, and in many examples that have been studied where warning time is ample, the required yields are quite modest (less than a megaton).

Before attempting to deflect an asteroid, we would almost certainly wish to study it in some detail with scientific flyby or orbiter spacecraft. It would be especially useful to characterize its size, shape, and spin vector as accurately as possible before attempting the deflection impulse. Given sufficient warning, such prior investigation should be possible, and indeed the deflection exercise itself could be undertaken in stages spread over several orbits.

If the time available for the deflection is relatively short, proportionately larger velocity changes may be required. Since it is important to avoid fracturing the asteroid, each impulsive velocity change should be kept smaller than the surface gravity of the object (Harris et al., 1994), leading to a strategy of multiple small nuclear impulses rather than a single multimegaton blast.

The alternative approach of destroying a projectile requires the application of much larger energy. In order to avoid making the situation worse by converting the incoming projectile from a cannon ball into a cluster bomb, it is necessary to do more than simply disrupt the object. Sufficient energy must be applied either to pulverize it (ensuring that no fragment is large enough to survive atmospheric entry) or to disperse all of the fragments so that none strikes the Earth. Destruction is the strategy of choice only when the warning time is so short that deflection is impractical. This is more likely to be the case for a cometary impactor than an asteroid.

9.5.3 Problems with Comets

Short-period comets will be discovered by the Spaceguard Survey and can be treated as equivalent to asteroids for purposes of the present discussion. However, long-period comets pose an entirely different class of problems. Since these comets do not reside in near-Earth space, no survey can hope to discover them decades before they approach the Earth. They might appear at any time, and models investigated for the Spaceguard Survey (Bowell and Muinonen, 1994) indicate that discovery cannot be expected more than a few months before the possible impact. Further, since no means exists to distinguish a faint object (either comet or asteroid) against the dense stellar background in the Milky Way, it is possible for a comet to "sneak up" on the Earth, escaping detection until it is only a few weeks from impact. Thus a perpetual survey is required to detect long-period comets, and even with such a survey we cannot be sure of success.

If only a few month's warning is available, it may not be possible to deflect an incoming comet. Since such a deflection impulse would have to be applied close to the Earth, a very large energy is required, risking unintentional disruption of the comet unless a series of smaller nuclear impulses is possible. Complicating matters further, there would be no time to precede the deflection mission with survey or reconnaissance missions to characterize the target. Such cases naturally lead toward a destruction strategy requiring very large bombs carried on high-energy rockets that can be launched on a few weeks warning.

Additional problems arise from the indeterminacy of the orbit for a long period comet entering the solar system. In the case of an asteroid, there is ample time to refine the orbit using both optical and radar astrometry in order to determine whether a future impact will take place. In contrast, an incoming comet can be observed over only a short orbital arc, and radar would not be available until it was very close to the Earth. Even if optical observations were sufficient to determine a high probability of impact, non-gravitational effects are likely to alter the orbit in unpredictable ways. It is probably impossible to provide reliable information on the trajectory of such a comet in advance of the decision to intercept it.

It is easy to imagine various nightmare scenarios for incoming comets. Suppose that a very large comet is discovered near the orbit of Jupiter, and that based on a few weeks of observations the Earth is found to lie within the error-ellipse of the projected orbit. But suppose also that the area of this error ellipse at the Earth is hundreds of times greater than the area of the Earth itself. What should be done? Should a crash program of defense be initiated even though the probability of impact is less than 1%? At what point is action justified? Suppose further that additional tracking does not shrink the ellipse greatly because of the onset of nongravitational forces. Perhaps the comet disappears behind the Sun for a few months and cannot be seen at all (the scenario described in the novel *Lucifer's*

Hammer (Niven and Pournelle, 1977)). How would a decision on countermeasures be made? Could a deflection be considered when its unintentional effect might be to *cause* an impact rather than to *avoid* one? What organization would take responsibility for dealing with these issues?

The situation described above is of concern for several reasons. Most important, it is the unexpected appearance of a large comet that has the most potential for truly catastrophic results, from which recovery may be impossible. Second, even though the probability of such an event is extremely small, the chances of a false alarm resulting from the indeterminant orbit of the comet are much larger. There are likely to be thousands of such false alarms for every real impact.

As a practical matter, we can expect that the first "threats" that society will have to face will be due to either a cometary false alarm or the detection of a relatively small asteroid (like Tunguska) on a collision course with the Earth. In either case the pressures for an active defense may be overwhelming, in spite of arguments that the probability of significant damage is quite low. The odds of either of these events happening within the lifetimes of ourselves or our grandchildren are fairly large.

9.6   Summary and Conclusions

The threat of impacts of asteroids and comets has existed since our planet's birth, but it has only recently been recognized as having practical consequences for modern life. The chances that civilization might be disrupted or destroyed by such an impact are very low, but they are not zero. By some measures, they are comparable to other hazards that society takes very seriously.

At the present time, only a small fraction of the potentially threatening objects have been discovered, but it is well within our capability to inventory most of the population of larger Earth-crossing asteroids and short-period comets. Unfortunately, there is no similar strategy that will ensure a long warning time for long-period or new comets. The asteroidal problem does not require the development and maintenance of expensive and inherently risky "space defenses" in advance of the discovery of a specific threat. A comprehensive sky survey would likely provide decades or more warning, permitting decisions about specific deflection schemes to be deferred until they really are required. However, a comet might be identified on an impact trajectory with a lead time of only a few months; should we prepare in advance to deal with such a contingency, probably at great expense, even though it is extremely unlikely?

The development of space defenses is an inherently hazardous undertaking. The existence of nuclear weapons and launch vehicles carries its own risk, since these might be used against targets on Earth rather than in space. In addition, as pointed out by Sagan and Ostro (1994a, b) and Harris et al. (1994), the technology developed to deflect threatening asteroids could be used as well to alter the course of a benign asteroid so that it would strike the Earth. The fact that such a suicidal act would require that a government be under the control of a madman does not make it impossible.

As public recognition of the impact hazard grows, such questions will increasingly fuel the hazard-mitigation debate. The July 1994 impact of P/Shoemaker–Levy 9 with Jupiter inevitable raises the issue of "could it happen here?" There may be demands for governmental action to protect the Earth from impacts. However, the impact hazard must be considered in parallel with, and balanced against, debates over society's priorities of dealing with other potential ecological disasters, many of which appear to be more severe than the impact hazard. Over the very long term, however, impacts must be dealt with to ensure the survival of civilization and the human species itself.

Acknowledgments: I am grateful to Clark Chapman, with whom I have collaborated on many of the results that are reviewed in this chapter; thanks also to Ted Bowell, Greg Canavan, Alan Harris, Steve Ostro, Carl Sagan, Gene Shoemaker, Duncan Steel, Ed Tagliaferri, Brian Toon, and Kevin Zahnle for numerous stimulating discussions of the issues discussed here, and to Kevin Zahnle and Chris McKay for their reviews of this paper.

9.7   References

Ahrens, T.J. and W.A. Harris (1992). Deflection and fragmentation of near-Earth asteroids. *Nature* **360**: 429–443.

Alvarez, L.W., W. Alvarez, F. Asaro, and H.V. Michel (1980). Extraterrestrial cause for the Cretaceous-Tertiary extinction. *Science* **208**: 1095–1108.

Baldwin, R.B. (1949). *The Face of the Moon*. University of Chicago Press, Chicago.

Bowell, E. and K. Muinonen (1994). Earth-crossing asteroids and comets: Groundbased search strategies. In *Hazards Due to Comets and Asteroids* (T. Gehrels, ed.). University of Arizona Press, Tucson, pp. 149–198.

Canavan, G.H., J. Solem, and G.H. Canavan (1994). Near-Earth object interception workshop. In *The Hazard of Impacts by Comets and Asteroids* (T. Gehrels, ed.). University of Arizona Press, Tucson, pp. 93–126.

Chapman, C.R. and D. Morrison (1989). *Cosmic Catastrophes*. Plenum Press, New York, p. 302.

Chapman, C.R. and D. Morrison (1994). Impacts on the Earth by asteroids and comets: Assessing the hazard. *Nature* **367**: 33–40.

Chyba, C.F. (1993). Explosions of small Spacewatch asteroids in the Earth's atmosphere. *Nature* **363**: 701–703.

Chyba, C.F., P.J. Thomas, and K.J. Zahnle (1993). The 1908 Tunguska explosion: Atmospheric disruption of a stony asteroid. *Nature* **361**: 40–44.

Covey, C., S.J. Ghan, J.J. Walton, and P.R. Weissman (1990). Global environmental effects of impact-generated aerosols: Results from a general circulation model. In it Global Catastrophes in Earth History (V.L. Sharpton and P.D. Ward, eds.), Geological Society of America Special Paper 247: 263–270.

Gehrels, T. (1991). Scanning with charge-coupled devices. *Space Sci. Rev.* **58**: 347–375.

Glasstone, S. and P.J. Dolan (1977). *The Effects of Nuclear Weapons*, 3rd edn. U.S. Government Printing Office, Washington, DC.

Harris, A.W., G.H. Canavan, C. Sagan, and S.J. Ostro (1994). The deflection dilemma: Use versus misuse of technologies for avoiding interplanetary collision hazards. In *The Hazard of Impacts by Comets and Asteroids* (T. Gehrels, ed.). University of Arizona Press, Tucson, pp. 1145–1156.

W.K. Hartmann, R.J. Phillips, and G.J. Taylor, editors (1986). *Origin of the Moon*. Lunar and Planetary Institute, Houston, TX.

Harwell, M.A. and T.C. Hutchinson (1989). *Environmental Consequences of Nuclear War II: Ecological and Agricultural Effects*, 2nd edn. Wiley, New York.

Hills, J.G. and M.P. Goda (1993). The fragmentation of small asteroids in the atmosphere. *Astron. J.* **105**: 1114–1144.

Hills, J.G., I.V. Nemtchinov, S.P. Popov, and A.V. Teterev (1994). Tsunami generated by small asteroid impacts. In *Hazards Due to Comets and Asteroids* (T. Gehrels, ed.). University of Arizona Press, Tucson, pp. 779–790.

Kleiman, L.A., ed. (1968). *Project Icarus*. MIT Press, Cambridge MA, p. 162.

Krinov, E.E. (1963). The Tunguska and Sikhote-Alin meteorites. In *The Moon, Meteorites, and Comets* (B.M. Middlehurst and G.P. Kuiper, eds.). University of Chicago Press, Chicago, pp. 208–234.

Morrison, D., ed. (1992). *The Spaceguard Survey: Report of the NASA International Near-Earth-Object Detection Workshop*. Unpublished NASA report.

Morrison, D. (1993). An international program to protect the Earth from impact catastrophe: Initial steps. *Acta Astronautica* **30**: 11–16.

Morrison, D., C.R. Chapman, and P. Slovic (1994). The impact hazard. In *Hazards Due to Comets and Asteroids* (T. Gehrels, ed.). University of Arizona Press, Tucson, pp. 59–92.

Pike, J. (1993). The big splash. Unpublished manuscript.

Niven, L. and J. Pournelle (1977). *Lucifer's Hammer*. Fawcett Crest Books, New York.

Rabinowitz, D.L., E. Bowell, E.M. Shoemaker, and K. Muinonen (1994). The population of Earth-crossing asteroids. In *Hazards Due to Comets and Asteroids* (T. Gehrels, ed.). University of Arizona Press, Tucson, pp. 285–312.

Sagan, C. and S.J. Ostro (1994a). Dangers of asteroid deflection. *Nature* **368**: 501.

Sagan, C. and S.J. Ostro (1994b). Long-range consequences of interplanetary collisions. *Issues in Sci. and Tech.* **20**: 67–72.

Sharpton V.I. and P.D. Ward, eds. (1990). *Global Catastrophes in Earth History: An Interdisciplinary Conference on Impacts, Volcanism, and Mass Mortality*. Geological Society of America Special Paper 247, Boulder.

Shoemaker, E.M. (1983). Asteroid and comet bombardment of the Earth. *Annual Rev. Earth Planet. Sci.* **11**: 461–494.

Shoemaker, E.M, R.F. Wolff, and C.S. Shoemaker (1990). Asteroid and comet flux in the neighborhood of the Earth. In *Global Catastrophes in Earth History* (V.L. Sharpton an P.D. Ward, eds.), Geological Society of America Special Paper 247: 155–170.

Slovic, P. (1987). Perception of risk. *Science* **236**: 280–285.

Steel, D.I., D.J. Asher, W.M. Napier, and S.V.M. Clube (1994). Are impacts correlated in time? In *Hazards Due to Comets and Asteroids* (T. Gehrels, ed.). University of Arizona Press, Tucson, pp. 463–478.

Tagliaferri, E., R. Spalding, C. Jacobs, S.P. Worden, and A. Erlich (1994). Detection of meteoroid impacts by optical sensors in Earth orbit. In *Hazards Due to Comets and Asteroids* (T. Gehrels, ed.). University of Arizona Press, Tucson, pp. 199–220.

Toon, O.B., J.B. Pollack, T.P. Ackerman, R.P. Turco, C.P. McKay, and M.S. Liu (1982).

Evolution of an impact-generated dust cloud and its effects on the atmosphere. In *Geological Implications of Impacts of Large Asteroids and Comets with the Earth* (L.T. Silver and P.H. Schultz, eds.), Geological Society of America Special Paper 190: 187–200.

Toon, O.B., K. Zahnle, R.P. Turco, and C. Covey (1994). Environmental perturbations caused by impacts. In *Hazards Due to Comets and Asteroids* (T. Gehrels, ed.). University of Arizona Press, Tucson, pp. 791–826

Turco, R.P. et al. (1991). Nuclear winter: Physics and physical mechanisms. *Annual Rev. Earth Planet. Sci.* 19: 383–422.

Weissman, P.R. (1982). Terrestrial impact rates for long and short-period comets. In *Geological Implications of Impacts of Large Asteroids and Comets with the Earth* (L.T. Silver and P.H. Schultz, eds.), Geological Society of America Special Paper 190: 15–24.

Weissman, P.R. (1991). The cometary impactor flux at the Earth. In *Global Catastrophes in Earth History* (V.L. Sharpton and P.D. Ward, eds.) Geological Society of America Special Paper 247: 171–180.

Wetherill, G.W. (1988). Where do the Apollo objects come from? *Icarus* **76**: 1–18.

Wetherill, G.W. (1989). Cratering of the terrestrial planets by Apollo objects. *Meteoritics* **24**: 15–22.

10

^{26}Al and Liquid Water Environments in Comets

M. Podolak and D. Prialnik

ABSTRACT Whether liquid water can exist in comets depends on the realtive rates of heating and cooling in the interior. Historically, as our understanding of the relevant processes has improved, conclusions have changed. We review the different computations performed in the past, and discuss the effects of heating by radioactives in general, ^{26}Al in particular, and the phase transition from amorphous to crystalline ice. Cooling mechanisms include heat conduction through the ice, and heat carried by the flow of gas through the porous nucleus. The most recent models show that conditions for producing liquid water can be achieved only under rather contrived circumstances.

10.1   Introduction

At first sight, comets seem to present environments that are unlikely to be conducive to the formation of life. They are formed far from the Sun, and spend most of their time in regions where the temperature is well below 100 K. Some years ago, however, Hoyle and Wickramasinghe (1978) pointed out that organic molecules are observed in interstellar clouds. They argued that as interstellar grains formed in these clouds, complex prebiotic molecules could form in these grains. If these grains are incorporated into comets, then the building blocks of life are incorporated with them. Each time a comet approaches the Sun, its surface is exposed to ultraviolet radiation which can alter these molecules. The results are quickly frozen and remain in storage until the next perihelion passage. In this way, they proposed, life developes on comets, and eventually finds its way to Earth.

These arguments are confined to the formation of life in the outer layers of the comet, because that is the region where the solar energy is deposited. It is for precisely this reason that the model has difficulties. In the first place the time of irradiation is relatively short, being limited to that part of the orbit when the comet is near the Sun. In addition, only a small fraction of a comet's life is spent in an orbit that brings it into the inner solar system. The majority of its life is spent in the Oort cloud, where the rate of energy deposition from the Sun (and other stars) is very low. Finally, at each perihelion passage large areas of the surface are lost, so that it is difficult to see how the irradiated material can be stored for more than one orbital period.

If it were possible to heat the *interior* of the comet to a high enough temperature, in particular, to the melting point of water, then an environment much more amenable to the formation of life could be imagined. At the time Hoyle and Wickramasinghe (1978) wrote their paper, however, this did not seem possible. The most likely mechanism for such heating was the decay of radioactve materials in the cometary interior. Comets are known to contain dust (e.g., Delsemme, 1982), and if this dust contains radioactive isotopes, then these will provide a source of heat even while the comet is in the Oort cloud. Whipple and Stefanic (1966) computed expected heating rates due to the radioactive decay of ^{40}K, ^{235}U, ^{238}U, and ^{232}Th, assuming that for each species the heating is given by

$$Q_i(t) = Q_{0i} e^{-\lambda_i t}, \tag{10-1}$$

where λ_i is the decay constant of the ith species and Q_{0i} is its initial rate of radioactive heat generation per gram of nucleus material. In such a case, for a mixture of n radioactive species in a nucleus that is initially at the uniform temperature, T_0, the central temperature, T_c is given by

$$T_c = \sum_{j=1}^{n} \frac{2\rho R^2 Q_{0j}}{\pi^2 K} \sum_{m=1}^{\infty} (-1)^m \frac{\exp(-\lambda_j t) - \exp\left(-\frac{m^2 \pi^2 K}{\rho c R^2} t\right)}{\left(\frac{\lambda_j \rho c R^2}{\pi^2 K} - m^2\right)}$$

$$+ 2T_0 \sum_{m=1}^{\infty} (-1)^{m+1} \exp\left(-\frac{m^2 \pi^2 K}{\rho c R^2} t\right). \tag{10-2}$$

Here K is the thermal conductivity of the nucleus, R its radius, ρ its density, and c its heat capacity. Whipple and Stefanic (1966) found that there is indeed a temperature rise. They assumed that the radioactive materials were present in the solar ratio to C, N, and O, and that these latter materials, in the form of CH_4, NH_3, and H_2O, made up the bulk of the comet. For comets of 5 km and 10 km radius, and an assumed initial temperature of 0 K, Whipple and Stefanic (1966) found that the temperature rose to a peak of about 90 K after about 10^8 yr, and began to cool quickly after about 10^9 yr. It should be noted that the abundances of radioactives used by Whipple and Stefanic (1966) are about a factor of 10 lower than the values used today, and their thermal diffusion coefficient, based on the conductivity of insulating powders, is about a factor of about 30 lower (Prialnik et al., 1987). These two effects oppose each other, but the net effect, in this case, is to actually *reduce* the expected central temperature. In any event, these computations showed that liquid water could not be expected in cometary interiors.

Larger bodies store heat more efficiently, so if the cometary nucleus were significantly larger than 10 km, higher central temperatures would be attained. Lewis (1971) computed the effect of radioactive heating in icy satellites from a simple model. He assumed an energy source similar to that of Whipple and Stefanic (1966), and a thermal diffusion coefficient about 10^3 higher (corresponding to solid crystalline ice). If the body has a radius R, and a density ρ, then the maximum surface temperature (neglecting solar heating) will be determined by a balance between heating and radiation to space

$$T_{\text{surf}} \leq \left(\frac{Q R \rho}{3 \sigma} \right)^{1/4}. \qquad (10\text{-}3)$$

This temperature is very weakly dependent on mass, varying as $M^{-1/12}$ for a given density, and is less than 10 K for cases of interest. If we assume the temperature gradient between the center and the surface to be constant, then for bodies less than 500 km in radius, Lewis (1971) found a difference between central and surface temperatures of less than 50 K. This strengthened the notion that sufficient heat to produce liquid water could not be generated in the cometary interior. Note, however, that because Lewis was considering icy satellites, he used a value of the conductivity consistent with crystalline ice. Comets are widely believed to be composed of amorphous ice (Prialnik and Bar-Nun, 1990, 1992) which has a conductivity lower by about a factor of at least 20 (and possibly more, see below), and therefore the temperature difference between the center and the surface should be higher by a corresponding factor. On the other hand, the assumption of constant gradient is too simple, and a more detailed model is necessary.

As it turns out, there are additional factors that were not included in the studies mentioned above. As our understanding of the cometary interior has improved, our assessment of the possibility of liquid water in the interior has changed with it. In what follows, we will show how the present view on the existence of liquid water evolved. The historical approach is deliberate because it carries a message: Our view of the cometary interior has changed as our understanding of the various processes has progressed. There is no reason to suppose that our present picture is the final one.

10.2 ^{26}Al as a Heat Source

In 1976 Lee et al. presented strong evidence that ^{26}Al had been present in the early solar nebula. This radioactive element is potentially a much more powerful heat source than any of the other species mentioned. The only difficulty is that as a result of its relatively short half-life (7.4×10^5 yr), the formation of comets would have had to occur shortly after the formation of the ^{26}Al in order for comets to contain any appreciable abundance of this material. Since the only known source of ^{26}Al at the time was a supernova explosion, this led some workers to suggest that a supernova had exploded near the time of formation of the solar system, and had, quite possibly, even triggered it (Cameron and Truran, 1977). Today, the idea of a supernova trigger for the formation of the solar system is no longer popular, largely because there is no convincing observational evidence for this kind of star formation (Cameron, 1985).

In recent years γ-ray observations from the *HEAO 3* and *Solar Maximum Mission* satellites have detected ^{26}Al dispersed throughout the galaxy (Mahoney et al., 1982; Share et al., 1985). A number of sources have been suggested including ordinary novae (Clayton, 1984), asymptotic giant branch stars (Cameron, 1985), and massive stars (Dearborn and Blake, 1985). In any event, it is now believed that the production rate in the galaxy is high enough so that a steady state abundance of ^{26}Al would have existed at the time of the formation of the solar system, and would have been incorporated into cometary nuclei. The detection of ^{26}Mg excesses in Ca–Al-rich inclusions found in the famous Allende meteorite (Wasserburg and Papanastassiou, 1982), indicating in situ decay of ^{26}Al during the early stages of the solar system formation, lends further support to this view. It remains to evaluate the effect of ^{26}Al on the behavior of the cometary nucleus.

10.3   Melting of Cometary Interiors—Simple Model

One of the earliest studies of the effect of ^{26}Al on cometary interiors was by Irvine et al. (1980). As we have noted above, the rate of heat loss depends on the radius of the body. Cometary nuclei are generally taken to be no more than a few kilometers in radius, and such bodies cool quickly. Irvine et al. argued that the comets we see are members of a population that includes larger bodies such as Chiron. In this case, radii as large as several 100 km must be considered. They assumed an ^{26}Al/^{27}Al ratio of $\sim 5 \times 10^{-5}$ in accordance with the determination of Lee et al. (1976) for Allende inclusions. This implies a total heat release of 8.4×10^{10} erg g^{-1}, as compared to the latent heat of melting of 3.3×10^9 erg g^{-1}.

Assuming that this is spread over two half-lives, gives an average heating rate of $Q \simeq 2 \times 10^{-3}$ erg g^{-1} s^{-1}. We can now make a simple estimate of the surface temperature using (10-1). Taking $\rho \sim 2$ g cm^{-3} and $R \sim 100$ km gives $T_{surf} \leq 120$ K. This is still well below the melting temperature of ice. Irvine et al. (1980), however point out that the heat does not get to the surface immediately, but must diffuse through the nucleus. If the conductivity is sufficiently low, enough heat may build up so that melting is possible. The timescale for transporting heat by conduction is

$$\tau = \frac{R^2}{\kappa}, \tag{10-4}$$

where R is the radius of the nucleus, and κ is the thermal diffusivity. They estimate a diffusivity of $\kappa = 0.004$, which gives a $\tau \simeq 10^7$ yr for a 10 km radius body, but $\tau \simeq 10^9$ yr for a 100 km body. These timescales could be made even longer if the surface of the comet were covered with an insulating dust layer. In such a case, it is possible to imagine an environment where liquid water is available for times long enough for life to develop.

These early estimates of the effect of ^{26}Al were extended by an analytic model developed by Wallis (1980). He took the heating rate by ^{26}Al to be like that in (10-1) and assumed the thermal diffusivity was constant and equal to about 4.6×10^{-3}. This is quite close to the value appropriate for solid amorphous ice. He then developed an approximate analytic solution to the heat diffusion equation for this case of the form

$$T(r, t) = T_0 + \frac{Q\rho}{6K}(R^2 - r^2)e^{-\lambda t}, \tag{10-5}$$

where r is the radial distance, Q is the heating rate of ^{26}Al in ergs per seconds per gram of nucleus, and T_0 is the surface temperature. Wallis found that the central temperature could reach the melting point of water for comets larger than about 10 km in radius.

Wallis pointed out that there is another condition that must be satisfied before liquid water can be sustained: the pressure must be higher than the pressure at the triple point, about 6×10^3 dynes cm^{-2}. The hydrostatic pressure due to self-gravity is barely sufficient. Although water vapor would supply additional pressure, it can escape through pores and cracks in the ice. On the other hand, refreezing of vapor would tend to seal these openings. Wallis concluded that comets larger than about 10 km–15 km should have had melted cores that remained liquid for about 10^6 yr. Upon refreezing a hollow ice core would remain. Wallis suggests that this explains why some comets split into fragments of comparable sizes. The fluid core would provide an "excellently-protected environment for the development of biological systems."

10.4   Numerical Models I: The Effect of Amorphous Ice

In 1987 Prialnik et al. used a numerical heat diffusion code to investigate the details of the thermal evolution of cometary nuclei. The details of the code are described by Prialnik and Bar-Nun (1987). Simply, it is a numerical scheme for following the heat diffusion in a one-dimensional, spherical, comet. The differencing scheme conserves energy explicitly. Heating by radioactive materials was allowed for as in (10-1), the thermal diffusion coefficient was taken to be $\kappa = 3 \times 10^{-3}$ for amorphous ice. One of the important differences between this model and previous studies is that Prialnik et al. (1987) allowed for the phase change from amorphous to crystalline ice with a resulting latent heat release of 9×10^8 erg g^{-1} (Klinger, 1981). At the time of this study, it was believed that the transition between these two forms of ice occurred sharply at a temperature of 137 K.

This phase transition has an immediate effect. If there is sufficient ^{26}Al, then the center of the comet will be heated to above 137 K, and the amorphous ice will spontaneously transform to crystalline ice and release latent heat. This latent heat is sufficient to transform the adjacent layer. This layer too releases latent heat, and the transformation continues out to very near the edge of the nucleus.

In particular, for a 2.5 km radius comet with an initial mass fraction of ^{26}Al, $X_0(^{26}\text{Al}) = 7.0 \times 10^{-8}$, there was a period of about 1.6×10^5 yr of heating, after which the center of the nucleus reached the transition temperature. Within 2.3×10^4 yr nearly the entire nucleus was transformed to crystalline ice with the exception of a thin outer layer where the heat could be conducted to the surface sufficiently rapidly to be radiated away. Toward the end of the crystallization phase, the entire comet had a temperature equal to, or slightly above, 180 K except for the outer few meters which were cooled by radiation. Because the thermal diffusivity of crystalline ice is almost an order of magnitude larger than that for amorphous ice, the rate of cooling is more rapid and the temperature throughout the nucleus dropped quickly.

This outcome is not very sensitive to the precise value of $X_0(^{26}\text{Al})$. Even for $X_0(^{26}\text{Al}) = 3.3 \times 10^{-7}$, corresponding to the $^{26}\text{Al}/^{27}\text{Al}$ ratio estimated for Allende ($^{26}\text{Al}/^{27}\text{Al} = 5 \times 10^{-5}$), the peak temperature attained was $\sim$200 K. Indeed, if the $^{26}\text{Al}/^{27}\text{Al}$ ratio were as high as 5×10^{-5}, then all bodies larger than several hundred meters in radius would have undergone the transition to crystalline ice. In addition, bodies larger than $\sim$6 km in radius would have melted in the center. On the basis of this result, Prialnik et al. (1987) argued that since only amorphous ice can trap sufficient gas to explain the observed gas/ice ratios in comets (Bar-Nun et al., 1985, 1987), comets that are observed must contain substantial masses of amorphous ice. This serves to set an upper limit on the ^{26}Al mass fraction at the time of the comet's formation based on the size of the largest body which still retains substantial amounts of amorphous ice. Clearly, if this upper limit is low enough, there may never be enough heat generated to allow for melting of the interior.

A very similar model was studied by Haruyama et al. (1993) with two important additions: First, they took the transition from amorphous to crystalline ice to proceed at all temperatures at a rate given by (Schmitt et al., 1989)

$$\frac{\partial \rho_a}{\partial t} = -\lambda(T)\rho_a, \qquad (10\text{-}6)$$

where ρ_a is the density of amorphous ice in a given region, and

$$\lambda(T) = 1.05 \times 10^{13} e^{-5370/T} \text{ s}^{-1}. \qquad (10\text{-}7)$$

Second, they took the conductivity of amorphous ice to be some four orders of magnitude lower than the value given by Klinger (1980), based on the work of Kouchi et al. (1992). Although this extremely low conductivity is raised somewhat by the presence of dust it still gives a thermal diffusivity much lower than that used in any of the other models in the literature. The few differences between this model and that of Prialnik et al. (1987) stem mainly from this source.

As a result of the very low thermal conductivity the heat produced by radioactive decay is retained by the nucleus, and much higher temperatures could, in principle, be attained in the center. As the temperature rises, however, the ice begins to transform more quickly, and a crystalline ice front begins to move toward the surface of the nucleus similar to that found by Prialnik et al. (1987). The difference is that the transformation can be induced without the presence of any ^{26}Al. This result depends, of course, on the radius of the nucleus, and Haruyama et al. (1993) only report results for comets with radii of 5 km or larger, but in view of the low thermal diffusivity, smaller bodies will also crystallize, especially if some ^{26}Al is present. Haruyama et al. (1993) estimate that amorphous ice will be preserved in bodies less than about 2 km.

Once the ice crystallizes, its thermal diffusivity rises sharply, and cooling proceeds rapidly. As a result the temperature drops sharply. Even for cometary nuclei with radii of 100 km the central temperature remains below 200 K. The conditions required for producing liquid water are therefore never achieved.

A different situation is encountered when larger icy bodies, including ^{26}Al, are considered. Prialnik and Bar-Nun (1990) have shown that even a relatively low initial abundance of ^{26}Al would be capable of partially melting icy bodies with radii of a few hundred kilometers. The transition from almost no melting to considerable melting with increasing radius is found to be quite sharp: thus, for example, an initial mass fraction of ^{26}Al of 3×10^{-8} would melt the center of an icy sphere of 200 km radius up to 20% of its radius, representing less than 1% of the mass. The same mass fraction of ^{26}Al would melt a sphere of twice this radius up to 80% of the radius or more than 50% of the mass.

It seems therefore that the main ingredients necessary to initiate melting in the center of a comet nucleus are a large radius, a low thermal conductivity and the presence of some ^{26}Al. Fortunately, the first two conditions appear to be favored by recent observations. The detailed observations of comet P/Halley have shown that the greater part of the nucleus surface is dark and inactive, the active area constituting only $\sim$10% of the surface. Thus the total surface and hence volume of the comet nucleus turns out to be much larger than previously thought. At the same time the bulk density is found to be much lower, implying a very high porosity. In porous materials the thermal conductivity is considerably diminished (Smoluchowski, 1981), due to the reduced contact surface between grains. Porosity, however, introduces new effects and raises new problems.

10.5   Numerical Models II: The Effect of Trapped Gas and Porosity

The difficulty in reaching central temperatures high enough to initiate melting has been demonstrated in recent work by Prialnik and Podolak (1994). In this work the comet was modeled as a porous medium (Mekler et al. 1990), and the effect of trapped gas on the heat transport was studied. Here too the phase change from amorphous to crystalline ice was taken to be temperature-dependent, in accordance with the work of Schmitt et al. (1989), although the thermal conductivity of amorphous ice was kept at the value given by Klinger (1980). In most of the models heating was restricted to that due to ^{40}K, ^{232}Th, ^{235}U, and ^{238}U, assumed to be contained in the dust in chondritic proportions (Prialnik et al., 1987). In some runs ^{26}Al was also included. The mass fraction of dust in the nucleus was taken to be 0.5, and the dust was assumed to be uniformly distributed throughout the nucleus. The density of the water ice and the dust were taken to be 0.9 g cm^{-3} and 2.65 g cm^{-3}, respectively. The density of the mixture was assumed to be 0.5 g cm^{-3}, corresponding to a porosity of 0.63. The amorphous ice was assumed to contain 10%, by mass, of occluded carbon monoxide (CO). Transport of dust through the pores, and the build-up of a dust mantle were not included, although the sealing of pores by recondensation of the gas was treated.

As the cometary interior is heated, crystallization proceeds. This transition releases occluded gases such as CO. This gas flows through the porous medium, and helps to move heat toward the surface. The effective thermal conductivity is high enough so that, in the absence of heating by ^{26}Al, the temperature of the nucleus remains below 100 K for bodies with radii less than about 20 km.

For larger bodies the temperature in the center can be high enough so that the transition from amorphous to crystalline ice proceeds rapidly and significant amounts of occluded gas are released. On the one hand, the crystallization of the ice releases heat; on the other hand, the gas released in the process of crystallization is an efficient cooling mechanism. The competition between these two processes determines the depth at which the crystallization front stops. In addition, a large fraction of the outflowing CO gas condenses in the cooler outer layers of the nucleus, just outside the region of crystalline ice, forming a shell with an enhanced CO content. The structure of such bodies can be therefore described as consisting of three layers: a core of crystalline ice, a shell of amorphous ice enhanced in CO, and an outer mantle of essentially pristine composition. In these bodies the central temperatures surpass 100 K. At these temperatures the thermal behavior is controlled by sublimation/condensation of ice from/on the pore walls. This is an even more efficient heat conducting mechanism than either conduction or advection by flowing gas (see Prialnik, 1992). As a result, a flat temperature profile is maintained over a large fraction of the nucleus. Therefore, instead of a crystallization front advancing toward the surface, crystallization takes place throughout an extended region, where the fraction of crystalline ice gradually increases. The ice temperature will thus remain much below the melting point everywhere at all times.

The presence of ^{26}Al can change these results due to its much higher rate of radiogenic heat release. In this case, the central temperature rises quickly, and crystallization sets in at an early stage, adding another heat source. Sublimation of ice from the pore walls occurs here as well, but, if the vapor is not removed sufficiently rapidly, enough pressure might build up to arrest further sublimation. The competition in this case is between the increasing rate of sublimation, as the temperature rises, and the rate of vapor flow toward the surface. Only if the former is larger than the latter, and if, in addition, the rate of heat release by radioactive decay is larger still, would it become possible to raise the internal temperature significantly. In terms of timescales these conditions become:

$$\tau_{^{26}\text{Al}} < \tau_{\text{sublim}} < \tau_{\text{diff}},$$

where $\tau_{^{26}\text{Al}}$ is the decay time of ^{26}Al, $\tau_{\text{sublim}} = \rho/Sq$ (where S is the surface to volume ratio of the porous material and q is the rate of sublimation from the pore walls) is the timescale of sublimation, and τ_{diff} is the characteristic diffusion time of the gas through the porous nucleus. The first term is a constant, the second is very strongly temperature-dependent, while the last is only weakly affected by the temperature, but increases as the square of the cometary radius. Fig. 10.1 shows

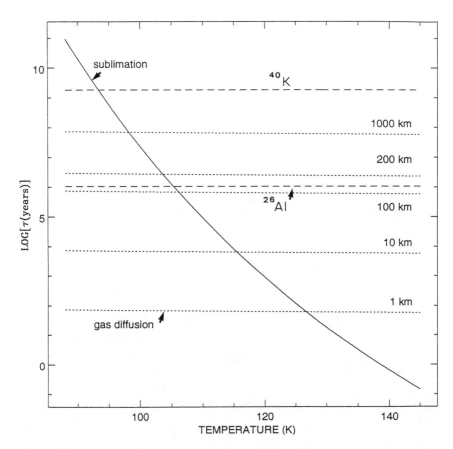

FIGURE 10.1. Comparison of relevant timescales. The horizontal dashed curves show the decay times of ^{26}Al and ^{40}K, while the dotted curves show the characteristic diffusion time of the gas through the porous nucleus for nuclei of different radii. The solid line is the sublimation timescale. Various parameters for this nuclear model are given in the text.

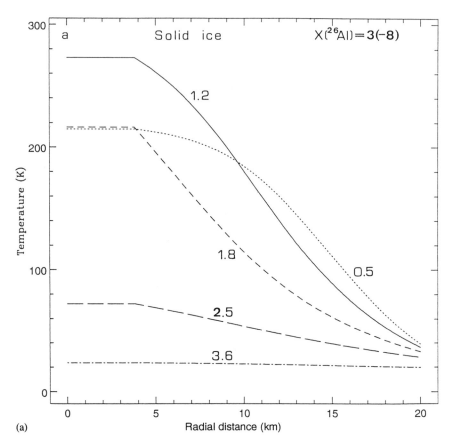

(a)

FIGURE 10.2. Evolution of the thermal profile throughout the nucleus for the case of (a) zero porosity. The various times (in millions of years) are indicated.

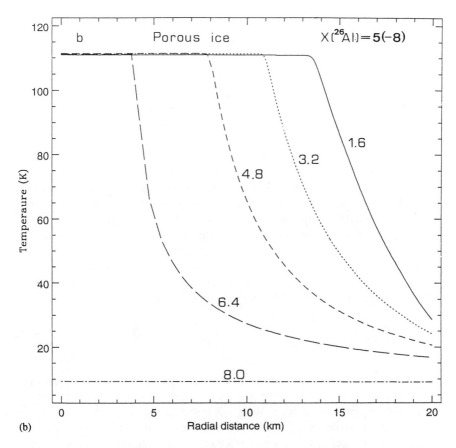

FIGURE 10.2. Evolution of the thermal profile throughout the nucleus for the case of (b) a porosity of 0.63. The various times (in millions of years) are indicated.

these timescales (on a logarithmic scale), as a function of temperature; τ_{diff} is plotted for different values of R for a porosity of 0.5. As can be seen from the figure, the above inequalities are satisfied only at temperatures well below the melting temperature of water. The presence of a more volatile ice such as CO will act to remove the heat more quickly than give by the simple timescale argument. In such a case, there would have to be sufficient ^{26}Al to supply the additional energy that is lost through this process. Fig. 10.2 shows the temperature profile for two typical cases for a 20 km nucleus. Fig. 10.2a shows the evolution of the temperature profile with time for the case of a crystalline ice nucleus with zero porosity. In such a case, with an ^{26}Al mass fraction of 3×10^{-8}, the temperature rises briefly above the melting point of water in the innermost few kilometers surrounding the center, and drops sharply with radius. This high temperature period lasts only of the order of a few million years. Fig. 10.2b shows the effect of a nonzero porosity. In this case cooling by vapor flow is so efficient, that the temperature never rises much above 100 K even for an ^{26}Al abundance of 5×10^{-8}, consistent with the expectations based on Fig. 10.1. Replacing crystalline with amorphous ice leads to more complex behavior (Prialnik and Podolak, 1994), but does not change that overall conclusion that liquid water can be formed in a comet only under conditions where the vapor is not free to flow through the medium.

10.6   Conclusions

As our understanding of the structure of the cometary nucleus has increased, our assessment of the possibility of the presence of liquid water has changed. At first it was thought that radioactive heating was insufficient. When ^{26}Al was discovered to be a viable heat source, the picture grew more optimistic. The exothermic transition from amorphous to crystalline ice provided yet another heat source. The most recent improvement in the models, the more careful modeling of the effects of a porous medium, seems to return a more pessimistic appraisal. The very act of heating the ice initiates an efficient cooling mechanism. If the old models indicated a (slim) possibility of water formation in the interior of comet nuclei and its preservation for significant periods of time, the new models seem to rule out such a possibility. Our understanding of the structure, composition and evolution of cometary nuclei is still rather poor, however, in spite of the rapidly accumulating observational data. In fact, recent data have only complicated the clear and simple picture of a comet nucleus that was beginning to emerge. If the short history of comet modeling has taught us anything, it is that any conclusions we draw from the models are subject to revision as our understanding of the processes involved improves. For a more definite conclusion, we may have to wait for yet another generation of comet simulation models and more comet exploration missions.

Acknowledgments: This work was supported by the Jack Adler Foundation, administered by the Israel Academy of Sciences and Humanities.

10.7   References

Bar-Nun, A., G. Herman, D. Laufer, and M.L. Rappaport (1985). Trapping and release of gases by water and implications for icy bodies. *Icarus* **63**, 317–332.

Bar-Nun, A., J. Dror, E. Kochavi, and D. Laufer (1987). Amorphous water ice and its ability to trap gases. *Physical Review B: Condensed Matter* **35**, 2427–2435.

Cameron, A. G. W. (1985). Formation and evolution of the primitive solar nebula. In *Protostars & Planets II* (D.C. Black and M.S. Matthews, eds.). University of Arizona Press, Tucson, pp. 1073–1099.

Cameron, A.G.W. and J.W. Truran (1977). The supernova trigger for the formation of the solar system. *Icarus* **30**, 447–461.

Clayton, D.D. (1984). [26]Al in the interstellar medium. *Astrophys. J.* **280**, 144–149.

Dearborn, D.S.P. and J.B. Blake (1985). On the source of the [26]Al observed in the interstellar medium. *Astrophys. J. (Letters)* **288**, L21–L24.

Delsemme, A.H. (1982). Chemical Composition of Cometary Nuclei. In *Comets*, (L.L. Wilkening, ed.). University of Arizona Press, Tucson, pp. 85–130.

Haruyama, J., T. Yamamoto, H. Mizutani, and J.M. Greenberg (1993). Thermal history of comets during residence in the Oort cloud: Effect of radiogenic heating in combination with very low thermal conductivity of amorphous ice. Submitted to *J. Geophys. Res.*

Hoyle, F. and N.C. Wickramasinghe (1978). Influenza from space. *New Scientist* **79**, 946–948.

Irvine, W.M., S.B. Leschine, and F.P. Schloerb (1980). Thermal history, chemical composition and relationship of comets to the origin of life. *Nature* **283**, 748–749.

Klinger, J. (1980). Influence of a phase transition of ice in the heat and mass balance of comets. *Science* **209**, 634–641.

Klinger, J. (1981). Some consequences of a phase transition of water ice on the balance of cometary nuclei. *Icarus* **47**, 320–324.

Kouchi, A., J.M. Greenberg, T. Yamamoto, and T. Mukai (1992). Extremely low thermal conductivity of amorphous ice: relevance to comet evolution. *Astrophys. J.* **388**, L73–L76.

Lee, T., D.A. Papanastassiou, and G.J. Wasserburg (1976). Demonstration of [26]Mg excess in Allende and evidence for [25]Al. *Geophys. Res. Let.* **3**, 109–112.

Lewis, J.S. (1971). Satellites of the outer planets: Their physical and chemical nature. *Icarus* **15** 174–185.

Mahoney, W.A., J.C. Ling, A.S. Jacobson, and R.E. Lingenfelter (1982). *HEAO 3* discovery of [26]Al in the interstellar medium. *Astrophys. J.* **262**, 578–585.

Mekler, Y., D. Prialnik, and M. Podolak (1990). Evaporation from a porous cometary nucleus. *Astrophys. J.* **356**, 682–686.

Prialnik, D. (1992). Crystallization, sublimation, and gas release in the interior of a porous comet nucleus. *Astrophys. J.* **388**, 196–202.

Prialnik, D. and A. Bar-Nun (1987). On the evolution of cometary nuclei. *Astrophys. J.* **313**, 893–905.

Prialnik, D. and A. Bar-Nun (1990). Gas release in comet nuclei. *Astrophys. J.* **363**, 274–282.

Prialnik, D., and A. Bar-Nun (1992). Crystallization of amorphous ice as the cause of comet P/Halley's outburst at 14 AU. *Astron. Astrophys.* **258**, L9–L12.

Prialnik, D., A. Bar-Nun, and M. Podolak (1987). Radiogenic heating of comets by [26]Al and implications for their time of formation. *Astrophys. J.* **319**, 993–1002.

Prialnik, D. and M. Podolak (1994). Radioactive heating of porous icy bodies. In preparation.

Schmitt, B., S. Espinasse, R.J.A. Grim, J.M. Greenberg, and J. Klinger (1989). Laboratory studies of cometary ice analogues. *ESA SP* **302**, 65–69.

Share, G.H., R.L. Kinzer, J.D. Kurfess, D.J. Forrest, and E.L. Chupp (1985). *Astrophys. J. (Letters)* **292**, L61.

Smoluchowski, R. (1981). Amorphous ice and the behavior of cometary nuclei. *Astrophys. J.* **244**, L31–L34.

Wallis, M.K. (1980). Radiogenic heating of primordial comet interiors. *Nature* **284** 431–433.

Wasserburg, G.J. and D.A. Papanastassiou (1982). Some short-lived nuclides in the early solar system—a connection with placental ISM. In *Essays in Nuclear Astrophysics* (C.A. Barnes, D.D. Clayton, and D.A. Schramm eds.). Cambridge University Press, Cambridge, pp. 77–135.

Whipple, F.L., and R.P. Stefanic (1966). On the physics and splitting of cometary nuclei. *Mem. Roy. Soc. Liege (Ser. 5)* **12**, 33–52.

11

Life in Comets

C.P. McKay

ABSTRACT Life appears early in the geological record of the Earth. This implies that either the origin of life was rapid or that life was carried to Earth from elsewhere. Comets have been suggested as the likely vectors for transporting life to Earth. The origin of life may have occurred in an initial phase of comet evolution when radioactive heating may have produced a liquid water core. However, a strong case cannot be made for the origin of life in comets. If life originated beyond the solar system and was carried along with interstellar organics to the solar nebula by unknown mechanisms, then comets are ideal for the collection of these lifeforms, as well as their storage and distribution to planetary surfaces. If comets were responsible for introducing life to Earth, then Earth-like life should be detectable in comets as well as in interplanetary dust particles originating from comets. The limited organic analyses of cometary material available from the missions to Comet Halley failed to detect amino acids and hence do not support the presence of Earth-type life in comets. Remote spectral analyses are virtually useless for this identification.

11.1   Introduction

Life appeared on Earth soon after the end of accretion. There is definite evidence for communities of micro-organisms which include photosynthesis at 3.5 Gyr ago (Schopf, 1993). Further back in time there is suggestive evidence for photosynthetic life as early as 3.8 Gyr ago. This evidence takes the form of sedimentary organic material in which the carbon isotopes have a distinctly biogenic signature (Schidlowski, 1988). The lack of firm evidence for life in the period from 4.0 Gyr to 3.5 Gyr ago is mostly due to the absence of well-preserved sedimentary rocks. The early appearance of life on Earth is even more surprising in light of the fact that the environment of Earth may have been inhospitable to life until the end of the accretionary period—perhaps as late as 3.8 Gyr ago. This inhospitality would have resulted from the heat released form large impacts; sterilizing impacts would have frustrated the origin of life (Maher and Stevenson, 1988; Sleep et al., 1989). Extremely large impacts—such as may have formed the moon—could have completely destroyed any complex organic molecules as well. The history of these events on the early Earth is shown in Fig. 11.1.

Comets enter into the origin of life on Earth in two ways. First, it is well understood that comets could have contributed to the formation of an environment on the Earth conducive to life prior to the first appearance of life. The accretion of comets could have provided water, carbon dioxide, nitrogen, and other compounds

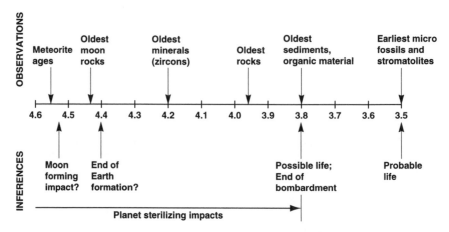

FIGURE 11.1. Events in the history of the Earth before the first appearance of life. Comets could have provided the compounds needed for a habitable environment, especially water. It has been suggested that they might have carried life to Earth—and continue to do so (from Davis and McKay, 1995.)

necessary for life (Chyba, 1987; Delsemme, 1992). In addition comets could have provided raw organic materials that could have been useful to life (Oró, 1961; Clark, 1988; Krueger and Kissel, 1989). The second way in which comets could have played a role in the origin of life on Earth is more direct. It has been suggested that life was carried to Earth by comets (Hoyle and Wickramasinghe, 1979; Wallis et al., 1992). This hypothesis obviates the need to have life originate on Earth in a geological short period of time but requires instead that life originate on comets or that comets carry lifeforms that originated elsewhere—presumably outside the solar system (Hoyle and Wickramasinghe, 1981).

In this chapter we consider the possibility that comets contain life. Since the essence of this hypothesis is that life on Earth first came from life within comets, it is logical to use the observed properties of life on Earth as a guide when we consider life in comets. The most relevant observation about life on Earth is that it requires liquid water. Liquid water is the quintessential environmental requirement for life on Earth (Kushner, 1981) and probably for the origin of life as well (Davis and McKay, 1995). There are no known organisms on the Earth that can survive on pure ice or that can extract liquid water from ice using metabolic energy (Kushner, 1981, p. 242). Organisms that thrive in snow and on glaciers do so only when there is a liquid water fraction present (e.g., Hoham, 1975; Hoham et al., 1983; Wharton et al., 1985). These organism are the passive recipients of liquid water produced environmentally.

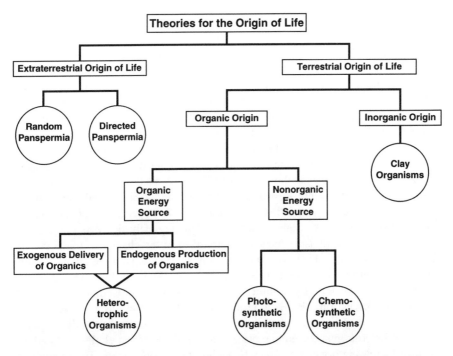

FIGURE 11.2. Diagram showing the range of theories for the origin of life adapted from Davis and McKay (1995). When applied to comets we assume that all the theories under the "terrestrial" branch could have occurred on comets as well. The "random panspermia" theories are consistent with life originating in comets or with comets merely transporting lifeforms present before the formation of the solar system.

11.2   The Origin of Life in Comets

Currently life on Earth is the only life was have available for study. Life has not been found on any other planet nor has life been created in the laboratory. For this reason the question of the origin of life has centered on the origin of life on Earth. Davis and McKay (1995) have reviewed the range of current scientific theories for the origin of life on Earth. Their diagrammatic representation of theories for the origin of life is shown in Fig. 11.2. Davis and McKay (1995) conclude that all theories for the origin of life on any object require the presence of liquid water on that object. This is the one feature that these theories have in common.

What might be called the standard theory for the origin of life posits that life arose on Earth composed from the start of organic material similar in detail to the organic components of life today. It is further assumed that the first life was heterotrophic, deriving energy from the consumption of pre-existing organic material. The prebiotic organic material could have been produced on the Earth itself via Miller–Urey reactions (Miller, 1992) or transported to the Earth by meteorites, dust particles, and comets (e.g., Anders, 1989; Chyba et al., 1990).

Alternate theories exist, however, that do not require prebiotic organic material as an energy source (see Fig. 11.2) and there are theories that posit an inorganic composition of the first organism. Finally, there are theories that do not require the origin of life on Earth itself but hold that life was carried here from elsewhere. In this case, the range of suggestions for the origin of life on Earth (the "terrestrial origin of life" branch in Fig. 11.2) would be applicable to whatever location or object on which life did originate. Comets are such a candidate location. Thus in considering the possibility of the origin of life on comets it is useful to consider the full range of suggestions put forward for the origin of life on Earth.

If life originated on comets the probable venue was deep inside the comet shortly after its formation. Many authors (Irvine et al., 1980; Wallis, 1980; Prialnik et al., 1987; Podolak and Prialnik, this volume) have considered the possibility that due to heat released by shortlived radioactive isotopes—[26]Al in particular—large comets would have maintained a liquid water core for geologically significant periods of time. The most recent assessment is summarized in the chapter in this book by Podolak and Prialnik. They conclude that comets probably did not enjoy a liquid core but admit that the uncertainties remain large. We keep this caveat in mind when we can consider the possibility that the life originated in comets.

Previous discussions (Bar-Nun et al., 1981; Marcus and Olsen, 1991) of the possible origin of life on comets have focused on the standard theory (organic origin of heterotrophs consuming organic material). Bar-Nun et al. (1981) argue against increasing organic complexity leading to life on comets. They point out that the surface of a comet is poorly suited to the origin of life due to the absence of an atmosphere, or discrete liquid water and solid surfaces. The low temperatures and the ablation of the surface on each perihelion passage further militate against the formation of complex organic molecules leading to a living system. As to the origin of life within a radiogenic liquid core, Bar-Nun et al. (1981) state that the radiation produced would destroy any organics present and sterilize any life that developed. Marcus and Olsen (1991) in a wide-ranging paper also discuss the possibility of life on comets. They provide an excellent summary of the standard theory for the origin of life updated to include the concept of the RNA phase in life. They do not attempt to decide if the origin of life in comets is possible or not.

We now turn to the other theories for the origin of life. The origin of life as photosynthetic organic organisms suffers from the unsuitability of the cometary surface as mentioned by Bar-Nun et al. (1981). As this is the only location receiving sunlight this theory does not favor the origin of life on comets. If, instead, life originated as a chemosynthetic system utilizing chemical energy, then this need not have occurred at the surface. A hypothetical liquid water core could serve nicely if there were suitable energy sources. Possible energy sources include those suggested for subsurface life on Mars (Boston et al., 1992) such as $CO_2 + 4H_2 \rightarrow CH_4 + 2H_2O$. The pyrite model for the origin of life (Wächtershäuser, 1988) uses the reaction $FeS + H_2S \rightarrow FeS_2 + H_2$ as an energy source in the production of biomass from CO_2. We note here that the chemosynthesis at deep-sea hydrothermal vents is not a suitable analog for the origin of life, within comets or elsewhere, since the reaction is the oxidation of H_2S using O_2 derived from plant photosynthesis on the surface.

If life originated in comets then it is possible that life originated many times in many different comets. In this case, one might expect that the nature of the lifeforms in each comet might be significantly different; for example, by the selection of D instead of L amino acids or perhaps by using a slightly different genetic sequence. Phylogenetic studies of life on Earth clearly indicate that there is only one type of life on this planet (Woese, 1987). All lifeforms on Earth are related implying descent from a common ancestor. This could be consistent with life originating on one comet but is more difficult to reconcile with life originating independently on several comets.

11.3   Comets as Vectors for Life

Panspermia theories for the origin of life (Fig. 11.2) are consistent with the hypothesis that life originated outside our solar system. In this case the presolar nebula would contain viable, presumably dormant, lifeforms intermixed with the interstellar grains of organic material. Most of the lifeforms and the organic material would have been decomposed during the heating that accompanied the accretion and formation of the solar system. However, there is clear evidence that interstellar organics survived in the outer reaches of the solar system (Allamandola et al., 1987; Simonelli et al., 1989; Ott, 1993) where temperatures never reached high values. Lifeforms could have survived as well and been incorporated into the materials from which comets, asteroids, the giant planets and their moons were subsequently formed. There is direct evidence that asteroids experienced higher temperatures and aqueous alteration (Shock and Schulte, 1990; Tingle et al., 1992) and it is possible that the lifeforms contained within them were destroyed. No such heatings episodes are thought to have affected comets.

The key uncertainty in panspermia theories is the ability of microorganisms to survive long—10^8 years—periods in space. There has been renewed interest in this question (Davies, 1988; Secker et al., 1994) and new data from space exposure experiments (Horneck et al., 1994). The results do not preclude the survival of bacteria over these long periods. There have also been studies of survival of micro-organisms in permafrost on the Earth for three million years at temperatures of $-10°$ C (Gilichinsky et al., 1992), as well as survival for 25 million years in amber in temperate latitudes (Cano and Borucki, 1995). Melosh (1985, 1988) has shown how impacts can launch micro-organisms from the surface a planet into solar orbit. How they escape from one star system to reach interstellar space has not yet been addressed. If life originated outside the solar system then it follows also that lifeforms on all comets would be identical.

A point to keep in mind here is that it would be difficult to discover life in fragments of asteroids and comets that reached the Earth. This is because the lifeforms on the extraterrestrial object would be similar to the lifeforms here on Earth since the latter are postulated to have evolved from the former. For example, it would be expected that only L amino acids would be found in any life extracted

TABLE 11.1. Atom-Percent in Comets and Bacteria.

Element	Comets[a]	Bacteria[b]
H	48%	63%
C	19	7.8
N	1	1.8
O	21	26
P	< 0.002[c]	0.17
S	1.7	0.06
Na	0.24	0.11
K	0.005	0..07
Ca	0.15	0.04

[a]From Jessberger et al. (1988).
[b]From Davies and Koch (1991) for *E. Coli.*
[c]Upper limit estimated from results of
Jessberger et al. (1988) and Kissel
and Krueger (1987).

from meteorites as is true for all Earth life. However, such samples would be dismissed as contamination (e.g., Engel et al., 1990). Furthermore, Marcus and Olsen (1991) point out that comets could contain real contamination from the Earth's biosphere injected into space by large impacts. Only the most fastidious of experimental protocols would allow for the determination on any extraterrestrial body—Mars, comets, or meteorites—of life identical to life on Earth.

The elemental composition of the dust from comet Halley (excluding the ice fraction) is compared to typical bacteria in Table 11.1. The alignment of the main biogenic elements—C, H, N, and O—is remarkable. However, as pointed out by Kissel and Krueger (1987; see also Marcus and Olsen, 1991), the analysis of comet Halley is not consistent with bacterial biomass composing the bulk of the grains; P, Na, K, sugars, and, most significantly, amino acids, were not observed in sufficient quantities. The absence of amino acids in the organic analysis (Kissel and Krueger, 1987) is particularly puzzling given the concentration of amino acids in meteorites as well as in the, possibly cometary, clay layer at the K/T boundary (Steel, 1992; Oberbeck and Aggarwal, 1992). The Halley organic analysis is uncertain due primarily to the high encounter speed, 70 km s^{-1}, and the difficulty of reconstructing the parent molecules from the fragments created by such a high speed impact. It is interesting to note here that purines and pyrimidines are reported as present in comet Halley (Kissel and Krueger, 1987).

Hoyle and Wickramasinghe (1979) use spectral comparison to conclude that there are bacteria, algae, and even viruses in comets and the interstellar medium. Hoover et al. (1986) follows the same technique to deduce the presence of diatom on comets, Europa, and interstellar space. (Hoover et al. (1986) are not dismayed by the fact that diatoms are aerobic eukaryotes and only evolved on Earth one hundred million years ago. In fact, they suggest that their sudden arrival on Earth is consistent with an extraterrestrial source. Why diatom laden comets would not have landed on the Earth earlier is not addressed.) It is clear that the spectra are well fit by invoking biological sources. However, the spectral features are of a broad nondescript nature and the fit cannot be shown to be unique to biological sources (see Davies et al., 1984; Chyba and Sagan, 1987).

Studies of the genetic record of the evolution of life on Earth contained within the vast genome of life argue against life originating in comets or for life originating long ago in another planetary system. Phylogenetic trees based on comparison of the ribosomal RNA for all life on Earth show the interesting feature that the organisms most similar to the common ancestor are thermophilic, sulfur metabolizing organisms (Woese, 1987). This could imply that the origin of life is somehow related to hot sulfur-rich environments—not an especially comet-like scenario. Alternatively there may have been a biological catastrophe, such as an ocean-boiling impact, after which only organisms in thermophilic sulfur-rich environments survived (Sleep et al., 1989). A final alternative is mere chance. Studies of the genome can also provide clues to the timing of the origin of life. Eigen et al. (1989) argue that the genetic code is not older (3.8 ± 0.6 Gyr) than but almost as old as the Earth—consistent with an origin on Earth.

11.4   Conclusion

There is a wide-ranging and vigorous debate on the role, if any, that comets played in the origin of life on Earth. The question of life in comets—either originating there or being held there—is the most interesting and speculative aspect of this debate. The issues cannot be settled with the current data set and no a priori considerations rule out one point of view or the other. The resolution will only come with detailed and careful biological and chemical analysis of comets.

11.5   References

Allamandolla, L.J., S.A. Sandford, and B. Wopenka (1987). Interstellar polycyclic aromatic hydrocarbons and carbon in interplanetary dust particles and meteorites. *Science* **237**, 56–59.

Anders, E. (1989). Prebiotic organic matter from comets and asteroids. *Nature* **342**, 255–257.

Bar-Nun, A., A. Lazcano-Araujo, and J. Oró (1981). Could life have evolved in cometary nuclei? *Origins Life* **11**, 387–394.

Boston, P.J., M.V. Ivanov, and C.P. McKay (1992). On the possibility of chemosynthetic ecosystems in subsurface habitats on Mars. *Icarus* **95**, 300–308.

R.J. Cano and M.K. Borucki (1995). Revival and identification of bacterial spores in 25- to 40-million-year-old Dominican amber. *Science* **268**, 1060–1064.

Chyba, C.F. (1987). The cometary contribution to the oceans of primitive Earth. *Nature* **330**, 632–635.

Chyba, C. and C. Sagan (1987). Cometary organics but no evidence for bacteria. *Nature* **329**, 208.

Chyba, C.F., P.J. Thomas, L. Brookshaw, and C. Sagan (1990). Cometary delivery of organic molecules to the early Earth. *Science* **249**, 366–373.

Clark, B.C. (1988). Primeval procreative comet pond. *Origins Life Evol. Biosphere* **18**, 209–238.

Davies, R.E. (1988). Panspermia: Unlikely, unsupported, but just possible. *Acta Astronaut.* **17** 129–135.

Davies, R.E. and R.H. Koch (1991). All the observed universe has contributed to life. *Phil. Trans. R. Soc. Lond.* B **334**, 391–403.

Davies, R.E., A.M. Delluva, and R.H. Koch (1984). Investigations of claims for interstellar organisms and complex organic molecules. *Nature* **311**, 748–750.

Davis, W.L. and C.P. McKay (1995). The origin of life: A survey of theories and application to Mars. submitted.

Delsemme, A.H. (1992). Cometary origin of carbon, nitrogen and water on the Earth. *Origins Life Evol. Biosphere* **21**, 279–298.

Eigen, M., Lindemann, B.F., Tietze, M., Winkler-Oswatitsch, R., Dress, A. and von Haeseler, A. (1989). How old is the genetic code? Statistical geometry provides an answer. *Science* **244**, 673–679.

Gilichinsky, D.A., E.A. Vorobyova, L.G. Erokhina, D.G. Fyordorov-Dayvdov, and N.R. Chaikovskaya (1992). Long-term preservation of microbial ecosystems in permafrost. *Adv. Space Res.* **12** (4) 255–263.

Hoham, R.W., J.E. Mullet, and S.C. Roemer (1983). The life history and ecology of the snow alga *Chloromonas polyptera* comb. nov. (Chlorophyta, Volvocales), *Canadian J. Botany* **61**, 2416–2429.

Hoham, R.W. (1975). Optimum temperatures and temperature ranges for growth of snow algae, *Arctic and Alpine Res* **7**, 13–24.

Hoover, R.B., F. Hoyle, N.C. Wickramasinghe, M.J. Hoover, and S. Al-Mufti (1986). Diatoms on Earth, comets, Europa and in interstellar space. *Earth, Moon, Planets*, **35**, 19–45.

Horneck, G., Brücker, H., and Reitz, G. (1994). Long-term survival of bacterial spores in space. *Adv. Space Sci.* **14**, (10) 41–45.

Hoyle, F. and C. Wickramasinghe (1979). On the nature of interstellar grains. *Astrophys. Space Sci.* **66**, 77–90.

Hoyle, F. and C. Wickramasinghe (1981). Comets—a vehicle for panspermia. In *Comets and the Origin of Life* (C. Ponnamperuma, Ed.). pp 227-239. Reidel, Dordrecht, Holland, p. 27.

Irvine, W.M., S.B. Leschine, and F.P. Schloerb (1980). Thermal history, chemical composition and relationship of comets to the origin of life. *Nature* **283**, 748–749.

Jessberger, E.K., A. Christoforidis, and J. Kissel (1988). Aspects of the major element composition of Halley's dust. *Nature*, **332**, 691–695.

Kissel, J. and F.R. Krueger (1987). The organic component in the dust from comet Halley as measured by the PUMA mass spectrometer on board Vega 1. *Nature* **326**, 755–760.

Krueger, F.R. and J. Kissel (1989). Biogenesis by cometary grains—thermodyamic apsects of self-organization. *Origins Life Evol. Biosphere* **19**, 87–93.

Kushner, D. (1981). Extreme environments: Are there any limits to life? In *Comets and the Origin of Life* (C. Ponnamperuma, ed.). Reidel, Dordrecht, Holland, pp. 241–248.

Maher, K. A., and D. J. Stevenson (1988). Impact frustration of the origin of life, *Nature* **331**, 612–614.

Marcus, J.N. amd M.A. Olsen (1991). Biological implications of organic compounds in comets. In *Comets in the Post-Hallye Era*, Vol. I (R.L. Newburn, Jr., ed.). Kluwer, Netherlands, pp. 439–462.

Melosh, H.J. 1985. Ejection of rock fragments from planetary bodies, *Geology* **13**, 144–148.

Melosh, H.J. (1988). The rocky road to paspermia. *Nature* **332**, 687–688.

Miller, S.L. (1992), The prebiotic synthesis of organic compounds as a step toward the origin of life. In *Major Events in the History of Life* (J.W. Schopf, ed.). Jones and Bartlett, Boston, MA., pp. 1–28.

Oberbeck, V.R. and H. Aggarwal (1992). Comet impacts and chemical evolution on the bombarded Earth. *Origins Life Evol. Biosphere* **21**, 317–338.

Oró, J. (1961). Comets and the formation of biochemical compounds on the primitive Earth. *Nature* **190**, 389–390.

Ott, U. (1993). Interstellar grains in meteorites. *Nature* **364**, 25–33.

Prialnik, D., A. Bar-Nun, and M. Podolak (1987). Radiogenic heating of comets by ^{26}Al and implications for their time of formation. *Astrophys. J.* **319** 993–1002.

Podolak M. and D. Prialnik (1996). ^{26}Al and liquid water environments in comets. This book.

Schidlowski, M. (1988). A 3,800-million-year isotopic record of life from carbon in sedimentary rocks. *Nature* **333**, 313–318.

Schopf, J.W. (1993). Microfossils of the early Archean apex chert: New evidence for the antiquity of life. *Science* **260**, 640–646.

Secker, J., J. Lepock, and P. Wesson (1994). Damage due to ultraviolet and ionizing radiaton during the ejection of shielded micro-organisms from the vicinity of 1 M$_\odot$ main sequence and red giant stars. *Astrophys. Space Sci.* **219**, 1–28.

Simonelli, D., J.B. Pollack, C.P. McKay, R.T. Reynolds, and A.L. Summers (1989). The carbon budget in the outer solar system. *Icarus*, **82**, 1–35.

Shock, E.L. and M.D. Schulte (1990). Amino-acid synthesis in carbonaceous meteorites by aqueous alteration of polycyclic aromatic hydrocarbons. *Nature* **343**, 728–731.

Sleep, N.H., Zahnle, K.J, and Kasting, J.F., and Morowitz, H.J. (1989). Annihilation of exosystems by large asteroid impacts on the early Earth. *Nature* **342**, 139–142.

Steel, D. (1992). Cometary supply of terrestrial organics: Lessons from the K/T and the present epoch. *Origins Life Evol. Biosphere* **21**, 2339–2357.

Tingle, T.N., J.A. Tyburczy, T.J. Ahrens, and C.H. Becker (1992). The fate or organic matter during planetary accretion: Preliminary studies of the organic chemistry of experimentally shocked Murchison meteorite. *Origins Life Evol. Biosphere* **21**, 385–397.

Wächtershäuser, G. (1988). Before enzymes and templates: Theory of surface methabolism, *Microbiological Reviews* **52**, 452–484.

Wallis, M.K. (1980). Radiogenic melting of primordial comet interiors. *Nature* **284**, 431–433.

Wallis M.K., F. Hoyle, and C. Wickramasinghe (1992). Cometary habitats for primitive life. *Adv. Space Res.* **12**(4), 281–285.

Wharton, R.A., Jr., C.P. McKay, G.M. Simmons, Jr., and B.C. Parker (1985). Cryoconite holes on glaciers, *BioScience* **35**, 499–503.

Woese, C. R. (1987). Bacterial evolution, *Microbiol. Rev.* **51**, 221–271.

12
Comets and Space Missions

C.P. McKay

ABSTRACT The exploration of comets by space missions will require several phases in both the types and complexity of the missions. Initial reconnaissance can be accomplished with flyby missions. More detailed study will require missions that rendezvous with comets, possibly attaching to them, and fly with them through perihelion passage, including a large segment of the orbit, so that changes in comet activity can be studied. Ultimately, a sample of cometary material must be returned to Earth (with due regard for planetary quarantine) for more detailed investigation of the complex organic chemistry in a comet. Missions to several comets will be needed to understand differences between comets, and target choice must reflect the diversity of comets. A particular challenge for comet missions is a rendezvous with a new comet as it enters the planetary region for the first time. The planned Rosetta mission with the surface landers Champollion and Roland promise to greatly expand our knowledge of comets.

12.1 Introduction

Ground-based observations of comets have indicated that comets are rich in volatiles and organic material, this being confirmed by the several missions to comet Halley. These compounds may date back to the early stages of the solar system, and perhaps even to the cloud of interstellar material from which the presolar nebula formed. Comets carrying these volatiles and organics to the inner solar system may have played a key role in the origin of life, making the basic building blocks available on the Earth (and, perhaps, other planets) soon after planetary agglomeration and cooling. Ultimately, a complete understanding of comets will require detailed study through a series of space missions. In the report, *An Integrated Strategy for the Planetary Sciences: 1995–2010*, published in 1994, the Space Studies Board (U.S. National Research Council) lists the following objectives for the study of primitive bodies, including comets.

- Describe the nature and provenance of carbonaceous materials in cometary nuclei, especially as they pertain to the origin of terrestrial life.

- Identify the sources of the extraterrestrial materials that are received on Earth.

- Delineate how asteroids and comets are related and how they differ.

- Determine the elemental, molecular, isotopic, and mineralogical compositions for a variety of samples of primitive bodies.

- Characterize the internal structure, geophysical attributes, and surface geology of a few comets and asteroids.

- Understand the range of activity of comets, including the causes of its onset and its evolution.

- Ascertain the early thermal evolution of primitive bodies, which led to the geochemical differentiation of these bodies.

The Space Studies Board appreciated that to achieve these goals will require that in the next few decades missions to comets be given the highest priority because "such an investigation would contribute so much to understanding how our solar system originated."

12.2   Missions

Table 12.1 lists the objectives and characteristics of comet missions that would be relevant to the question of the origin of life. For flyby missions such as the P/Halley missions the objectives include imaging the nucleus and characterizing its activity. This includes any outflowing dust and gas. Since the surface of a comet is not uniform these missions can catalog the distribution of jets and dark features on the surface as well as the structure of the coma. As demonstrated by the spacecraft sent to P/Halley, even flyby missions can collect and analyze organic (and other) material. Detailed studies of collected dust and organic grains with activation and microscopic techniques are possible if flybys can be achieved at low speeds. Because P/Halley is in a retrograde orbit, the relative velocity at encounter was about 70 km s^{-1}, too high to collect material intact.

In rendezvous mission the spacecraft would remain with a comet as it approaches the Sun even through perihelion. This would allow for the imaging of the nucleus during its closest approach to the Sun, and thus investigations of how the activity varies with time. This type of data would be needed to characterize how the outflowing of dust and gas changes with increasing insolation. It is likely also that the structure of the surface mantle, and the distribution and activity of jets, change with temperature while the temperature, in turn, is buffered by the energy required to sublime the volatiles. These effects may result in changes in organic species which are evaporated from the surface at different heliocentric distances, and the relative rates of loss for each. Refractory organics—like kerogens—that remained on the surface, perhaps even acting as a cement in the mantle material, may sublime as surface temperatures increase.

Missions which place a lander on the comet surface allow another step up in the complexity of analysis and instrumentation that can be applied. In situ analysis can

TABLE 12.1. Objectives and Characteristics of Comet Missions Relevant to the Origin of Life.

Type	Objectives
Flyby	• Image the nucleus • Characterize outflowing dust and gas, particularly organics • Observe the surface and jet structure • Detailed studies of dust and organic grains with neutron activation, X-ray fluorescence, and microscopic techniques • Coma structure
Rendezvous	• Image the nucleus over time through perihelion passage • Characterize outflowing dust and gas with increasing temperature, look for changes in organic content of coma • Observe the surface and jet structure evolution throughout orbit • Coma structure near perihelion
Lander	• In situ analysis of nuclei composition and organic structure: – detection of organics present, especially polymers – search for optical activity – isotopic analysis – experiments liquid water added to cometary material • Measurement of the physical properties of the crust • Direct measurement of radiative energy balance and subsurface heat flow • In situ measurement of ionizing radiation
Sample Return	• Detailed organic analysis and isotopes • Experiments on catalytic properties of grains • Experiments of chemical evolution with cometary organics • Search for life

determine the nucleus composition and organic structure free from any fractionation effects that occur during the outflow. Presumably, the organic materials seen in the coma are at least in part the fragments (daughter products) of more complex structures on the surface. Analysis would include the detection of organics present, especially polymers, searches for optical activity, isotopic analyses, and also simple experiments such as adding water to a sample. Measurements of the physical properties of the surface could elucidate the role that organics play in the stabilization of the mantle. Direct measurements of radiative energy balance and ionizing radiation would be possible. Both are relevant to the state and evolution of organics in the comet.

Finally, a key mission would be a sample return of a piece of a comet nucleus to Earth transported under conditions as close as possible to those where it was collected. This would allow for a wide range of analytical techniques in Earth-based laboratories to be used in the study of the material. These techniques have been extensively developed in the study of organic-rich meteorites. Important analyses would include the detailed organic and isotopic characterization. Direct experiments on the catalytic properties of grains and the efficacy of chemical evolution with cometary organics could be conducted. Last, but not least, the direct search for life in the comet sample could be conducted. It is probable that determining the biological activity of the sample would be a planetary quarantine requirement, and this would need to be accomplished prior to delivery to the terrestrial surface.

Taking comets as a general class, we know little yet about the diversity of their nature. Differences could arise due to initial conditions as well as evolutionary effects. Thus, to develop an understanding of comets will require that we study several in detail. A key question (as mentioned in the Space Science Board objectives listed above) is the determination of the relationship between inactive (dormant/moribund/extinct) comets and asteroids. It may be that a key comet mission will be a visit to one of the several asteroids that are thought to be "comets" rendered inactive by thick mantles. Another interesting type of comet to visit would be one of those entering the inner solar system for the first time. There is some observational evidence that these comets differ in several characteristics from periodic comets which have experienced repeated perihelion passages. From an operations point of view this poses severe challenges since the time from discovery until perihelion for a new comet can be less than a year or two. A mission to a new comet could be built and held ready for launch as telescopes searched for a suitable target. Failing that, a list of back-up target comets would be available. However, an extensive Spaceguard program would likely discover a new comet within a year.

12.3   Conclusion

The prospects for upcoming comet missions are positive. There is a broad appreciation in the scientific community of the importance and potential of studying comets. The field is ripe for future developments which must come from missions that allow for direct analysis. Current plans include the ESA/NASA Rosetta mission which will carry the Champollion and Roland landers to a comet surface. If comets are omens, they now portend an increase in our understanding of the solar system and life in it.

Acknowledgments: The author's participation in comet missions was supported as an Exobiology Interdisciplinary Scientist during planning for the CRAF mission.

Index